METHODS IN MOLECULAR BIOLOGY

For further volumes:
http://www.springer.com/series/7651

The Bacterial Flagellum

Methods and Protocols

Edited by

Tohru Minamino

Graduate School of Frontier Biosciences, Osaka University, Suita, Osaka, Japan

Keiichi Namba

Graduate School of Frontier Biosciences, Osaka Univeristy, Suita, Osaka, Japan

Quantitative Biology Center RIKEN, Suita, Osaka, Japan

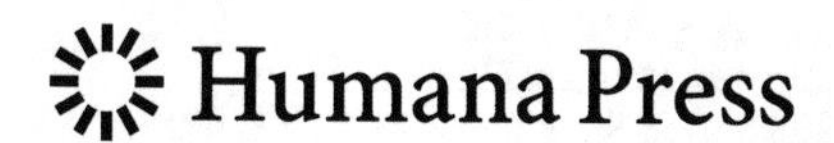

Editors
Tohru Minamino
Graduate School of Frontier Biosciences
Osaka University
Suita, Osaka, Japan

Keiichi Namba
Graduate School of Frontier Biosciences
Osaka Univeristy
Suita, Osaka, Japan

Quantitative Biology Center RIKEN
Suita, Osaka, Japan

ISSN 1064-3745 ISSN 1940-6029 (electronic)
Methods in Molecular Biology
ISBN 978-1-4939-8341-4 ISBN 978-1-4939-6927-2 (eBook)
DOI 10.1007/978-1-4939-6927-2

Printed on acid-free paper

This Humana Press imprint is published by Springer Nature
The registered company is Springer Science+Business Media LLC
The registered company address is: 233 Spring Street, New York, NY 10013, U.S.A.

Preface

Many motile bacteria can swim in liquid environments and move on semi-solid surface by rotating flagella. The bacterial flagellum is a supramolecular assembly composed of 30 different proteins and consists of at least three parts: the basal body, the hook, and the filament. The basal body is embedded within the cell membranes and acts as a bidirectional rotary motor. The energy for motor rotation is supplied by cation influx driven by an electrochemical potential difference of specific ions, such as H^+ and Na^+, across the cytoplasmic membrane, i.e., the ion motive force. The hook and filament extend outwards in the cell exterior. The filament works as a helical propeller to produce thrust. The hook connects the filament with the basal body and functions as a universal joint to smoothly transmit torque produced by the motor to the helical filament. *Escherichia coli* and *Salmonella enterica* produce several flagella per cell. The *E. coli* and *S. enterica* flagellar motor can operate in either counterclockwise (CCW, viewed from filament to motor) or clockwise (CW) direction. When most of the motors rotate in the CCW direction, the filaments form a bundle and propel the cell smoothly. When one or more motors spin in the CW direction, the bundle is disrupted and hence the cell tumbles and changes the swimming direction. *E. coli* and *S. enterica* can move towards more favorable conditions and escape from undesirable ones for their survival by sensing temporal variations of environmental stimuli such as chemical attractants and repellents, temperature, and pH via methyl-accepting chemotaxis proteins (MCP). MCPs are transmembrane proteins with a large cytoplasmic domain involved in interactions with a histidine kinase CheA and an adaptor protein CheW. The MCPs control CheA autophosphorylation. Phosphorylated CheA transfers its phosphate group to a response regulator CheY, and then phosphorylated CheY (CheY-P) binds to the cytoplasmic face of the flagellar motor, letting the motor spin in the CW direction.

In this volume we have brought together a set of cutting-edge research protocols to study the structure and dynamics of the bacterial flagellum using bacterial genetics, molecular biology, biochemistry, structural biology, biophysics, cell biology, and molecular dynamics simulation. Our aim is to provide a pathway to the investigation of the bacterial flagellum derived from various bacterial species through techniques that can be applied. The protocols are generally applicable to other supramolecular motility machinery, such as gliding machinery of bacteria. Since the principal goal of the book is to provide researchers with a comprehensive account of the practical steps of each protocol, the Methods section contains detailed step-by-step descriptions of every protocol. The Notes section complements the Methods to get the hang of each experiment based on the authors' experiences and to figure out the best way to solve any problem and difficulty that might arise during the experiment.

Flagellar assembly begins with the basal body, followed by the hook and finally the filament. The flagellar transcriptional hierarchy is coupled to the assembly process. A remarkable feature is that the flagellar type III protein export apparatus, which is required for flagellar assembly beyond the cellular membranes, coordinates flagellar gene expression with assembly. *E. coli* and *S. enterica* are model organisms that have provided detailed insights into the structure, assembly, and function of the bacterial flagellum. Chapters in

Part I describe flagellar type III protein export (Chapters 1 and 2), assembly (Chapter 3), and gene regulation (Chapters 4 and 5) in *S. enterica.*

The flagellar motor consists of a rotor and a dozen stators. The rotor consists of several rings called the C, M-S, and P-L rings, and a drive shaft. The stator is composed of two transmembrane proteins, commonly referred to as MotA and MotB in the H^+-driven motor of *E. coli* and *S. enterica* and PomA and PomB in the Na^+-driven motor of marine *Vibrio*. The stator acts as an ion channel to couple the ion flow through the channel with torque generation by electrostatic interactions of MotA or PomA with a rotor protein FliG. The stator is anchored to the peptidoglycan layer through the C-terminal periplasmic domain of MotB ($MotB_C$) or PomB ($PomB_C$). $MotB_C$ and $PomB_C$ coordinate stator assembly around the rotor with ion channel formation, thereby suppressing undesirable ion flow through the channel when the stator is not installed into the motor. Chapters in Part II describe how to isolate the flagella from the bacterial cell bodies (Chapter 6) and how to carry out high-resolution structural and functional analyses of the flagellar motor (Chapters 7, 8, 9, and 11). In silico modeling of the MotAB proton channel complex (Chapter 10) is included in Part II.

Torque is produced by electrostatic interactions of MotA or PomA with a rotor protein FliG. Ion translocation through the ion channel is coupled with cyclic conformational changes of MotA or PomA for torque generation. CheY-P binds to FliM and FliN in the C ring, resulting in switching of flagellar motor rotation from the CCW to CW directions without changing the direction of the ion flow. Direct observations of flagellar motor rotation by nanophotometry with high spatial and temporal resolutions have revealed that the elementary process of torque generation by stator-rotor interactions is symmetric in CCW and CW rotation. Single molecule imaging techniques have shown that both stator and rotor are highly dynamic structures, thereby showing rapid exchanges between localized and freely diffusing forms even during motor rotation. Chapters in Part III describe how to measure flagellar motor rotation over a wide range of external load (Chapters 12, 13, and 14), how to measure ion motive force across the cytoplasmic membrane (Chapters 15), and how to measure the dynamic properties of the flagellar motor proteins by fluorescence microscopy with single molecule precision (Chapters 16 and 17).

Intact flagellar motor structures derived from different bacteria species have been visualized. Most components of the core structure of the basal body and their organization are well conserved among bacteria species. Recently, novel and divergent structures with different symmetries have been observed to surround the conserved core structure in different species. Chapters in Part IV describe the structure and function of Spirochetal (Chapters 18 and 19), *Vibrio* (Chapters 20 and 21), *Rhodobacter* (Chapter 22), *Shewanella* (Chapter 23), Alkaliphilic *Bacillus* (Chapter 24), and *Magnetococcus* flagellar motors (Chapter 25).

All the contributors are leading researchers in the bacterial flagellar field, and we would like to acknowledge them for providing their comprehensive protocols and techniques for this volume. We would like to thank Dr. John Walker, the Editor-in-Chief of the Methods in Molecular Biology series, for giving us a great opportunity to edit this volume and his continuous support and encouragement.

We hope you all have lots of fun with and benefit from this volume of Methods in Molecular Biology.

Osaka, Japan

Tohru Minamino
Keiichi Namba

Contents

Contributors

Shin-Ichi Aizawa • *Department of Life Sciences, Prefectural University of Hiroshima, Shobara, Hiroshima, Japan*

Yoshiyuki Arai • *The Institute of Scientific and Industrial Research, Osaka University, Osaka, Japan*

Paul M. Bergen • *Department of Pathology, University of Cambridge, Cambridge, UK*

Susanne Brenzinger • *Institute for Microbiology and Molecular Biology, Justus-Liebig-Universität Gießen, Gießen, Germany*

Owain J. Bryant • *Department of Pathology, University of Cambridge, Cambridge, UK*

Laura Camarena • *Instituto de Investigaciones Biomédicas, Universidad Nacional Autónoma de México, Mexico City, Mexico*

Fabienne F.V. Chevance • *Department of Biology, University of Utah, Salt Lake City, UT, USA*

Georges Dreyfus • *Instituto de Fisiología Celular, Universidad Nacional Autónoma de México, Mexico City, Mexico*

Marc Erhardt • *Helmholtz Centre for Infection Research, Braunschweig, Germany*

Lewis D.B. Evans • *Department of Pathology, University of Cambridge, Cambridge, UK*

Gillian M. Fraser • *Department of Pathology, University of Cambridge, Cambridge, UK*

Hajime Fukuoka • *Graduate School of Frontier Biosciences, Osaka University, Osaka, Japan*

Michio Homma • *Division of Biological Science, Graduate School of Science, Nagoya University, Nagoya, Japan*

Kelly T. Hughes • *Department of Biology, University of Utah, Salt Lake City, UT, USA*

Katsumi Imada • *Department of Macromolecular Science, Graduate School of Science, Osaka University, Toyonaka, Osaka, Japan*

Yuichi Inoue • *Institute of Multidisciplinary Research for Advanced Materials, Tohoku University, Sendai, Miyagi, Japan; Sigmakoki Co., Ltd., Tokyo Head Office, Tokyo, Japan*

Md. Shafiqul Islam • *Department of Applied Physics, Graduate School of Engineering, Tohoku University, Sendai, Miyagi, Japan; Department of Microbiology and Hygiene, Faculty of Veterinary Science, Bangladesh Agricultural University, Mymensingh, Bangladesh*

Masahiro Ito • *Bio-Nano Electronics Research Centre, Toyo University, Saitama, Japan; Graduate School of Life Sciences, Toyo University, Itakura, Gunma, Japan*

Taishi Kasai • *Department of Frontier Bioscience, Hosei University, Tokyo, Japan; Research Center for Micro-Nano Technology, Hosei University, Tokyo, Japan*

Ikuro Kawagishi • *Department of Frontier Bioscience, Hosei University, Tokyo, Japan; Research Center for Micro-Nano Technology, Hosei University, Tokyo, Japan*

Akihoro Kawamoto • *Graduate School of Frontier Biosciences, Osaka University, Osaka, Japan*

Miki Kinoshita • *Graduate School of Frontier Biosciences, Osaka University, Osaka, Japan*

Akio Kitao • *Institute of Molecular and Cellular Biosciences, The University of Tokyo, Tokyo, Japan*

SANTOSH KOIRALA • *Department of Chemical and Biomolecular Engineering, University of Illinois at Urbana-Champaign, Urbana, IL, USA*
SEIJI KOJIMA • *Division of Biological Science, Graduate School of Science, Nagoya University, Nagoya, Japan*
LAWRENCE K. LEE • *European Molecular Biology Laboratory Australia Node for Single Molecule Science, School of Medical Sciences, University of New South Wales, Sydney, NSW, Australia; Structural and Computational Biology Division, Victor Chang Cardiac Research Institute, Darlinghurst, NSW, Australia*
TSAI-SHUN LIN • *Department of Physics, National Central University, Taiwan, Republic of China*
JUN LIU • *Department of Pathology and Laboratory Medicine, McGovern Medical School at UTHealth, Houston, TX, USA*
CHIEN-JUNG LO • *Department of Physics, National Central University, Taiwan, Republic of China; Graduate Institute of Biophysics, National Central University, Taiwan, Republic of China*
TOHRU MINAMINO • *Graduate School of Frontier Biosciences, Osaka University, Osaka, Japan*
JAVIER DE LA MORA • *Instituto de Fisiología Celular, Universidad Nacional Autónoma de México, Mexico City, Mexico*
DUSTIN R. MORADO • *Department of Pathology and Laboratory Medicine, McGovern Medical School at UTHealth, Houston, TX, USA*
YUSUKE V. MORIMOTO • *Quantitative Biology Center, RIKEN, Osaka, Japan*
SHUICHI NAKAMURA • *Department of Applied Physics, Graduate School of Engineering, Tohoku University, Sendai, Miyagi, Japan*
KEIICHI NAMBA • *Graduate School of Frontier Biosciences, Osaka University, Osaka, Japan; Quantitative Biology Center, RIKEN, Osaka, Japan*
YASUTAKA NISHIHARA • *Institute of Molecular and Cellular Biosciences, The University of Tokyo, Tokyo, Japan*
MASAYOSHI NISHIYAMA • *The Hakubi Center for Advanced Research/Institute for Integrated Cell-Material Sciences, Kyoto University, Kyoto, Japan*
SO-ICHIRO NISHIYAMA • *Department of Frontier Bioscience, Hosei University, Tokyo, Japan; Research Center for Micro-Nano Technology, Hosei University, Tokyo, Japan*
YASUHIRO ONOUE • *Division of Biological Science, Graduate School of Science, Nagoya University, Nagoya, Japan*
ZHUAN QIN • *Department of Pathology and Laboratory Medicine, McGovern Medical School at UTHealth, Houston, TX, USA*
CHRISTOPHER V. RAO • *Department of Chemical and Biomolecular Engineering, University of Illinois at Urbana-Champaign, Urbana, IL, USA*
YOSHIYUKI SOWA • *Department of Frontier Bioscience, Hosei University, Tokyo, Japan; Research Center for Micro-Nano Technology, Hosei University, Tokyo, Japan*
YI-REN SUN • *Department of Physics, National Central University, Taiwan, Republic of China*
YUKA TAKAHASHI • *Bio-Nano Electronics Research Centre, Toyo University, Saitama, Japan*
KAI M. THORMANN • *Institute for Microbiology and Molecular Biology, Justus-Liebig-Universität Gießen, Gießen, Germany*
JUYU WANG • *Department of Pathology and Laboratory Medicine, McGovern Medical School at UTHealth, Houston, TX, USA*

LONG-FEI WU • *Aix Marseille Univerité, CNRS, LCB, Marseille, France; Laboratoire International Associé de la Bio-Minéralisation et Nano-Structures, Centre National de la Recherche Scientifique, Marseille, France*
SHENG-DA ZHANG • *Laboratory of Deep-Sea Microbial Cell Biology, Sanya Institute of Deep-Sea Science and Engineering, Chinese Academy of Sciences, Sanya, China*
WEI-JIA ZHANG • *Laboratory of Deep-Sea Microbial Cell Biology, Sanya Institute of Deep-Sea Science and Engineering, Chinese Academy of Sciences, Sanya, China; Laboratoire International Associé de la Bio-Minéralisation et Nano-Structures (LIA-BioMNSL), Centre National de la Recherche Scientifique, Marseille cedex, France*
SHIWEI ZHU • *Department of Pathology and Laboratory Medicine, McGovern Medical School at UTHealth, Houston, TX, USA*

Part I

Flagellar Type III Protein Export, Assembly and Gene Regulation in *Salmonella enterica*

Chapter 1

Fuel of the Bacterial Flagellar Type III Protein Export Apparatus

Tohru Minamino, Miki Kinoshita, and Keiichi Namba

Abstract

The flagellar type III export apparatus utilizes ATP and proton motive force (PMF) across the cytoplasmic membrane as the energy sources and transports flagellar component proteins from the cytoplasm to the distal growing end of the growing structure to construct the bacterial flagellum beyond the cellular membranes. The flagellar type III export apparatus coordinates flagellar protein export with assembly by ordered export of substrates to parallel with their order of the assembly. The export apparatus is composed of a PMF-driven transmembrane export gate complex and a cytoplasmic ATPase complex. Since the ATPase complex is dispensable for flagellar protein export, PMF is the primary fuel for protein unfolding and translocation. Interestingly, the export gate complex can also use sodium motive force across the cytoplasmic membrane in addition to PMF when the ATPase complex does not work properly. Here, we describe experimental protocols, which have allowed us to identify the export substrate class and the primary fuel of the flagellar type III protein export apparatus in *Salmonella enterica* serovar Typhimurium.

Key words ATPase, Flagellar assembly, Proton motive force, Sodium motive force, Substrate specificity switching, Type III protein export

1 Introduction

The bacterial flagellum is a supramolecular motility machine consisting of basal body rings and an axial structure. The axial structure is composed of the rod, the hook, the hook-filament junction zone, the filament, and the filament cap. Axial component proteins are exported via the flagellar type III export apparatus from the cytoplasm to the distal end of the growing structure where their assembly occurs. The export apparatus is composed of a transmembrane export gate complex made of FlhA, FlhB, FliO, FliP, FliQ, and FliR and a cytoplasmic ATPase ring complex consisting of FliH, FliI, and FliJ (Fig. 1). The MS ring of the flagellar basal body, which is made of 26 copies of FliF, is a housing for the flagellar type III export apparatus as well as a mounting platform for the C ring, which is made of FliG, FliM, and FliN. The C ring acts as

Tohru Minamino and Keiichi Namba (eds.), *The Bacterial Flagellum: Methods and Protocols*, Methods in Molecular Biology, vol. 1593, DOI 10.1007/978-1-4939-6927-2_1,

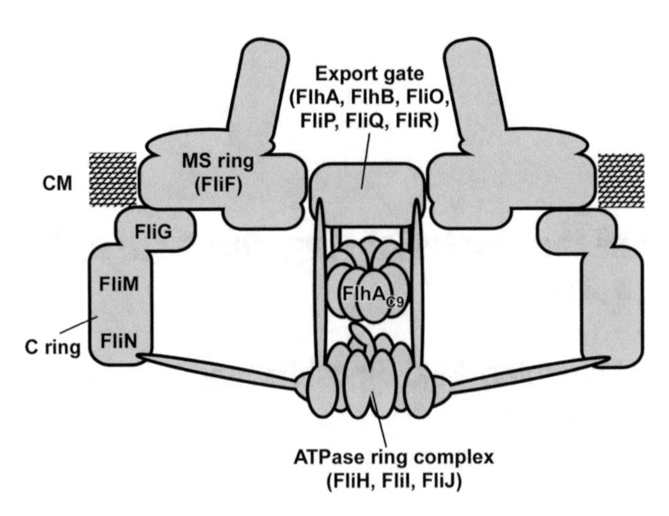

Fig. 1 Schematic diagram of the flagellar type III export apparatus. The export gates made of FlhA, FlhB, FliO, FliP, FliQ, and FliR are located within the central pore of the MS ring. The C-terminal cytoplasmic domain of FlhA ($FlhA_C$) forms a nonameric ring structure and projects into the cavity of the C ring formed by FliG, FliM, and FliN. FliI and FliJ form a $FliI_6FliJ$ ring complex. The $FliI_6FliJ$ ring complex associates with the FBB through interactions of FliH with both FliN and FlhA. CM, cytoplasmic membrane

a sorting platform for efficient assembly of the $FliH_{12}FliI_6FliJ$ ring complex at the flagellar base. These components are highly homologous components of the type III secretion system of pathogenic Gram-negative bacteria, which directly transport virulence effector proteins into host cells for their invasion [1–4].

Assembly of the axial structure proceeds successively from most cell-proximal structures to most cell-distal ones; it begins with the rod, followed by the hook and finally the filament. The flagellar type III export apparatus monitors the state of rod-hook assembly and switches its substrate specificity upon completion of the hook structure [5–7]. The export substrates are divided into two classes: the rod/hook-type export class, comprising proteins needed for the structure and assembly of the rod (FliE, FlgB, FlgC, FlgF, FlgG, and FlgJ) and hook (FlgD, FlgE, FliK) and the filament-type export class, comprising proteins responsible for filament formation (FlgK, FlgL, FlgM, FliC, and FliD). During rod and hook assembly, the export apparatus transports the rod/hook-type substrates but not the filament-type substrates. Upon completion of hook assembly, the export apparatus terminates the export of the rod/hook-type proteins and initiates the export of the filament-type proteins. At least, five flagellar proteins, FliK, FlhB, FlhA, FlhE, and Flk (RflH) are involved in the export specificity switching mechanism [8–10]. FlgN, FliS, and FliT, which function as flagellar type III export chaperones specific for FlgK and FlgL,

FliC and FliD, respectively, bind to FlhA and coordinate the export of their cognate substrates with the assembly of the flagellar filament [11, 12].

The flagellar type III export apparatus utilizes both ATP and proton motive force (PMF) across the cytoplasmic membrane to drive protein translocation into the central channel of the growing structure [13, 14]. Since FliH, FliI, and FliJ are dispensable for flagellar protein export, PMF is the primary fuel for unfolding and translocation of export substrates [13, 14]. Since the export apparatus processively transports flagellar proteins to grow flagella even by the E211D mutation resulting in an extremely low ATPase activity, the role of ATP hydrolysis by FliI ATPase appears to activate the export gate complex, allowing the gate to transport flagellar axial proteins in a PMF-dependent manner [15]. Interestingly, the export apparatus can also use a Na^+ gradient across the cytoplasmic membrane in addition to a H^+ gradient when FliH, FliI, and FliJ do not work under certain conditions [16, 17].

This book chapter describes the protocols we used for the identification of the export substrate class and the energy source for flagellar protein export in *Salmonella*.

2 Materials

Prepare all solutions using Milli-Q water and analytical grade reagents and then autoclave them at 121 °C for 20 min.

2.1 *Salmonella enterica*

1. SJW1103 (wild type for motility and chemotaxis) [18].
2. SJW1353 (*flgE*) [19].
3. SJW2177 (*flgK*) [20].
4. MMHI0117 (Δ*fliH-fliI flhB(P28T)*) (*see* **Note 1**) [13].

2.2 Plasmids

1. pTrc99AFF4 (Cloning vector) [21].
2. pMM1001 (pTrc99AFF4/FliE) [22].
3. pMM1101 (pTrc99AFF4/FlgB) [22].
4. pMM1201 (pTrc99AFF4/FlgC) [7].
5. pRCD001 (pTrc99AFF4/FlgD) [7].
6. pMM1301 (pTrc99AFF4/FlgF) [7].
7. pMM202 (pTrc99AFF4/FlgG) [7].
8. pMM1903 (pTrc99AFF4/FlgL) [7].

2.3 Culture Media

1. Luria broth (LB): 1% (w/v) Bacto tryptone, 0.5% (w/v) Yeast extract, 0.5% (w/v) NaCl.
2. LB agar plate (LA): 1% (w/v) Bacto tryptone, 0.5% (w/v) Yeast extract, 0.5% (w/v) NaCl, 1.5% (w/v) Bacto agar.

3. T-broth (TB): 1% (w/v) Bacto tryptone, 10 mM potassium phosphate pH 7.5.
4. Ampicillin.

2.4 Transformation

1. 0.1 M $MgCl_2$.
2. 0.1 M $CaCl_2$.
3. 50%(w/v) glycerol.
4. 37 °C shaker.
5. Spectrophotometer.
6. Refrigerated centrifuge.
7. 50 mL centrifugation tubes.
8. 1.5 mL Eppendorf tubes.
9. Ice.
10. 42 °C heating block.
11. 37 °C incubator.

2.5 Flagellar Protein Export Assays

1. 30 °C shaker.
2. Spectrophotometer.
3. Refrigerated microcentrifuge.
4. Isopropyl-β-D-thiogalactopyranoside (IPTG).
5. 50 mM carbonyl cyanide *m*-chlorophenylhydrazone (CCCP) in dimethyl sulfoxide.
6. 5 mL Eppendorf tubes.
7. 1.5 mL Eppendorf tubes.
8. Ice.
9. Trichloroacetic acid (TCA).
10. 95 °C heating block.
11. SDS-loading buffer (2×): 125 mM Tris–HCl, pH 6.8, 4% (w/v) sodium dodecyl sulfate (SDS), 20% (w/v) glycerol, 0.002% (w/v) bromophenol blue.
12. 1 M Tris.
13. 12.5% SDS-polyacrylamide gel and gel apparatus for SDS-PAGE.
14. Running buffer for SDS-PAGE: 30 g Tris, 144 g Glycine, 10 g SDS per liter.
15. Nitrocellulose membranes.
16. Blotting apparatus.
17. Transfer buffer for immunoblotting: 25 mM Tris, 250 mM glycine, 20% (v/v) methanol.

18. TBS contacting Tween-20 (TBS-T): 20 mM Tris–HCl, pH 7.5, 500 mM NaCl, 0.1% (v/v) Tween-20.
19. Blocking buffer: 5% skim milk in TBS-T.
20. Polyclonal antibodies against flagellar proteins were produced by MBL (Nagoya, Japan).
21. Goat anti-rabbit IgG-HRP.
22. Chemiluminescence reagents (e.g., ECL Prime immunoblotting detection kit).
23. Chemiluminescence detection system.

2.6 Measurements of Intracellular ATP Levels

1. 30 °C shaker.
2. Spectrophotometer.
3. Refrigerated microcentrifuge.
4. 1.5 mL Eppendorf tubes.
5. Ice.
6. 96-well microplates.
7. ATP bioluminescence assay reagent.
8. 100 °C heating block.
9. Microplate reader.

2.7 ATPase Activity Measurements

1. FliI ATPase freshly purified on the day of the experiment (*see* **Note 2**).
2. X 10 reaction buffer: 300 mM HEPES-NaOH, pH 8.0, 300 mM KCl, 300 mM NH_4Cl, 50 mM $Mg(CH_3COO)_2$, 10 mM DTT.
3. 5 mg/mL BSA.
4. 100 mM ATP freshly dissolved in 1 M Tris–HCl, pH 7.5 on the day of the experiment.
5. 34% (w/v) citric acid.
6. 1 mM KH_2PO_4 in 0.01 N H_2SO_4.
7. Freshly prepared malachite green-ammonium molybdate (MGAM) reagent: three volumes of 0.045% (w/v) malachite green hydrochloride, one volume of 4.2% (w/v) ammpnium molybdate tetrahydrate in 4 N HCl, 1/50th volume of 1% (w/v) Triton X-100 (*see* **Note 3**).

2.8 In Vitro Reconstitution of the $FliI_6$ Ring

1. FliI ATPase freshly purified on the day of the experiment.
2. 50 mM Tris–HCl, pH 8.0, 150 mM NaCl.
3. 1 mg/mL *E. coli* acidic phospholipids freshly suspended in Milli-Q water on the day of the experiment.
4. 300 mM NaF.

5. 100 mM $AlCl_3$.
6. 100 mM $MgCl_2$.
7. 100 mM DTT freshly dissolved in Milli-Q water on the day of the experiment.
8. 100 mM ADP freshly dissolved in 1 M Tris–HCl, pH 7.5 on the day of the experiment.
9. 37 °C heating block.
10. Carbon-coated copper grids.
11. 2% (w/v) uranyl acetate.
12. Electron microscope.

3 Methods

3.1 Transformation

1. Inoculate 0.3 mL of overnight culture of *Salmonella* cells into 30 mL of fresh LB and incubate at 37 °C with shaking until the cell density has reached an OD_{600} of 0.6–0.8.
2. Measure OD_{600} of each culture using a spectrophotometer.
3. Transfer the cultures into 50 mL centrifuge tubes.
4. Centrifuge the tubes (8000 × *g*, 5 min, 4 °C).
5. Discard supernatant and suspend cell pellets in 30 mL of cold 0.1 M $MgCl_2$.
6. Centrifuge the tubes (8000 × *g*, 5 min, 4 °C).
7. Discard supernatants and suspend the cell pellets in 15 mL of cold 0.1 M $CaCl_2$.
8. Leave the tubes on ice for more than 30 min.
9. Discard supernatants and suspend the cell pellets in 1.5 mL of cold 0.1 M $CaCl_2$ and 1.5 mL of cold 50% (w/v) glycerol.
10. Competent cells are either used immediately or stored at −80 °C.
11. Add 100 μL of the competent cells into 1.5 mL Eppendorf tubes containing 1–2 μL of plasmid DNA prepared from the *Salmonella* JR501 strain (*see* **Note 4**).
12. Leave the tubes on ice for more than 30 min.
13. Heat the tubes at 42 °C for 2 min.
14. Leave the tubes on ice for 5 min.
15. Add fresh 1.0 mL of LB to the tubes.
16. Incubates the tubes at 37 °C for 1 h.
17. Centrifuge the tubes (8000 × *g*, 5 min, 4 °C).
18. Resuspend cell pellets in 100 μL of LB.
19. Streak the cell suspensions on LA containing 50 μg/mL ampicillin using glass beads and incubate overnight at 37 °C.

3.2 Flagellar Type III Protein Export Assays

3.2.1 Analysis of the Export Properties of Flagellar Axial Proteins

1. Construct plasmids encoding flagellar axial proteins on the pTrc99FF4 vector plasmid (*see* **Note 5**).
2. Transform *Salmonella* SJW1353 (*flgE*) and SJW2177 (*flgK*) with the pTrc99AFF4-based plasmids (*see* **Note 6**).
3. Inoculate fresh transformants into 5 mL of LB containing 100 μg/mL ampicillin and incubate at 30 °C with shaking until the cell density has reached an OD_{600} of 0.6–0.8.
4. Measure OD_{600} of each culture using a spectrophotometer.
5. Transfer 3 mL of each culture into a 5 mL Eppendorf tube.
6. Centrifuge the tubes (8000 × *g*, 5 min, 4 °C).
7. Discard supernatant and suspend cell pellet in 1 mL of fresh LB containing 100 μg/mL ampicillin.
8. Repeat **steps 6** and 7 twice.
9. Discard supernatant and resuspend the cell pellet in 3 mL of fresh LB containing 100 μg/mL ampicillin and 1 mM IPTG.
10. Transfer the cell suspensions into test tubes and incubate at 30 °C for 1 h with shaking.
11. Measure OD_{600} of each culture using the spectrophotometer.
12. Transfer 1.5 ml of each culture into a 1.5 mL Eppendorf tube.
13. Centrifuge (12,000 × *g*, 5 min, 4 °C) to collect the cell pellet and culture supernatant, separately.
14. Suspend cell pellets in OD_{600} × 250 μL of 1× SDS-loading buffer containing 1 μL of 2-mercaptoethanol and then boil at 95 °C for 3 min.
15. Transfer 900 μL of each culture supernatant to a 1.5 ml Eppendorf tube.
16. Add 100 μL TCA to the culture supernatant and vortex well.
17. Leave on ice for 1 h.
18. Centrifuge the tubes (20,000 × *g*, 20 min, 4 °C).
19. Discard supernatants completely and suspend the pellets in OD_{600} × 25 μL of Tris–SDS-loading buffer (one volume of 1 M Tris, nine volumes of 1× SDS-loading buffer) containing 1 μL of 2-mercaptoethanol.
20. Vortex well and then boil at 95 °C for 3 min.
21. Run proteins in the whole cell and culture supernatant fractions on SDS-PAGE and analyzed by CBB staining (Fig. 2).

3.2.2 Effect of Depletion of Total PMF on Flagellar Protein Export

1. Inoculate 50 μL of overnight culture of *Salmonella* SJW1103 (wild-type) and MMHI0117 [*ΔfliH-fliI flhB(P28T)*] strains into 5 mL of fresh LB and incubate at 30 °C with shaking until the cell density has reached an OD_{600} of 0.8–1.0.
2. Measure OD_{600} of each culture using a spectrophotometer.

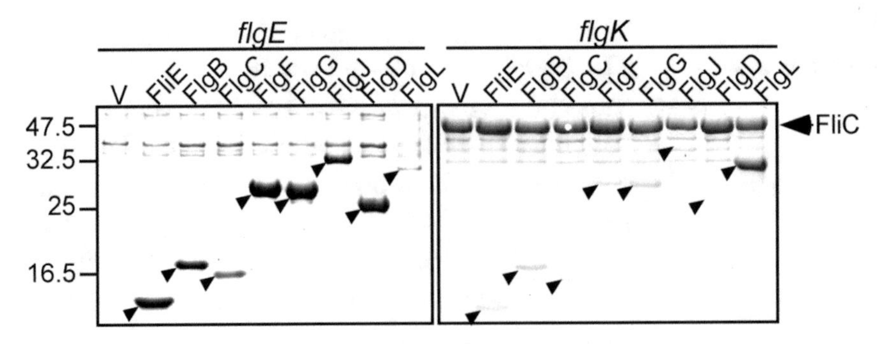

Fig. 2 Export assays of flagellar axial proteins before and after hook completion. Coomassie-stained gels of the culture supernatants of (*left panel*) SJW1353 (*flgE*) and (*right panel*) SJW2177 (*flgK*) mutants carrying pTrc99AFF4-based plasmids producing various flagellar proteins under induction with 1 mM IPTG. The proteins being produced are identified above each lane, and their positions are indicated by *arrowheads*. *v*, vector alone. Molecular mass markers (in kDa) are shown to the *left*

3. Transfer 3 mL of each culture into a 5 ml Eppendorf tube.
4. Centrifuge the tubes (8000 × *g*, 5 min, 4 °C).
5. Discard supernatants and suspend cell pellets in 1 mL of fresh LB.
6. Repeat **steps 4** and **5** twice.
7. Discard the supernatants and resuspend the cell pellets in 3 mL of fresh LB with 0, 5, 10, and 25 μM CCCP (*see* **Note 7**).
8. Transfer the cell suspensions into test tubes and measure OD_{600} using a spectrophotometer.
9. Incubate at 30 °C for 1 h with shaking.
10. Measure OD_{600} of each culture using the spectrophotometer.
11. Transfer 1.5 ml of each culture into a 1.5 mL Eppendorf tube.
12. Centrifuge (12,000 × *g*, 5 min, 4 °C) to collect cell pellets and culture supernatants, separately.
13. Suspend the cell pellets in OD_{600} × 250 μL of 1× SDS-loading buffer containing 1 μL of 2-mercaptoethanol and then boil at 95 °C for 3 min.
14. Transfer 900 μL of each culture supernatant to a 1.5 ml Eppendorf tube.
15. Add 100 μL TCA to the culture supernatant and leave on ice for 1 h.
16. Centrifuge the tubes (20,000 × *g*, 20 min, 4 °C).
17. Discard supernatants completely and suspend the pellets in OD_{600} × 25 μL of Tris–SDS-loading buffer (one volume of 1 M Tris, nine volumes of 1× SDS-loading buffer) containing 1 μL of 2-mercaptoethanol.
18. Vortex well and then boil at 95 °C for 3 min.

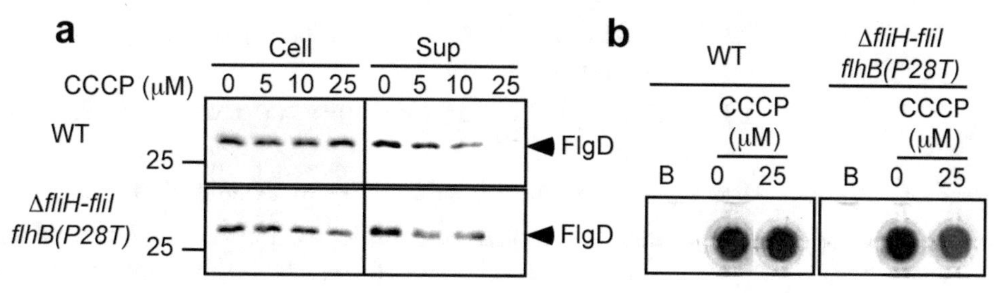

Fig. 3 Effect of depletion of proton motive force on flagellar protein export. (**a**) Immunoblotting, using polyclonal anti-FlgD antibody, of whole cell proteins (Cell) and culture supernatant fractions (Sup) prepared from SJW1103 (WT) and MMHI0117 [Δ*fliH-fliI flhB(P28T)*] grown at 30 °C in LB containing 5, 10, and 25 μM CCCP. DMSO (solvent for CCCP) is added as a control (0 μM CCCP). The position of FlgD is indicated on the *right*. (**b**) Effect of CCCP on the cellular ATP level. SJW1103 (WT) and MMHI0117 [Δ*fliH-fliI flhB(P28T)*] were grown at 30 °C in LB with or without 25 μM CCCP. The cultures were centrifuged and then cell pellets were resuspended in 100 mM Tris–HCl, pH 7.8, 4 mM EDTA to adjust the cell density to an OD_{600} of 1.0. The cell suspensions were boiled for 2 min. Samples were centrifuged, and 100 μL of each supernatant was transferred to a 96-well microtiter plate. 100 μL of luciferase reagent was injected to each well, and then luminescence was detected by a microplate reader. B indicates blank wells with the buffer only

19. Run proteins in the whole cellular and culture supernatant fractions on SDS-PAGE and analyzed by immunoblotting with polyclonal anti-FlgD antibody (Fig. 3a).

3.2.3 Effect of Depletion of a Na^+ Gradient on Flagellar Protein Export

1. Inoculate 50 μL of overnight culture of the *Salmonella* SJW1103 (wild-type) and MMHI0117 [Δ*fliH-fliI flhB(P28T)*] strain into 5 mL of fresh TB (pH 7.5) with 100 mM NaCl and incubate at 30 °C with shaking until the cell density has reached an OD_{600} of 0.8–1.0 (*see* **Note 8**).
2. Measure OD_{600} of each culture using a spectrophotometer.
3. Transfer 3 mL of each culture into a 5 mL centrifuge tube.
4. Centrifuge the tubes (8000 × *g*, 5 min, 4 °C).
5. Discard supernatants and suspend cell pellets in 1 mL of fresh TB (pH 7.5).
6. Repeat **steps 4** and **5** twice.
7. Discard the supernatants and resuspend the cell pellets in 3 mL of fresh TB (pH 7.5) with or without 100 mM NaCl.
8. Transfer the cell suspensions into test tubes and incubate at 30 °C for 1 h with shaking.
9. Measure OD_{600} of each culture using the spectrophotometer.
10. Centrifuge 1.5 ml of the cultures to collect cell pellets and culture supernatants (12,000 × *g*, 5 min, 4 °C).
11. Suspend the cell pellets in OD_{600} × 250 μL of 1× SDS-loading buffer containing 1 μL of 2-mercaptoethanol and then boil at 95 °C for 3 min.

12. Transfer 900 μL of the culture supernatant to a 1.5 mL Eppendorf tube.
13. Add 100 μL TCA to the culture supernatant and leave on ice for 1 h.
14. Centrifuge the tubes (20,000 × *g*, 20 min, 4 °C).
15. Discard supernatants completely and suspend the pellets in OD_{600} × 25 μL of Tris-SDS-loading buffer containing 1 μL of 2-mercaptoethanol.
16. Vortex well and then boil at 95 °C for 3 min.
17. Run proteins in the whole cellular and culture supernatant fractions on SDS-PAGE and analyzed by immunoblotting with polyclonal anti-FlgD antibody.

3.3 Measurements of Intracellular ATP Level

1. Inoculate 50 μL of overnight culture of the *Salmonella* SJW1103 (wild-type) and MMHI0117 [Δ*fliH-fliI flhB(P28T)*] strain into 5 mL of fresh LB and incubate at 30 °C with shaking until the cell density has reached an OD_{600} of 0.8–1.0.
2. Transfer 3 mL of culture into a 5 mL centrifuge tube.
3. Centrifuge the tubes (8000 × *g*, 5 min, 4 °C).
4. Discard supernatants and suspend cell pellets in 1 mL of fresh LB.
5. Repeat **steps 3** and **4** twice.
6. Discard the supernatants and resuspend the cell pellets in 3 mL of fresh LB with or without 25 μM CCCP.
7. Transfer the cell suspensions into fresh test tubes and incubate at 30 °C for 1 h with shaking.
8. Measure OD_{600} of each culture using a spectrophotometer.
9. Centrifuge 1.5 mL of cultures (8000 × *g*, 5 min, 4 °C).
10. Discard supernatants completely and suspend the cell pellets in 100 mM Tris–HCl, pH 7.75, 4 mM EDTA, with adjustment of the optical density at 600 nm (OD_{600}) of the cell suspension to 1.0.
11. Boil 300 μL of the cell suspensions for 2 min at 100 °C and then centrifuge (8000 × *g*, 5 min, 4 °C).
12. Transfer 100 μL of each supernatant to a 96-well microplate that is kept on ice until measurement.
13. Inject 100 μL luciferase reagent to each well and then read bioluminescence by a microplate reader at 20 °C (Fig. 3b).

3.4 Measurements of FliI ATPase Activity

1. Create a standard curve of phosphate with various concentrations using freshly prepared MGAM reagent.
2. Mix 90 μL of 10× reaction buffer, 90 μL of 5 mg/mL BSA, 36 μL of 100 mM ATP and 669 μL of Milli-Q water in a 1.5 mL Eppendorf tube.

3. Add 15 μL of FliI with various concentrations and vortex (*see* **Note 9**).
4. Incubate at 37 °C.
5. Take 100 μL of the reaction mixture at various time points (0, 5, 10, 20, 40, 60, 90, and 120 min) and then transfer to the 1.5 mL Eppendorf tube containing 800 μL of MGAM reagent.
6. Vortex well.
7. Leave for 1 min at room temperature.
8. Add 100 μL of 34% citric acid to the tube to stop color development and then leave for 20 min at room temperature.
9. Measure OD_{660} of each reaction mixture using a spectrophotometer (Fig. 4a).

3.5 In Vitro Reconstitution of the $FliI_6$ Ring

1. Prepare 30 μL of a reaction mixture of 35 mM Tris–HCl, pH 8.0, 113 mM NaCl, 1 mM DTT, 5 mM ADP, 5 mM $AlCl_3$, 5 mM NaF, 5 mM $MgCl_2$, and 100 μg/mL acidic phospholipids in a 1.5 mL Eppendorf tube (*see* **Note 10**).
2. Add 70 μL of freshly purified FliI sample at a final concentration of 1 μM.
3. Incubate at 37 °C for a few minutes.

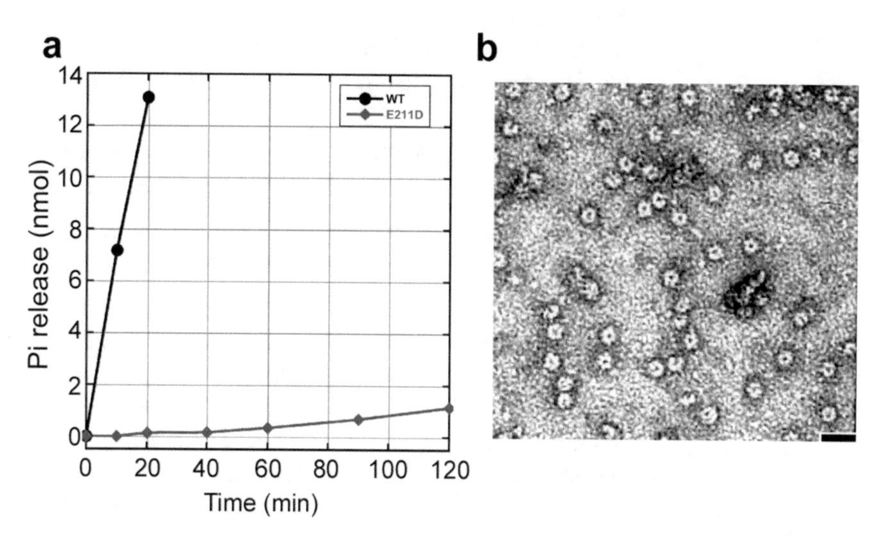

Fig. 4 FliI ATPase activity and its hexameric ring formation. (**a**) Measurements of the ATPase activity of wild-type FliI (WT) and its mutant variant (E211D) using malachite green assay. The carboxyl group of Glu-211 in FliI polarizes a water molecule for the nucleophilic attack to the γ-phosphate of ATP [23] and hence the E211D substitution reduces the ATPase activity of FliI by about 100-fold [15]. (**b**) Electron micrograph of negatively stained samples of purified FliI preincubated with 5 mM $MgCl_2$, 5 mM ADP, 5 mM $AlCl_3$, 5 mM NaF, and 100 μg/mL of acidic phospholipids. Electron micrograph was recorded at a magnification of X 50,000. The scale bar represents 200 Å

4. Apply the reaction mixture to a carbon-coated copper grid and then stain the sample with 2% (w/v) uranyl acetate.
5. Observe negative stained samples of FliI by an electron microscope operated at 100 kV (Fig. 4b).

4 Notes

1. The MMHI0117 [*ΔfliH-fliI flhB(P28T)*] strain has been isolated as a gain-of-function mutant from a *Salmonella ΔfliH-fliI* mutant. The FlhB(P28T) bypass mutation increases the export efficiency of flagellar axial proteins to considerable degrees, allowing the bypass mutant cells to form a couple of flagella in the absence of FliH and FliI [13].
2. The pTrc99AFF4 vector is a modified form of pTrc99A in which the *Nde*I site within the vector is removed and the *Nco*I site in the multiple cloning sites is replaced by the *Nde*I site [21]. A fragment containing *Nde*I and *Bam*HI restriction sites is generated by PCR with the chromosomal DNA prepared from SJW1103 as a template. The purified DNA fragment is digested with *Nde*I and *Bam*HI and then is inserted into the *Nde*I and *Bam*HI sites of pTrc99AFF4.
3. For rapid and efficient purification of FliI, a Histidine tag derived from the pET19b vector (Novagen) is attached to the N-terminus of FliI. His-FliI is purified by Ni-NTA affinity chromatography as described [23, 24]. The ATPase activity of FliI is significantly decreased in several days during storage at 4 °C.
4. Freshly prepared MGAM reagent must be passed through a Whatman Grade 597 Qualitative Filter Paper prior to its usage.
5. The *Salmonella* JR501 strain is used for conversion of *E. coli*-derived plasmids to compatibility with *Salmonella* [25].
6. SJW1353 (*flgE*) is used as an example of a strain blocked in hook assembly and therefore the flagellar type III export apparatus remains in the rod- and hook-type specificity state. SJW2177 (*flgK*) is used as an example of a strain where hook assembly has been completed and therefore the switch to the filament-type specificity has occurred.
7. The cell growth rate decreases when CCCP concentration increases and 25 μM CCCP treatment immediately results in the growth arrest [13]. Since the *Salmonella* flagellar motor rotation is driven by PMF across the cytoplasmic membrane [3, 26], free-swimming motility of *Salmonella* cells must be observed under an optical microscope to check whether the PMF is significantly reduced.

8. PMF consists of two components: the electric potential difference ($\Delta\psi$) and the proton concentration difference (ΔpH). $\Delta\psi$ alone is sufficient for flagellar protein export but the export gate alone, in the absence of FliH and FliI, requires the ΔpH component of PMF in addition to $\Delta\psi$ [16]. The intracellular pH is maintained at around 7.5 in *Salmonella* cells and so an external pH of 7.5 diminishes the ΔpH component of the PMF [17].
9. FliI ATPase shows the positive cooperativity in its ATPase activity and hence the FliI ATPase activity is stimulated by an increase in the protein concentration [27, 28].
10. FliI forms a homo-hexamer in the presence of a non-hydrolyzable ATP analog, Mg^{2+}-ADP-AlF_4, much more efficiently than in the presence of Mg^{2+}-ATP because ATP binding induces FliI hexamerization and the release of ADP and Pi destabilizes the ring structure in vitro [23, 27, 28].

Acknowledgments

This research has been supported in part by JSPS KAKENHI Grant Numbers JP26293097 (to T.M.) and JP25000013 (to K.N.) and MEXT KAKENHI Grant Numbers JP25121718 and JP15H01640 (to T.M.).

References

1. Macnab RM (2003) How bacteria assemble flagella. Annu Rev Microbiol 57:77–100
2. Minamino T, Imada K, Namba K (2008) Mechanisms of type III protein export for bacterial flagellar assembly. Mol Biosyst 4:1105–1115
3. Minamino T, Imada K, Namba K (2008) Molecular motors of the bacterial flagella. Curr Opin Struct Biol 18:693–701
4. Minamino T (2014) Protein export through the bacterial flagellar type III export pathway. Biochim Biophys Acta 1843:1642–1648
5. Minamino T, Macnab RM (1999) Components of the *Salmonella* flagellar export apparatus and classification of export substrates. J Bacteriol 181:1388–1394
6. Minamino T, Doi H, Kutsukak K (1999) Substrate specificity switching of the flagellum-specific export apparatus during flagellar morphogenesis in Salmonella typhimurium. Biosci Biotechnol Biochem 63:1301–1303
7. Hirano T, Minamino T, Namba K, Macnab RM (2003) Substrate specificity class and the recognition signal for *Salmonella* type III flagellar export. J Bacteriol 185:2485–2492
8. Kutsukake K, Minamino T, Yokoseki T (1994) Isolation and characterization of FliK-independent flagellation mutants from *Salmonella typhimurium*. J Bacteriol 176: 7625–7629
9. Williams AW, Yamaguchi S, Togashi F, Aizawa S, Kawagishi I, Macnab RM (1996) Mutations in *fliK* and *flhB* affecting flagellar hook and filament assembly in *Salmonella typhimurium*. J Bacteriol 178:2960–2970
10. Hirano T, Mizuno S, Aizawa S, Hughes KT (2009) Mutations in Flk, FlgG, FlhA, and FlhE that affect the flagellar type III secretion specificity switch in Salmonella enterica. J Bacteriol 181:3938–3949
11. Bange G, Kümmerer N, Engel C, Bozkurt G, Wild K, Sinning I (2010) FlhA provides the adaptor for coordinated delivery of late flagella building blocks to the type III secretion system. Proc Natl Acad Sci U S A 107: 11295–11300
12. Kinoshita M, Hara N, Imada K, Namba K, Minamino T (2013) Interactions of bacterial chaperone-substrate complexes with FlhA con-

tribute to co-ordinating assembly of the flagellar filament. Mol Microbiol 90:1249–1261
13. Minamino T, Namba K (2008) Distinct roles of the FliI ATPase and proton motive force in bacterial flagellar protein export. Nature 451:485–488
14. Paul K, Erhardt M, Hirano T, Blair DF, Hughes KT (2008) Energy source of flagellar type III secretion. Nature 451:489–492
15. Minamino T, Morimoto YV, Kinoshita M, Aldridge PD, Namba K (2014) The bacterial flagellar protein export apparatus processively transports flagellar proteins even with extremely infrequent ATP hydrolysis. Sci Rep 4:7579
16. Minamino T, Morimoto YV, Hara N, Namba K (2011) An energy transduction mechanism used in bacterial type III protein export. Nat Commun 2:475
17. Minamino T, Morimoto YV, Hara N, Aldridge PD, Namba K (2016) The bacterial flagellar type III export gate complex is a dual fuel engine that can use both H^+ and Na^+ for flagellar protein export. PLoS Pathog 12:e1005495
18. Yamaguchi S, Fujita H, Sugata K, Taira T, Iino T (1984) Genetic analysis of H2, the structural gene for phase-2 flagellin in *Salmonella*. J Gen Microbiol 130:255–265
19. Ohnishi K, Ohto Y, Aizawa S, Macnab RM, Iino T (1994) FlgD is a scaffolding protein needed for flagellar hook assembly in *Salmonella typhimurium*. J Bacteriol 176: 2272–2281
20. Homma M, Fujita H, Yamaguchi S, Iino T (1984) Excretion of unassembled flagellin by *Salmonella typhimurium* mutants deficient in the hook-associated proteins. J Bacteriol 159:1056–1059
21. Ohnishi K, Fan F, Schoenhals GJ, Kihara M, Macnab RM (1997) The FliO, FliP, FliQ, and FliR proteins of *Salmonella typhimurium*: putative components for flagellar assembly. J Bacteriol 179:6092–6099
22. Minamino T, Yamaguchi S, Macnab RM (2000) Interaction between FliE and FlgB, a proximal rod component of the flagellar basal body of *Salmonella*. J Bacteriol 182:3029–3036
23. Kazetani K, Minamino T, Miyata T, Kato T, Namba K (2009) ATP-induced FliI hexamerization facilitates bacterial flagellar protein export. Biochem Biophys Res Commun 388:323–327
24. Fan F, Macnab RM (1996) Enzymatic characterization of FliI: an ATPase involved in flagellar assembly in *Salmonella typhimurium*. J Biol Chem 271:31981–31988
25. Ryu J, Hartin RJ (1990) Quick transformation in *Salmonella typhimurium* LT2. Biotechniques 8:43–45
26. Minamino T, Imada K (2015) The bacterial flagellar motor and its structural diversity. Trends Microbiol 23:267–274
27. Claret L, Calder SR, Higgins M, Hughes C (2003) Oligomerisation and activation of the FliI ATPase central to bacterial flagellum assembly. Mol Microbiol 48:1349–1355
28. Minamino T, Kazetani K, Tahara A, Suzuki H, Furukawa Y, Kihara M, Namba K (2006) Oligomerization of the bacterial flagellar ATPase FliI is controlled by its extreme N-terminal region. J Mol Biol 360:510–519

Chapter 2

Interactions of Flagellar Structural Subunits with the Membrane Export Machinery

Lewis D.B. Evans, Paul M. Bergen, Owain J. Bryant, and Gillian M. Fraser

Abstract

During assembly of the bacterial flagellum, structural subunits synthesized inside the cell must be exported across the cytoplasmic membrane before they can crystallize into the nascent flagellar structure. This export process is facilitated by a specialized Flagellar Type III Secretion System (fT3SS) located at the base of each flagellum. Here, we describe three methods—isothermal titration calorimetry, photo-crosslinking using unnatural amino acids, and a subunit capture assay—used to investigate the interactions of flagellar structural subunits with the membrane export machinery component FlhB.

Key words *Salmonella*, Flagella, Flagellar type III secretion system (fT3SS), Protein export, Protein-protein interactions, Isothermal titration calorimetry (ITC), *p*-benzoyl-L-phenylalanine (p*Bpa*) photo-crosslinking, Capture assay

1 Introduction

1.1 Background

A striking feature of the bacterial flagellum is that it self-assembles, aided by a dedicated Type III export machinery located at each flagellum base. The export machinery unfolds nascent structural subunits and delivers them across the cell membrane into a central 2 nm diameter channel that runs the entire length of the growing flagellum [1]. The unfolded subunits must then transit through this channel in the external flagellum to its tip where they crystallize beneath cap foldases [2]. In this way, the flagellar rod that spans the cell envelope is built first, followed by assembly of the hook and then the filament, both of which extend from the cell surface.

Structural subunits are thought to dock initially at the FliI export ATPase [3, 4]. Subunits of the rod and hook go on to bind a surface exposed hydrophobic pocket on the cytosolic domain of the FlhB export gate ($FlhB_C$, residues 212–383) via a conserved gate recognition motif (GRM sequence FxxxΦ, where Φ is any hydrophobic residue) near the subunit N-terminus [5]. Subunits

Tohru Minamino and Keiichi Namba (eds.), *The Bacterial Flagellum: Methods and Protocols*, Methods in Molecular Biology, vol. 1593, DOI 10.1007/978-1-4939-6927-2_2, © Springer Science+Business Media LLC 2017

then move across the cytoplasmic membrane into the central channel of the external flagellum, where there is no conventional biological energy source, and transit to the assembly site at the distal tip.

There are a number of alternative theoretical models for the mechanism of subunit transit through the flagellar central channel [6–8], however, there is experimental evidence to suggest that the energy for transit is intrinsic to the unfolded subunits themselves [5]. It is proposed that transit could be achieved by linking of the subunit docked at the FlhB export gate to the free C-terminus of the preceding subunit that has already partly crossed the membrane into the channel [5]. The juxtaposed N- and C-terminal helices of successive subunits link as parallel coiled-coils and the newly linked subunit is then pulled from the export gate into the channel by the thermal motion of the unfolded subunit chain, which is anchored at its other end to the tip of the growing flagellum. Repeated crystallization of subunits at the tip causes the chain to shorten, stretch, and exert an increasing force on the next subunit docked at the export gate, eventually pulling this subunit into the channel. In this way, linking of consecutive subunits at the membrane export machinery is coupled to subunit folding at the flagellum tip to produce directional subunit transit [5].

1.2 Overview of Methods

Here, we describe the experimental techniques we used to investigate the mechanisms underlying the transfer of nascent flagellar subunits from the cytosol into the growing flagellum [5]. Specifically, interactions between the export gate component FlhB and subunits were analyzed using (1) Isothermal Titration Calorimetry (ITC) [9] to determine the binding affinities of subunits for $FlhB_C$ (Subheading 3.7), (2) a photo-crosslinking assay based on the incorporation of unnatural amino acids [10] to identify subunit residues that directly bind $FlhB_C$ (Subheading 3.8), and (3) a capture assay to examine linking and capture of $FlhB_C$-docked-subunit by free subunit (Subheading 3.9). These techniques rely on a common set of molecular microbiology procedures to maintain bacterial strains and plasmids (Subheading 3.1), construct recombinant plasmids harboring flagellar genes (Subheading 3.2), express recombinant flagellar proteins (Subheading 3.3), generate clarified bacterial cell lysates (Subheading 3.4), and purify proteins and/or protein complexes (Subheadings 3.5 and 3.6).

The biophysical technique ITC was used to measure the thermodynamics of (GST)$FlhB_C$ binding to the hook cap subunit FlgD in solution (Subheading 3.7), generating quantitative data on the binding constant (K_D = 39 μM), stoichiometry (1:1), and enthalpy of binding (ΔH_b -1×10^3). Recombinant FlgD subunits in which the GRM (FlgD residues 36–40) was intact (wild-type) or deleted (gate-blind subunit, FlgDΔ36–40) were engineered to introduce a C-terminal hexa-histidine tag to enable purification using Ni^{2+}-affinity chromatography. An N-terminal translational fusion of Glutathione S-transferase (GST) to $FlhB_C$ and a GST control were purified using Glutathione affinity chromatography with Glutathione Sepharose resin (GSH-resin). Exhaustive dialysis of purified flagellar proteins into analysis buffer was carried out prior to performing ITC to prevent heats of dilution (ΔH_d; background heats) masking observations.

Direct interactions between the subunit FlgD GRM (residues 36–40) and $FlhB_C$ were identified using photo-cross-linkable unnatural amino acids (Uaa; Subheading 3.8) incorporated at specific sites in FlgD, either in the GRM (L_{39}Uaa or L_{40}Uaa) or elsewhere (L_5Uaa). This was achieved by mutating the corresponding codons in *flgD* to the amber codon (UAG). In the presence of an orthogonal tRNA and recombinant mutated aminoacyl-tRNA synthetase, the photo-cross-linkable Uaa *p*-benzoyl-L-phenylalanine (p*Bpa*) was site-specifically incorporated into the polypeptide chain at the position encoded by the amber codon. Binding of the subunit GRM to $FlhB_C$ was demonstrated by covalent linkage of FlgD to (GST)$FlhB_C$ via the engineered ultraviolet (UV)-activated site-specific crosslinking residues. Soluble extracts of *E. coli* individually expressing FlgD wild-type or photo-cross-linkable derivatives were incubated with purified (GST)$FlhB_C$, exposed to UV light, separated by sodium dodecyl sulfate-polyacrylamide gel electrophoresis (SDS-PAGE), and analyzed by immunoblotting using anti-FlgD sera to identify cross-linked protein complexes. In addition to assessing subunit binding to (GST)$FlhB_C$, our study also used a GST fusion to the autocleavage-defective $FlhB_CN_{269}A$ [11] to enhance the shift in the migration of the cross-linked protein adducts when analyzed by SDS-PAGE. Binding of FlgD to (GST) $FlhB_C$ and GST-$FlhB_CN_{269}A$ was shown to be comparable [5].

The N-terminus of a subunit docked at the $FlhB_C$ export gate can be captured by the free C-terminus of the preceding subunit already in the channel. It is thought that the docked subunit can be pulled into the central channel by the entropic force generated by the thermal motion of the chain of unfolded subunits in the channel. To characterize the interactions between $FlhB_C$-docked subunits and free subunits, and to demonstrate subunit release from $FlhB_C$, we developed an in vitro capture assay (Subheading 3.9). In this assay, subunits docked at the $FlhB_C$ export gate are captured and released from $FlhB_C$ by free gate-blind subunits, which are deleted for the GRM and cannot interact with $FlhB_C$ [5]. Below, we describe the protocol used to show that the flagellar hook subunit FlgE can, in a concentration-dependent manner, capture and release FlgD subunit from a preformed (GST)$FlhB_C$-FlgD complex.

2 Materials

Prepare all solutions using sterile Milli-Q water.

2.1 Materials for the Maintenance of Bacterial Strains and Plasmids

1. Bacterial strains for molecular biology and protein expression (Table 1).
2. 37 °C static incubator for culture plates.
3. 37 °C shaking incubator for liquid culture tubes and flasks.
4. Luria-Bertani (LB) broth.
5. Plastic tubes with lids (e.g., Falcon©), 5 mL, 14 mL, and 50 mL for bacterial cell culture.
6. Conical flasks for bacterial cell culture.
7. Luria-Bertani (LB) 1.5% agar plates.
8. Spectrophotometer to measure absorbance from ultraviolet to visible light or densitometer to measure bacterial culture density.
9. Floor-standing centrifuge with rotors (e.g., Avanti, Beckman Coulter) and thick-wall polypropylene tubes (e.g., 25 × 89 mm Beckman Coulter) to collect bacterial cell pellets.
10. Plastic microcentrifuge tubes, 0.5 mL or 1.5 mL for storage of electrocompetent cells.
11. Ice-cold sterile Milli-Q water.
12. 10% (v/v) glycerol.
13. Liquid nitrogen in a Dewar flask.
14. Ultra-low temperature (−80 °C) freezer.
15. Ice.
16. Electroporation apparatus (e.g., Bio Rad Gene Pulser™ set at 12.5 kVcm^{-1} field strength, 25 μFD and 200 Ω at 2.5 kV).
17. Electroporation cuvette, 0.1 cm gap.

Table 1
Bacterial strains

Bacterial strain	Description	Source
Salmonella enterica serovar Typhimurium SJW1103	Wild-type	Yamaguchi et al. [12]
Escherichia coli DH5α	F$^-$*endA1 glnV44 thi-1 recA1 relA1 gyrA96 deoR nupG purB20* φ80d*lacZ*ΔM15 Δ(*lacZYA-argF*)U169, *hsd*R17(*rK*$^-$*mK*$^+$), λ$^-$	Hanahan [13]
Escherichia coli C41 (DE3)	F$^-$*ompT gal dcm hsdS*$_B$(r_B^- m_B^-) DE3	Miroux and Walker [14]

18. Super Optimal broth with Catabolite repression (SOC) medium: 2% (w/v) tryptone, 0.5% (w/v) yeast extract, 10 mM NaCl, 2.5 mM KCl, 10 mM $MgCl_2$, 20 mM Glucose (add filter-sterilized $MgCl_2$ and glucose after sterilization of broth at 121 °C).
19. Antibiotics for plasmid selection (Table 2): ampicillin (100 μg/mL), chloramphenicol (25 μg/mL), tetracycline (12.5 μg/mL).

2.2 Materials for the Construction of Recombinant Plasmids Harbouring Flagellar Genes

1. Oligonucleotide primers (Table 3) for amplification of flagellar genes using the polymerase chain reaction (PCR).
2. High-fidelity DNA polymerase, template genomic DNA, deoxynucleotide triphosphates (dNTPs), and buffers for PCR.
3. Thermal cycler.
4. Equipment for agarose gel electrophoresis (separation of nucleic acids).
5. Plasmid vectors (Table 2).
6. Enzymes (restriction endonucleases, DNA ligase) for the construction of recombinant plasmids.

2.3 Materials for the Expression of Recombinant Flagellar Proteins

1. 2× Tryptone-Yeast Extract (2× TY) broth medium.
2. Two-liter conical flasks for bacterial cell culture.
3. Isopropyl β-D-1-thiogalactopyranoside (IPTG), 1 M stock solution.
4. 20% (w/v) L-arabinose.
5. *p*-benzoyl-L-phenylalanine (*p*Bpa).
6. Sodium hydroxide (1 M).
7. Equipment for sodium dodecyl sulfate-polyacrylamide gel electrophoresis (SDS-PAGE).
8. SDS-polyacrylamide gels for SDS-PAGE (10%, 12.5%, or 15% acrylamide).

Table 2
Plasmids

Plasmid (antibiotic resistance)	Origin of replication	Restriction sites	Transcription promoter	Source
pGEX-4T-3 (Amp^R)	pBR322	BamHI, XhoI	T7	Kaelin et al. [16]
pDULE (Tet^R)	p15A	n/a	lpp	Farrell et al. [10]
pBAD18 (Amp^R)	pBR322	XbaI, HindIII/SalI	P_{BAD}	Guzman et al.[17]
pET20b (Amp^R)	pBR322	NdeI, HindIII	T7	Studier and Moffatt [18]
pACT7 (Cm^R)	p15A	NdeI, BamHI	T7	Evans et al. [19]

Table 3
Primers for amplification of recombinant *Salmonella* flagellar genes by PCR

Gene	Description	Primer sequence (5′ to 3′)
flhBC	BamHI site; (GST) fusion; residue 219; forward; for the construction of pGEX-4 T-3-*flhBC*	CCGCGTGGATCCGTGGCAGAAGAGAGCGACGACGA
flhBC	Residue 383 Stop; XhoI site; reverse; for the construction of pGEX-4 T-3-*flhBC*	GTCAGCCTCGAGTTAGCCATCAGTATTCTT
flhB	$N_{269}A$; forward; for the construction of pGEX-4 T-3- *flhBC*$N_{269}A$	CGGACGTCATTGTCACTGCCCCGACGCACT
flhB	$N_{269}A$; reverse; for the construction of pGEX-4 T-3- *flhBC*$N_{269}A$	GAATAGTGCGTCGGGGCAGTGACAATGACGT
flgD	XbaI site; RBS, residue 1; forward for the construction of pBAD18 flgD derivatives	GCTCCTTCTAGAAGGAGAGCCCAAATGTCTATTGCCGT AAATATGAATG
flgD	Residue 232 stop; HindIII; reverse for the construction of pBAD18 *flgD* derivatives	GCATGCAAGCTTGATTATTTGCCGAACT TCGTCGAGTGT
flgD	Deletion of GRM residues 36–40; forward for the construction of "gate-blind" flgD derivatives	GATCTGCAAAGCAGTGTCGCGCAATTGAA
flgD	Deletion of GRM residues 36–40; reverse for the construction of "gate-blind" flgD derivatives	CTTCAATTGCGCGACACTGCTTTGCAGATC
flgD	V_5Uaa; forward for the construction of amber codon containing *flgD* derivative	GCTCCTCTCTAGGAGATATACCATGTCTATTGCCTAGA ATATGAATGACCCGA
flgD	L_{39}Uaa; forward for the construction of amber codon containing *flgD* derivative	GCAGTTTCCTGACCTAGCTGGTCGCGCAATTGA
flgD	L_{39}Uaa; reverse for the construction of amber codon containing *flgD* derivative	CAATTGCGCGACCAGCTAGGTCAGGAAACTGCT

(continued)

Table 3 (continued)

Gene	Description	Primer sequence (5′ to 3′)
flgD	L_{40}Uaa; forward for the construction of amber codon containing *flgD* derivative	CAGTTTCCTGACCTTATAGGTCGCGCAATTGAA
flgD	L_{40}Uaa; forward for the construction of amber codon containing *flgD* derivative	CAATTGCGCGACCAGCTAGGTCAGGAAACTGCT
flgD	NdeI site; residue 11; forward for the construction of N-terminal truncate of FlgD	CGCCTGCATATGACCAACACGGGCGTCAAAACGACG ACCGGCA
flgD	FlgD residue 232; FLAG3; stop; SalI site; reverse for the construction of full-length *flgD* with C-terminal FLAG × 3 tag	CGTAGTGTCGACTTACTTGTCATCGTCATCTTTATAAT CAATATCATGATCTTTATAGTCGCCGTCATGATCTTT ATAATCGATTATTTGCCGAACTTCGTCGAGTGTGG
Flag tag	FLAG × 3; stop; BamHI site; reverse for the construction of C-terminally tagged flagellar gene derivatives	CGTAGTGGATCCTTACTAGTCATCGTCATCTTTAT
flgE	NdeI site; Residue 1; forward	CGTAGTCATATGTCTTTTTCTCAAGCGGTTAGC
flgE	HindIII site; 403, no stop; reverse for the construction of full-length *flgE* C-terminally FLAG × 3 tagged	GCTATCCCGTCAAGCTTGCGCAGGTTA
flgE	HindIII site; 359, no stop; reverse for the construction of C-terminally truncated and FLAG × 3 tagged *flgE*	CGTAGTAAGCTTGCCGTTCGTCAGCTTACCGAAGTT
flgE	Deletion of GRM residues 39–43; forward for the construction of "gate-blind" *flgE* derivative	GTCCGGTACGGCATCAGCCGGTTCCAAAGTGGGGCT
flgE	Deletion of GRM residues 39–43; reverse for the construction of "gate-blind" *flgE* derivative	CACTTTGGAACCGGCTGATGCCGTACCGGACTTAAA

Table 4
Buffers for purification of recombinant proteins from *E. coli*

Protein	Resin	Lysis buffer	Wash buffer	Elution buffer
His-tagged proteins	Nickel	50 mM Tris–HCl, pH 7.4 400 mM NaCl, 10 mM imidazole Protease inhibitor (one tablet in 20 mL of lysis buffer) DNase I (10 μg/mL)	50 mM Tris–HCl, pH 7.4 400 mM NaCl 10 mM imidazole	50 mM Tris–HCl, pH 7.4 200 mM NaCl 600 mM imidazole
GST-fusion proteins	GSH	50 mM Tris–HCl, pH 7.4 200 mM NaCl Protease inhibitor (one tablet in 20 mL of lysis buffer) DNase I (10 μg/mL)	50 mM Tris–HCl, pH 7.4 200 mM NaCl	50 mM Tris–HCl, pH 7.4 200 mM NaCl 10 mM glutathione
Untagged proteins	Q HP	50 mM sodium phosphate buffer, pH 7.0–7.5 5–50 mM NaCl Protease inhibitor (one tablet in 20 mL of lysis buffer) DNase I (10 μg/mL)	50 mM sodium phosphate buffer, pH 7.0–7.5 5–50 mM NaCl	50 mM sodium phosphate buffer, pH 7.0–7.5 1 M NaCl

9. SDS-PAGE running buffer: 25 mM Tris base, 192 mM glycine, 0.1% SDS.
10. SDS-PAGE sample buffer: 100 mM Tris–HCl, pH 6.8, 4% (w/v) SDS, 0.2% (w/v) bromophenol blue, 20% (v/v) glycerol, 200 mM dithiothreitol (DTT).
11. Prestained protein molecular weight marker for SDS-PAGE.
12. Coomassie Brilliant Blue G-250 stain: 0.1% (w/v) Coomassie Brilliant Blue G-250, 50% (v/v) methanol, 10% (v/v) glacial acetic acid.

2.4 Materials for the Production of Clarified Bacterial Cell Lysates

1. Cells lysis buffers (Table 4) for purification of recombinant proteins from *E. coli*.
2. Plastic tubes with lids (e.g., Falcon©), 5 mL, 14 mL, and 50 mL.
3. Cell disruptor or French© Pressure Cell.
4. cOmplete© protease inhibitor cocktail, EDTA-free (Roche).
5. DNase I (Thermo Scientific).

2.5 Materials for the Purification of Recombinant Proteins and Recombinant Protein Complexes

1. Buffers for purification of recombinant proteins (Table 4) and recombinant subunit/export gate complexes (Table 5) from *E. coli*.
2. Plastic tubes with lids (e.g., Falcon©), 5 mL, 14 mL, and 50 mL.
3. Ni^{2+}-Sepharose HisTrap© Excel column (GE Healthcare).
4. Ni^{2+}-nitrilotriacetic acid (NTA) resin.

Table 5
Buffers for purification of recombinant subunit/export gate complexes

Buffer	Components
Lysis buffer	50 mM sodium phosphate, pH 7.4, 150 mM NaCl, 1 mM β-mercaptoethanol, protease inhibitor, DNase 1
Wash buffer	50 mM sodium phosphate, pH 7.4, 150 mM NaCl, 1 mM β-mercaptoethanol
Elution buffer	50 mM sodium phosphate, pH 7.4, 150 mM NaCl, 1 mM β-mercaptoethanol, 10 mM reduced glutathione

5. Glutathione Sepharose 4B resin (GSH-resin; GE Healthcare).
6. Anion exchange resin (Q HP; GE Healthcare).
7. Bench-top tube rotator.
8. Fast protein liquid chromatography (FPLC) system.
9. Peristaltic pump.
10. Plastic beaker, 5 L.
11. Dialysis tubing, 6–8 kDa, 10 kDa, and 25 kDa molecular weight cutoff (MWCO).
12. Pasteur pipettes.

2.6 Additional Materials for ITC

1. Vacuum pump for degassing solutions.
2. ITC assay buffer: 50 mM Tris–HCl pH 7.4, 100 mM NaCl.
3. Spin concentrators, 3.5–5 kDa membrane cutoff.
4. DC™ Protein Assay (Bio-Rad).
5. Heating block.
6. Degassed sterile Milli-Q water for cleaning calorimeter.
7. VP-Isothermal Titration Calorimeter (MicroCal) with loading needle.
8. Plastic syringes, 2 mL, 5 mL, and 10 mL.
9. Computer with Origin® analysis software.

2.7 Additional Materials for Photo-Crosslinking

1. Phosphate buffered saline (PBS).
2. Polystyrene 24-well plate.
3. Shallow box.
4. Lamp that emits 350 nm light and cooling fan.

5. Trichloroacetic acid (TCA), 100%.
6. Acetone, 4 °C.
7. Bench top microcentrifuge.
8. Wet blotting transfer equipment (e.g., Hoefer TE22).
9. Nitrocellulose membrane for immunoblotting.
10. Blotting transfer buffer: 10 mM CAPS pH 11.0, 10% (v/v) methanol.
11. Phosphate buffered saline (PBS), pH 7.4: 137 mM NaCl, 2.7 mM KCl, 10 mM Na_2HPO_4, 1.8 mM KH_2PO_4.
12. PBS-Triton: 137 mM NaCl, 2.7 mM KCl, 10 mM Na_2HPO_4, 1.8 mM KH_2PO_4, 0.05% (v/v) Triton X100.
13. Immunoblotting blocking buffer: PBS-Triton, 5% (w/v) skimmed milk powder.
14. Specific antisera (rabbit) raised against purified *Salmonella* FlhB or FlgD, and commercial antibodies raised against glutathione S-transferase (GST).
15. IRDye-conjugated goat-anti-rabbit secondary antibodies (Licor).
16. Licor Odyssey® CLx imaging system.

2.8 Additional Materials for Capture Assay

1. Ni^{2+}-nitrilotriacetic acid (NTA) Resin (Qiagen).
2. Bovine serum albumin (BSA).
3. Specific antisera (rabbit) raised against purified *Salmonella* FlhB or FlgE, and commercial antibodies raised against the FLAG epitope.

3 Methods

3.1 Maintenance of Bacterial Strains and Plasmids

1. Grow bacterial strains *Salmonella enterica* serovar Typhimurium SJW1103 [12], *Escherichia coli* DH5α [13], and *Escherichia coli* C41 [14] (Table 1) at 37 °C in Luria-Bertani (LB) broth with vigorous shaking (200 rpm) or on LB agar (1.5%) plates.
2. To prepare electrocompetent bacteria cells [15] for plasmid propagation and maintenance (*E. coli* DH5α) and for protein expression (*E. coli* C41), inoculate LB broth (500 mL) using an overnight bacterial broth culture to a starting density of approximately A_{600} 0.005 and grow at 37 °C, with shaking (200 rpm), to mid-exponential phase (A_{600} 0.6–0.8). Carry out all subsequent steps at 4 °C or on ice. Harvest cells by centrifugation (10 min, 5000 × *g*). Resuspend cells in 200 mL ice-cold sterile Milli-Q water, and repeat harvest and wash steps twice. After final harvest, resuspend cells in 100 mL ice-cold 10% (v/v) glycerol and pellet cells by centrifugation (10 min, 5000 × *g*). Finally, resuspend cells in 0.5 mL 10% (v/v) glycerol

and snap freeze 50 μL aliquots of the cell suspension in plastic microcentrifuge tubes (e.g., 1.5 mL tubes) by immersing in liquid nitrogen in a Dewar flask. Store electrocompetent cells at −80 °C.

3. Isolate and purify plasmids from *E. coli* DH5α (or comparable strain, deleted for *endA* and *recA1*) using standard laboratory techniques [15].
4. Transform *E. coli* with plasmids harboring recombinant flagellar genes by mixing 0.5 μg of plasmid DNA with 50 μL electrocompetent cells that have been thawed on ice. Transfer mixture to a chilled electroporation cuvette (0.1 cm gap) and electroporate the bacterial cells using a Gene Pulser™ (Bio Rad) set at 12.5 $kVcm^{-1}$ field strength, 25 μFD and 200 Ω at 2.5 kV. Immediately add 1 mL SOC medium to the electroporation cuvette, resuspend cells, and transfer to a 5 mL plastic tube with lid. Incubate transformed cells at 37 °C for 1 h with shaking (200 rpm). Select cells carrying appropriate plasmid(s) by plating onto LB agar (1.5%) plates containing, where appropriate (Table 2), ampicillin (100 μg/mL), tetracycline (12.5 μg/mL), and/or chloramphenicol (25 μg/mL).

3.2 Generation of Recombinant Plasmids Harbouring Flagellar Genes

1. Isolate genomic DNA from *Salmonella enterica* serovar Typhimurium SJW1103 (*Salmonella*) using standard laboratory techniques [15].
2. To construct recombinant plasmids, amplify *Salmonella* flagellar genes (e.g., *flhB*, *flgD*, or *flgE*) using the polymerase chain reaction (PCR; [15]) with specific oligonucleotide primers (Table 3), high fidelity DNA polymerase, dNTPs, *Salmonella* genomic DNA as template and recommended buffer conditions in a thermal cycler. Use overlap-extension PCR [15] to introduce site-specific point mutations (e.g., in the gene-encoding $FlhBN_{269}A$), the amber codon UAG for photo-crosslinking (*see* **Note 1** on mutagenic primer design) or deletions (e.g., in the gene-encoding FlgDΔ36–40). Perform restriction digestions and ligations using standard laboratory techniques [15].
3. Validate recombinant plasmids by DNA sequencing.

3.3 Protein Production in E. coli Cells Carrying Expression Plasmids Harbouring Flagellar Genes

1. Following transformation of *E. coli* C41 with appropriate expression plasmid(s) (Subheading 3.1, **step 4**), select fresh colonies to inoculate 50 mL 2TY medium containing appropriate antibiotics (Table 2) and grow at 37 °C with shaking (200 rpm) to late stationary phase ($A_{600} > 2.0$).
2. For plasmids harboring *flgD* derivatives with amber codons for incorporation of photo-cross-linkable p*Bpa*, transform into competent *E. coli* cells that already carry pDULE.

3. Pre-warm (30 °C) 500 mL 2TY medium containing appropriate antibiotics (Table 2) in a 2 L conical flask.
4. Inoculate the pre-warmed 2TY with the late stationary phase ($A_{600} > 2.0$) *E. coli* C41 culture (**step 1**) to a starting density of A_{600} 0.005.
5. Grow cells to mid-exponential phase (A_{600} 0.6–0.8) and induce protein expression using 0.8 mM IPTG for pGEX-4 T3 and pDULE or 0.2% (w/v) L-arabinose for pBAD18 (Table 2).
6. For expression of photo-cross-linkable FlgD derivatives, dissolve 108 mg of *p*-benzoyl-L-phenylalanine (*p*Bpa) in 440 μL of 1 M sodium hydroxide, then add to 400 mL of pre-warmed 2TY medium to give a final concentration of 1 mM *p*Bpa (*see* **Note 2**).
7. Incubate expression cultures at 30–37 °C for 4–6 h, with shaking (200 rpm). Harvest cells by centrifugation at 8000 × *g* for 10 min. Store bacterial cell pellets at −80 °C.
8. Assess protein expression by analyzing preinduced and postinduced whole cell samples by SDS-(15%)PAGE followed by staining with Coomassie Brilliant Blue G-250 stain.

3.4 Preparation of Cleared Cell Lysates

Prepare cleared lysates for purification of proteins and/or protein complexes (Subheadings 3.5 and 3.6) or to use directly in the capture assay (Subheadings 3.9). Preparation of cleared lysates is typically performed in parallel and, to minimize degradation, all steps are carried out on ice.

1. Resuspend *E. coli* C41 cells producing flagellar proteins in the appropriate Lysis Buffer (Table 4) containing protease inhibitors (1 tablet in 20 mL of lysis buffer) and 10 μg/mL DNAse I.
2. For small-scale *E. coli* C41 cultures (<5 g wet weight cells, culture volume < 500 mL) lyse cells using a precooled French© pressure cell at 30 kpsi.
3. For larger-scale *E. coli* C41 cultures (> 5 g wet weight, culture volume > 500 mL) lyse using a cell disruptor (Constant Systems) at 30 kpsi.
4. Centrifuge (40,000 × *g*, 1 h) cell lysates to remove insoluble cell debris and then decant the soluble fraction, which is the cleared cell lysate, into a plastic tube with lid.
5. Store the cleared cell lysate on ice for later use.

3.5 Purification of Flagellar Proteins

1. Equilibrate Ni^{2+}-sepharose resin, Ni^{2+}-nitrilotriacetic acid (NTA) resin, GSH-resin, or anion exchange resin (e.g., Q HP, GE Healthcare) in 50 volumes of the appropriate Lysis Buffer (Table 4), where one volume equals the volume of resin used.
2. If performing batch purification, equilibrate the resin [3 mL resin slurry, 50% (v/v)] in a 50 mL plastic tube with lid, allow

the resin to settle to the bottom of the tube, and then carefully remove the Lysis Buffer with a Pasteur pipette to leave only the resin in the tube. Carefully add the cleared cell lysate (*see* Subheading 3.4) to the resin and incubate for 1 h at room temperature on a bench-top tube rotator.

3. If using resin in a column, pass the cleared cell lysate over the equilibrated resin using a peristaltic pump or FPLC at 1 mL/min.
4. Wash the resin with 10 volumes of the appropriate Wash Buffer (*see* Table 4 and **Note 3**).
5. Elute bound protein from the resin using the appropriate Elution Buffer (Table 4). It is optimal to elute in multiple fractions over 3–5 volumes (*see* **Note 4**).
6. In a 5 L plastic beaker, dialyze (using, e.g., dialysis tubing, 8–10 kDa MWCO) the eluted protein at 4 °C in 4 L of the buffer appropriate for the downstream experiment (*see* **Note 5**).

3.6 Purification of Recombinant Subunit/Export Gate Protein Complexes

1. Place 1 mL of GSH-resin slurry [50% (v/v)] into a 50 mL plastic tube with lid and wash with 5 mL of sterile Milli-Q water followed by 5 mL of the appropriate Lysis Buffer (Table 5). Allow the washed resin to settle to the bottom of the plastic tube and then carefully remove the Lysis Buffer with a Pasteur pipette to leave only the resin.
2. Decant the cleared cell lysate containing (GST)FlhB$_C$ into the 50 mL plastic tube containing the washed GSH-resin.
3. Incubate tube-containing lysate and resin on a bench-top tube rotator at room temperature for 1 h.
4. Place the 50 mL plastic tube containing lysate and resin on ice for 10 min to allow the resin to settle and then carefully remove the supernatant with a Pasteur pipette to leave only the resin.
5. Wash the resin with 10 mL of Wash Buffer (Table 5).
6. Apply the cleared cell lysate containing recombinant FlgD (*see* **Notes 6** and **7**) to the washed resin and incubate on a bench-top tube rotator at 4 °C for 2 h.
7. Repeat **step 4**.
8. Wash the resin with 2 mL of wash buffer (*see* **Note 8**).
9. Repeat **step 4**.
10. For the capture assay protocol (Subheading 3.9), store the resin on ice and carry out the capture assay on the same day. For the photo-crosslinking protocol proceed to **step 11**.
11. Elute the protein complex by adding 1 mL Elution Buffer (Table 5) per 1 mL of resin and incubate in a 50 mL plastic tube with lid on a bench-top tube rotator at room temperature for 10 min.

12. Place the 50 mL plastic tube with lid on ice to allow the resin to settle to the bottom of the tube, collect the supernatant containing the eluted protein, and store in a plastic tube with lid on ice.
13. Repeat **steps 11** and **12** two more times.
14. Store the elution fractions on ice.
15. Confirm protein purity by analyzing the elution fractions by SDS-(15%)PAGE and Coomassie Brilliant Blue G-250 staining.

3.7 ITC

All buffers and samples used in ITC are degassed under a vacuum prior to use.

1. Proteins to be analyzed by ITC should be dialyzed (dialysis tubing, 6–8 kDa MWCO) into ITC assay buffer (50 mM Tris–HCl, pH 7.4, 100 mM NaCl) at 4 °C (*see* **Note 5**).
2. Concentrate solutions of purified and dialyzed GST or (GST) $FlhB_C$ to a concentration of 50 μM using a spin concentrator (3.5–5 kDa MWCO). Ligand (in this case untagged FlgD or untagged FlgDΔGRM) should be concentrated to 1.6 mM (*see* **Note 9**).
3. For each ITC experiment, use a heating block to pre-warm (to 20 °C) 2.0 mL of ITC assay buffer (50 mM Tris–HCl, pH 7.4, 100 mM NaCl), 2 mL of 50 μM GST or (GST)$FlhB_C$ in ITC assay buffer, and 200 μL 1.6 mM protein ligand (untagged FlgD or untagged FlgDΔGRM) in ITC assay buffer.
4. Pass 100 mL degassed, sterile Milli-Q water through the sample cell of the calorimeter using a vacuum pump.
5. Fill the reference cell with ITC assay buffer.
6. Equilibrate the sample cell of the calorimeter by passing 10 mL ITC assay buffer through the cell using a 5 mL syringe and the loading needle provided with the calorimeter. The loading needle should be as straight as possible to avoid touching the sample cell walls (*see* **Note 10**).
7. Load 1.8 mL degassed and pre-warmed GST, (GST)$FlhB_C$ or ITC assay buffer control into the sample cell (*see* **Note 11**).
8. Load the calorimeter syringe with 200 μL protein ligand in ITC assay buffer (in this case untagged FlgD or untagged FlgDΔGRM). Use the system software to slowly push the plunger until the air is expelled from the syringe (*see* **Note 11**).
9. Set experimental parameters on the calorimeter's control software. Input the protein and ligand concentrations. Sequentially inject the protein sample until it is fully titrated against the ligand. The parameters provided below are specific for (GST) $FlhB_C$ and FlgD measurements, but may need to be modified for other proteins and ligands.

(a) Initial injection of ligand: 0.5 μL for 200 s.

(b) Remaining injections (29): 4 μL, 200 s between injections. The final titration volume is 116.5 μL (*see* **Note 12**).

10. Using Origin® software, find the integral of the calorimetric signal after each injection and apply a best-fit line. The baseline, where Origin® calculates the top of each data point, can be altered by the investigator but this is not advised (*see* **Note 13**).
11. Using Origin®, produce a final figure from the experimental data. This will show both the μcal/s of each injection (top plot) and kcal/mol of injectant against the molar ratio of protein and ligand (bottom plot) with a best (single) fit line. Origin® will provide an association constant (K_a) that can be converted into the dissociation constant using $K_d = 1/K_a$.
12. Other data provided by Origin® include the Chi-squared value, the R score, the stoichiometry (n), and the enthalpy of binding (ΔH_b). These are automatically calculated by Origin® and provided with the K_a.

3.8 Photo-Crosslinking of Purified Subunit-Export Gate Complexes

1. Prepare co-purified protein complexes (as described under Subheading 3.6) of (GST)FlhB$_C$N$_{269}$A with FlgD subunit derivatives containing the Uaa p*Bpa* in either the GRM (L$_{39}$Uaa or L$_{40}$Uaa) or elsewhere (L$_5$Uaa). A 1 mL elution fraction from 1 mL GSH-resin yields ~100 μM of FlgD-(GST)FlhB$_C$N$_{269}$A complex.
2. Dialyze the FlgD-(GST)FlhB$_C$N$_{269}$A complexes for 2 h against 2 L of PBS in a 5 L plastic beaker at 4 °C.
3. Transfer 200 μL aliquots of the co-purified complexes (100 μM) of FlgD-(GST)FlhB$_C$N$_{269}$A into the wells of a polystyrene 24-well plate mounted in a shallow box containing ice.
4. Set aside equivalent amounts of protein complexes as non-irradiated control samples.
5. Place the samples in the polystyrene 24-well plate mounted in a shallow box containing ice 5 cm beneath a lamp that emits 350 nm light.
6. With the cooling fan on, UV irradiate (350 nm) the samples for 5 min, and then remove the samples in the polystyrene 24-well plate mounted in a shallow box containing ice and allow to cool for 5 min. Repeat this process eight times, replacing the ice surrounding the plate as necessary.
7. Precipitate 200 μL of each sample with 10% (v/v) TCA at 4 °C for 1 h, then centrifuge at 16,000 × *g* for 15 min to pellet the precipitated proteins. Decant supernatant and wash precipitated protein pellet with two volumes of ice-cold 100% acetone, centrifuge at 16,000 × *g* for 15 min, decant the supernatant, and air-dry the resulting protein pellet. Resuspend

the dried protein pellet in 100 μL of SDS-PAGE sample buffer and heat at 100 °C for 10 min. Cool the samples on ice and load 15 μL directly onto SDS-PAGE gels (15% and 12.5% polyacrylamide). Run the electrophoresis at 200 V until the loading dye migrates out of the gel.

8. The formation of cross-linked protein adducts is assessed by immunoblotting with specific antisera against FlhB, FlgD, or GST. Briefly, transfer separated protein complexes from SDS-PAGE gels to nitrocellulose membranes using wet blotting transfer equipment and blotting transfer buffer at 400 mA for 1 h. Transfer nitrocellulose membranes into 20 mL immunoblotting blocking buffer in a square plastic dish and incubate overnight at 4 °C. Wash blots with PBS and then incubate blots in 20 mL PBS containing the appropriate antisera. Wash blots with 20 mL PBS-Triton for 15 min. Repeat wash step five times. Incubate blots in 20 mL PBS containing the appropriate IRDye-conjugated goat-anti-rabbit secondary antibodies (Licor) for 1 h in the dark. Wash blots with 20 mL PBS-Triton for 15 min in the dark. Repeat wash step five times. Image blots using a Licor Odyssey® CLx imaging system.
9. Compare irradiated samples to non-irradiated samples. Determine the percentage of cross-linking by integrating the band density corresponding to the cross-linked and non-cross-linked material. Typically, efficiency of cross-linking is relatively low (>30%).

3.9 Capture Assay

1. Purify FlgEΔ39–43 (ΔGRM) and FlgEΔ39–43, Δ360–403 (ΔGRM, ΔCt) (*see* Subheading 3.5 and **Note 14**) and store on ice.
2. Prepare a co-purified protein complex of (GST)FlhB$_C$-FlgD$_{FLAG}$ (*see* Subheading 3.6).
3. Split the GSH-resin-bound complex of (GST)FlhB$_C$-FlgD$_{FLAG}$ into 50 μL aliquots in 2 mL microcentrifuge tubes.
4. Prepare 1 mL solutions of increasing concentrations (0, 0.1, 1, and 10 μM) of purified FlgEΔ39–43 (ΔGRM) and FlgEΔ39–43, Δ360–403 (ΔGRM, ΔCt).
5. In 2 mL microcentrifuge tubes, mix each 1 mL aliquot of the FlgE variants described in **step 4** with a 50 μL aliquot of resin-bound complex of (GST)FlhB$_C$-FlgD$_{FLAG}$ and incubate on a bench-top tube rotator at room temperature for 20 min.
6. Place the samples in the 2 mL microcentrifuge tubes on ice for 10 min to allow the resin to settle.
7. Collect the supernatant fractions containing unbound/captured proteins using a pipette and take a sample (900 μL) for analysis. Precipitate the proteins in the analysis samples, as described under Subheading 3.8, **step 11**.

8. Wash the resin with five volumes of wash buffer (for GST-fusion proteins, Table 4).
9. Repeat **step 6** and discard the supernatant.
10. Elute remaining resin-bound proteins ([GST]FlhB$_C$-FlgD$_{FLAG}$) by adding one volume of SDS-PAGE loading buffer to the resin and heating at 80 °C for 10 min.
11. Adjust the 10 μM subunit-challenged unbound/captured fractions (from **step 7**) to contain 5 mM imidazole and 0.5% BSA (w/v).
12. Add 80 μL of 50% suspension of Ni-NTA resin prewashed in buffer (50 mM sodium phosphate pH 7.4, 200 mM NaCl, 1 mM β-mercaptoethanol) to the 10 μM subunit-challenged unbound/captured fractions and incubate on a bench-top tube rotator at room temperature for 1 h.
13. Repeat **step 6** and discard the supernatant.
14. Elute resin-bound (subunit-captured) proteins by adding one volume of SDS-PAGE loading buffer to the resin and heating at 80 °C for 10 min.
15. For the unbound/captured samples (from **step 7**) and the subunit-captured samples eluted from the Ni-NTA (from **step 14**), load 5 μL onto a SDS-(15%)PAGE gel. Run the electrophoresis at 200 V until the loading dye migrates out of the gel and analyze by immunoblotting with antisera against FlhB$_C$, FlgE, and the FLAG epitope (as described under Subheading 3.8, **step 12**).

4 Notes

1. When selecting codons for substitution to amber codons, avoid codons for highly conserved residues as substitution may alter protein function or expression.
2. Always freshly prepare the *p*Bpa sodium hydroxide solution. We occasionally observe *p*Bpa precipitating out of solution on addition, but this has no effect on protein expression.
3. It is best to perform the wash in two or three steps. When purifying a single protein, you can use more than ten volumes of wash buffer; however, you may lose some protein with additional washes.
4. For untagged proteins, use an increasing gradient (usually over five to ten volumes) of NaCl to elute protein. Collect 0.5 mL elution fractions.
5. For some proteins, it is possible to dialyze overnight with 100–300 times more buffer than sample with little loss of protein. Other proteins might precipitate over this time, in which case

use shorter dialysis times, e.g., 2–3 h, with a buffer change every hour. It is important to be sure that all proteins are in the same buffer before beginning an experiment (such as ITC).

6. To prepare the cleared cell lysate containing $FlgD_{FLAG}$, lyse the resuspended cells expressing recombinant $FlgD_{FLAG}$ using a French© pressure cell. Aim to use a cell wet weight-to-buffer ratio of 1:1 to obtain a high $FlgD_{FLAG}$ concentration.
7. Ensure that you saturate the (GST)$FlhB_C$-bound resin with $FlgD_{FLAG}$. The binding capacity of glutathione sepharose 4B resin is >25 mg horse liver GST/mL resin. Aim to saturate the GST-$FlhB_C$-bound resin with at least twice the molar ratio of $FlgD_{FLAG}$ to (GST)$FlhB_C$.
8. The dissociation constant of FlgD for $FlhB_C$ is 39 μM [5]. As this is a low affinity interaction, it is important to wash the subunit-gate complex (GST)-$FlhB_C$)-$FlgD_{FLAG}$ with a low volume of wash buffer to avoid dissociating the complex.
9. You can use a spectrophotometer, such as a Nanodrop (Thermo), or a commercial Bradford assay (e.g., DC™ Protein Assay, Bio-Rad) to measure protein concentration. It is critical to use a technique that gives a precise and reliable measure of protein concentration for use in ITC.
10. Be sure not to scratch the sides of the sample cell as this can damage the experiment and lead to poor data.
11. The MicroCal control software allows control of the plunger with 1 μL increments. This is handy to lower the plunger until only protein is left in the syringe. Be sure to purge and refill once the air has been expelled from the syringe. The plunger should sit at the liquid interface and there should not be any empty space between it and your protein. If there is, the volume injected into the sample cell may be incorrect.
12. You may need to adjust the length of each injection to allow for a return to a temperature baseline, depending on the protein injected into the sample cell.
13. The heats of reactions of the buffer with (GST)$FlhB_C$ and the buffer with ligand should be subtracted from the heat of reaction of (GST)$FlhB_C$ with ligand. This is accomplished by collecting that experimental data and adding it to the spreadsheets containing the $FlhB_C$-ligand reaction data. A final column with the recalculated data should be used to analyze the association of $FlhB_C$ and ligand (e.g., FlgD).
14. FlgE has a short motif comprising a conserved phenylalanine followed by three residues and a hydrophobic residue (FxxxΦ) termed the gate recognition motif (GRM). The two FlgE derivatives used in the capture assay lack the GRM and can no longer interact with $FlhB_C$ and hence are unable to compete with $FlgD_{FLAG}$ docked on GST-$FlhB_C$.

Acknowledgments

Research in the Fraser laboratory is supported by grant BB/M007197/1 from the Biotechnology and Biological Sciences Research Council. Paul Bergen is supported by a postgraduate scholarship from Gates Cambridge.

References

1. Evans LD, Hughes C, Fraser GM (2014) Building a flagellum outside the bacterial cell. Trends Microbiol 22:566–572
2. Yonekura K, Maki S, Morgan DG, DeRosier DJ, Vonderviszt F, Imada K, Namba K (2000) The bacterial flagellar cap as the rotary promoter of flagellin self-assembly. Science 290:2148–2152
3. Thomas J, Stafford GP, Hughes C (2004) Docking of cytosolic chaperone-substrate complexes at the membrane ATPase during flagellar type III protein export. Proc Natl Acad Sci U S A 101:3945–3950
4. Stafford GP, Evans LD, Krumscheid R, Dhillon P, Fraser GM, Hughes C (2007) Sorting of early and late flagellar subunits after docking at the membrane ATPase of the type III export pathway. J Mol Biol 374:877
5. Evans LD, Poulter S, Terentjev EM, Hughes C, Fraser GM (2013) A chain mechanism for flagellum growth. Nature 504:287–290
6. Keener JP (2006) How *Salmonella Typhimurium* measures the length of flagellar filaments. Bull Math Biol 68:1761–1778
7. Tanner DE, Ma W, Chen Z, Schulten K (2011) Theoretical and computational investigation of flagellin translocation and bacterial flagellum growth. Biophys J 100:2548–2556
8. Stern A, Berg H (2013) Single-file diffusion of flagellin in flagellar filaments. Biophys J 105:182–184
9. Pierce MM, Raman CS, Nall BT (1999) Isothermal titration calorimetry of protein–protein interactions. Methods 19:213–221
10. Farrell IS, Toroney R, Hazen JL, Mehl RA, Chin JW (2005) Photo-cross-linking interacting proteins with a genetically encoded benzophenone. Nat Methods 2:377–384
11. Fraser GM, Hirano T, Ferris HU, Devgan L, Kihara M, Macnab RM (2003) Substrate specificity of type III flagellar protein export in Salmonella is controlled by subdomain interactions in FlhB. Mol Microbiol 48:1043–1057
12. Yamaguchi S, Fujita H, Taira T, Kutsukake K, Homma M, Iino T (1984) Genetic analysis of three additional fla genes in *Salmonella typhimurium*. J Gen Microbiol 130:3339–3342
13. Hanahan D (1985) In: Glover DM (ed) DNA cloning: a practical approach, vol 1. IRL Press, McLean, VA, pp 109.
14. Miroux B, Walker JE (1996) Over-production of proteins in *Escherichia coli*: mutant hosts that allow synthesis of some membrane proteins and globular proteins at high levels. J Mol Biol 260:289–298
15. Sambrook J, Russell DW (2001) Molecular cloning: a laboratory manual. Cold Spring Harbor Laboratory Press, Cold Spring Harbor, NY
16. Kaelin WG Jr, Krek W, Sellers WR, DeCaprio JA, Ajchenbaum F, Fuchs CS, Chittenden T, Li Y, Farnham PJ, Blanar MA et al (1992) Expression cloning of a cDNA encoding a retinoblastoma-binding protein with E2F-like properties. Cell 70:351–364
17. Guzman LM, Belin D, Carson MJ, Beckwith J (1995) Tight regulation, modulation, and high-level expression by vectors containing the arabinose pBAD promoter. J Bacteriol 177:4121–4130
18. Studier FW, Moffatt BA (1986) Use of bacteriophage T7 RNA polymerase to direct selective high-level expression of cloned genes. J Mol Biol 189:113–130
19. Evans L, Stafford GP, Ahmed S, Fraser GM, Hughes C (2006) An escort mechanism for cycling export chaperones. Proc Natl Acad Sci U S A 103:17474–17479

Chapter 3

Fluorescent Microscopy Techniques to Study Hook Length Control and Flagella Formation

Marc Erhardt

Abstract

The bacterial flagellum is a sophisticated motility device made of about 30 different proteins and consists of three main structural parts: (1) a membrane-embedded basal body, (2) a flexible linking structure (the hook) that connects the basal body to, (3) the rigid filament that extends up to 10 μm from the cell surface. In *Salmonella enterica* serovar Typhimurium, the hook structure is controlled to a length of 55 nm by a molecular ruler protein, FliK. Only upon hook completion, FliK induces a switch in substrate specificity of the flagellar export apparatus, which allows secretion of filament-type substrates, such as flagellin. Up to 20,000 subunits of flagellin assemble one flagellar filament that extends several micrometers beyond the cell surface. The formation of hook and filament structures as hallmarks of the hook length control mechanism can be monitored by immunofluorescence microscopy as described in this chapter.

Key words Bacterial flagellum, Hook length control, Flagellar assembly, Fluorescent microscopy, Immunostaining

1 Introduction

The structure of the bacterial flagellum is composed of three main parts: a membrane-embedded basal body; a long filament that functions as a propeller and extends several micrometers beyond the cell surface; and the hook, a universal joint that connects the basal body and the rigid filament [1]. The basal body complex includes rotor and stator components embedded in the cytoplasmic membrane, a flagellum-specific export apparatus, and a rod structure that spans the periplasmic space. Assembly of the hook starts upon completion of the P- and L-rings, which polymerize around the distal rod and form a pore in the outer membrane [2, 3]. The extracellular, flexible hook structure allows for transmission of torque generated by the cytoplasmic motor to rotational energy of the flagellar filament [4].

The lengths of the rod and the hook structures are controlled by different length control mechanisms. The rod extends from the

Tohru Minamino and Keiichi Namba (eds.), *The Bacterial Flagellum: Methods and Protocols*, Methods in Molecular Biology, vol. 1593, DOI 10.1007/978-1-4939-6927-2_3, © Springer Science+Business Media LLC 2017

cytoplasmic membrane through the periplasmic space to the outer membrane and covers a distance of 21.5 nm [5]. The length of the rod structure is determined by self-limited polymerization of the distal rod protein [2]. The hook extends 55 nm from the outer membrane and is made of approximately 120 subunits of FlgE [4, 6]. Upon completion of the hook structure, a signal is transduced to the export apparatus embedded within the basal body complex and results in a switch in secretion substrate specificity from rod-hook-type ("early") to filament-type ("late") substrates [7]. By this mechanism, the cell ensures that a functional hook-basal-body complex is present on the top of which the long flagellar filament made of several tens of thousands flagellin molecules can assemble.

The FliK protein is responsible for both hook length measurement and signal transmission to the export apparatus, after which the switch in substrate specificity occurs. In strains defective in the *fliK* gene, the length of the hook is not controlled and results in a so-called polyhook phenotype, where the hook structure reaches lengths of more than a micron [8]. FliK is secreted in random intervals and thereby determines the physiological length of the hook during hook-basal-body assembly. During secretion, FliK measures the length of the rod-hook structure as a molecular ruler [9, 10]. In this infrequent ruler mechanism, the hook length is determined by a stochastic process, where the probability of termination of hook polymerization is an increasing function of hook length [11]. In a hook of physiological length or longer, the molecular ruler protein will reside within the export channel for a sufficiently long time to allow for a productive interaction of the C-terminal domain of the molecular ruler with the export apparatus, which subsequently induces a switch in secretion substrate specificity.

Here, I describe the techniques to analyze termination of hook growth and initiation of filament formation as hallmarks for the substrate specificity switch using anti-hook and anti-filament immunofluorescence microscopy.

2 Materials

Standard chemicals are purchased in analytical quality from established commercial suppliers. Prepare all solutions using ultrapure water unless indicated otherwise.

1. *Salmonella enterica* serovar Typhimurium strains:
   (a) TH3730: P*flhDC*5451::Tn*10*dTc[del-25].
   (b) TH15801: P*flhDC*5451::Tn*10*dTc[del-25] Δ*hin*-5717::FCF (FRT–Chloramphenicol acetyl transferase–FRT).
   (c) TH16941: Δ*araBAD*7606::*fliK*⁺*flgE*7742::3 × HA Δ*fliK*6140 P*flhDC*5451::Tn*10*dTc[del-25] (*see* **Note 1**).

2. Lysogeny broth (LB): 10 g tryptone, 5 g yeast extract, 5 g NaCl. Tryptone broth (TB): 10 g tryptone, 5 g NaCl. Add 12 g agar for LB agar plates. Add water to a volume of 1 L and autoclave.
3. Shaking incubator (*see* **Note 2**).
4. Spectrophotometer for OD_{600} determination.
5. 0.2 mg/mL anhydrotetracycline stock solution in H_2O (*see* **Note 3**).
6. 20% (w/v) L-arabinose stock solution in H_2O (*see* **Note 3**).
7. Primary antibodies:
   (a) anti-FliC: Difco *Salmonella* H Antiserum I, dilution 1:5000–1:10,000.
   (b) anti-FljB: Difco *Salmonella* H Antiserum Single Factor 2, dilution 1:5000–1:10,000.
8. Anti-hemagglutinin antibodies conjugated to Alexa Fluor®594, dilution 1:1000.
9. Anti-rabbit polyclonal antibodies conjugated to Alexa Fluor®488, dilution 1:1000.
10. FM® 4–64 membrane staining dye (*N*-(3-Triethylammoniumpropyl)-4-(6-(4-(Diethylamino) Phenyl) Hexatrienyl) Pyridinium Dibromide).
11. Fluoroshield™ mounting solution with DAPI (4′,6-diamidino-2-phenylindole).
12. 0.1% (w/v) poly-L-lysine solution.
13. Polysine microscope adhesion slides (*see* **Note 4**).
14. Microscope coverslips and microscope slides.
15. High-vacuum silicone grease.
16. 36.5–38% Formaldehyde solution in H_2O.
17. 25% Gluteraldehyde solution in H_2O.
18. 10× Phosphate Buffered Saline (PBS): 80 g NaCl, 2 g KCl, 14.4 g Na_2HPO_4, 2.4 g KH_2PO_4, pH 7.4. Sterilize by autoclaving.
19. 10% (w/v) Bovine Serum Albumin in 1× PBS and store at 4 °C until use.
20. Epifluorescence microscope equipped with a motorized microscope stage, a digital camera for image acquisition, and appropriate filter sets for DAPI (absorption maximum 358 nm, emission maximum 461 nm), Alexa Fluor®488 (absorption maximum 490 nm, emission maximum 525 nm), Alexa Fluor®594 (absorption maximum 590 nm, emission maximum 617 nm) and FM® 4–64 dye (absorption maximum 515 nm, emission maximum 640 nm).

3 Methods

Perform experiments at room temperature unless indicated otherwise.

3.1 Culture Conditions

1. Streak *S.* Typhimurium strains TH16941 (Δ*araBAD*7606::*fliK*[+]*flgE*7742::3×HAΔ*fliK*6140P*flhDC*5451::Tn*10*dTc[del-25]) for hook immunostaining or TH15801 P*flhDC*5451::Tn*10*dTc[del-25] Δ*hin*-5717::FCF for anti-FliC immunostaining (*see* **Note 1**) for single colonies on fresh LB plates. Incubate overnight at 37 °C.
2. Inoculate a single colony of strain TH16941 or strain TH15801 into 1 mL LB and incubate overnight at 37 °C in a water bath incubator, shaking at 200 rpm (*see* **Note 5**).
3. Dilute overnight culture of *S.* Typhimurium strains TH16941 or TH15801 1:100 in 10 mL LB or TB and incubate at 37 °C or 30 °C (*see* **Note 6**) in a water bath incubator, shaking at 200 rpm. Grow approximately 2 h until optical density (OD_{600}) of 0.5.
4. Induce flagellar gene expression by the addition of 100 ng/mL anhydrotetracycline and resume incubation in a water bath incubator for an additional 60 min. To induce expression of *fliK* in case of strain TH16941, additionally supplement the culture with 0.2% L-arabinose to induce expression of *fliK* for 60 min at 37 °C.

3.2 Flow Cell Preparation

1. To coat a cover slip with poly-L-lysine, pipette approximately 10 μL of 0.1% (w/v) poly-L-lysine solution onto a microscope slide and place a cover slip on the top of the drop of poly-L-lysine (*see* **Note 4**).
2. Incubate for 10 min at room temperature and remove the microscope slide. Dry the cover slip for approximately 15 min at room temperature.
3. Use a fresh microscope slide and the poly-L-lysine coated cover slip to prepare a flow cell. Apply two layers of double-sided sticky tape or plastic paraffin film (Parafilm, *see* **Note 7**) to the microscope slide and seal with a poly-L-lysine-coated cover slip, forming a chamber of approximately 50 μL volume (Fig. 1)

antibodies. In case of co-immunostaining of flagellin and the 3× hemagglutinin epitope tagged hook, add a mixture of anti-hemagglutinin antibodies conjugated to Alexa Fluor®594 and anti-rabbit Alexa Fluor®488. *Step 5*: Wash samples with 1× PBS and add aqueous mounting solution and membrane stain FM® 4–64 dye (optional). *Step 6*: Perform fluorescent microscopy at 100× magnification using an epifluorescence microscope equipped with a motorized microscope stage

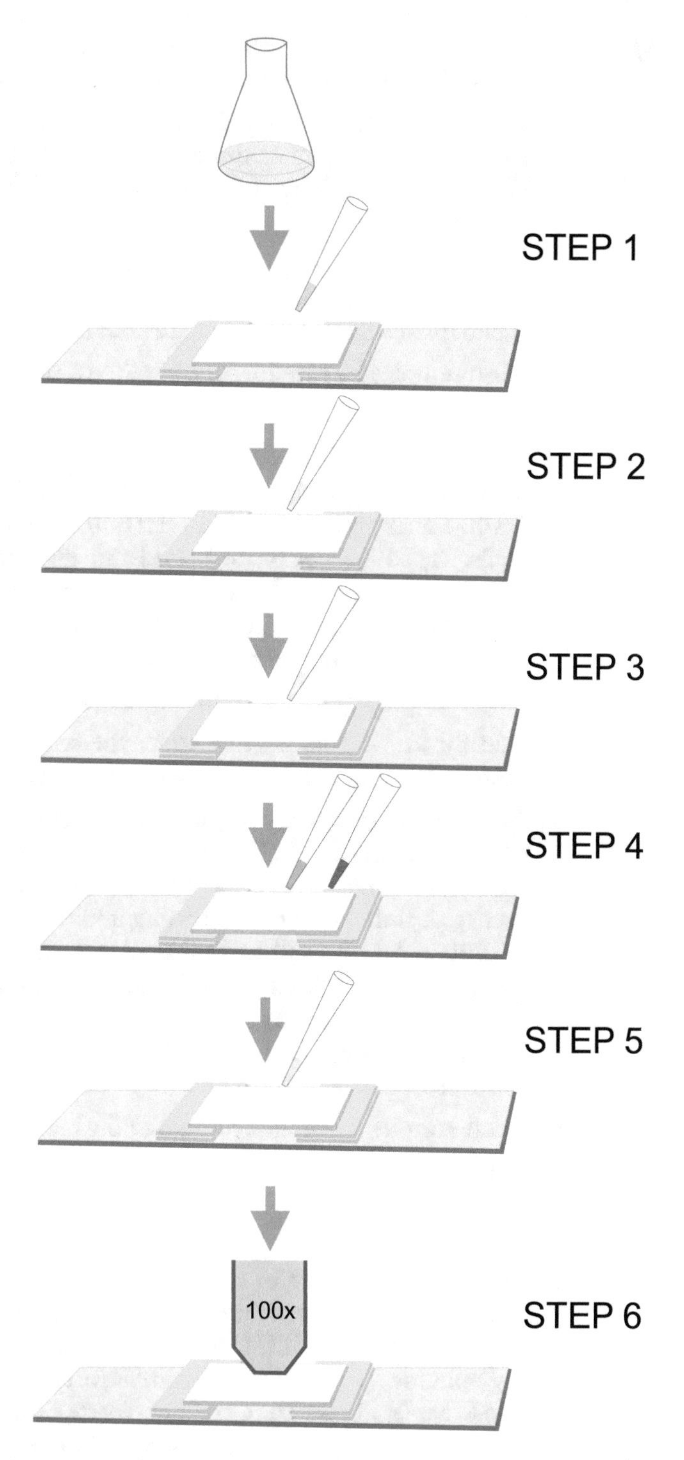

Fig. 1 Cartoon depiction of flow cell assembly and schematic workflow illustrating the most important steps to perform anti-flagellin immunostaining. *Step 1*: Grow culture to mid-log phase and apply 50 μL into the flow cell. *Step 2*: Fix bacteria by applying 50 μL of a mixture of 2% formaldehyde and 0.2% gluteraldehyde into the flow cell. *Step 3*: Wash samples with 1× PBS, block with 10% BSA, and add primary antibodies. *Step 4*: Wash samples with 1× PBS, block with 10% BSA, and **Fig. 1** (continued) add secondary anti-rabbit Alexa Fluor®488 coupled

3.3 Immunostaining

1. Carefully pipet about 50 μL of the mid-log phase culture of strains TH16941 or TH15801 into the flow cell.
2. Let the flow cell sit coverslip-side down at room temperature for 5 min in a humid chamber (*see* **Note 8**). (Optional: repeat filling the flow cell by addition of 50 μL of the mid-log phase culture.)
3. Carefully pipette 50 μL of a mixture of 2% formaldehyde and 0.2% gluteraldehyde into the flow cell to fix the samples. Incubate in a humid chamber at room temperature for 10 min.
4. Wash samples by carefully adding 50 μL of 1× PBS.
5. Add 50 μL of 10% BSA to block the samples. Incubate in a humid chamber at room temperature for 10 min.
6. Add 50 μL of primary antibody [mixture of anti-FliC, rabbit, 1:1000 and anti-FljB, rabbit, 1:1000 in 2% BSA in case of strain TH16941 (*see* **Note 1**) and anti-FliC, rabbit, 1:1000 in 2% BSA in case of TH15801]. Incubate in a humid chamber at room temperature for 60 min (Optional: incubate in a humid chamber overnight at 4 °C).
7. Wash samples by carefully adding 50 μL of 1× PBS.
8. Add 50 μL of 10% BSA to block the samples. Incubate in a humid chamber at room temperature for 10 min.
9. Add 50 μL of secondary antibody. In case of co-immunostaining of flagellin and the 3× hemagglutinin epitope tagged hook of strain TH16941, add a mixture of monoclonal anti-hemagglutinin antibodies conjugated to Alexa Fluor®594, 1:1000 in 1× PBS and anti-rabbit Alexa Fluor®488, 1:1000 in 1× PBS. For flagellin immunostaining of strain TH15801, add anti-rabbit antibodies conjugated to Alexa Fluor®488, 1:1000 in 1× PBS. Incubate in a humid chamber in the dark at room temperature for 60 min.
10. Wash samples by carefully adding 50 μL of 1× PBS.
11. Repeat washing step by the addition of 50 μL of 1× PBS.
12. Carefully pipette 50 μL of aqueous mounting solution (Fluoroshield™ with DAPI) into the flow cell. Store the flow cell in a humid chamber in the dark and proceed to fluorescence microscopy. (Optional: In case of anti-flagellin immunostaining of strain TH3730 or TH15801, prepare a mixture of the mounting solution (Fluoroshield™ with DAPI) and 5 μg/mL FM® 4–64 dye (*see* **Note 9**). Carefully pipette 50 μL of the mixture into the flow cell and store flow cell in a humid chamber in the dark until performing fluorescence microscopy analysis).
13. Optional: seal flow cell by the addition of nail polish or silicone grease. Store sealed flow cell in the dark at 4 °C.

3.4 Fluorescent Microscopy

1. Use an epifluorescence microscope equipped with a motorized microscope stage, a digital camera, and appropriate filter sets (DAPI, Alexa Fluor®488, Alexa Fluor®594, and FM® 4–64 dye) to perform fluorescent microscopy analyses of the samples fixed in the flow cell.
2. Collect images at 100× magnification. It is recommended to collect optical Z sections every 200–300 nm using the motorized microscope stage.

3.5 Image Processing

1. Use ImageJ software version 1.48 (National Institutes of Health) to analyze fluorescent images and project individual Z sections on a single plane (using settings for maximal intensity). Figure 2 illustrates representative co-immunostaining images of strain TH16941 for samples representing physiological hook length control, polyhook conditions in the absence of *fliK*, and late induction of the molecular ruler FliK. (Optional: If available, use deconvolution software such as softWoRx v.3.4.2 (Applied Precision) to deconvolve the fluorescent images).

4 Notes

1. Use strain TH3730 (P*flhDC*5451::Tn*10*dTc[del-25]) to induce flagellar gene expression by the addition of 100 ng/mL anhdyrotetracyline. Strains TH3730 and TH16941 are not phase-locked and will stochastically express both flagellins FliC and FljB. Thus, it is recommended to use a mixture of anti-FliC and anti-FljB antibodies for immunostaining of the flagellar filament. Alternatively, use strain TH15801 (P*flhDC*5451::Tn*10*dTc[del-25] Δ*hin*-5717::FCF), which is locked into the FliC^{ON} phase. Strain TH16941 (Δ*araBAD*7606::*fliK*$^{+}$ *flgE*7742::3 × HA Δ*fliK*6140 P*flhDC*5451::Tn*10*dTc[del-25]) is used to additionally induce the molecular ruler protein FliK by the addition of 0.2% arabinose. In the absence of *fliK* induction, hook length is uncontrolled and the switch to filament-type substrates does not occur. A late induction of *fliK* results in elongated hook structures, but retains the ability to form filaments, thus demonstrating that FliK functions as an infrequent molecular ruler (Fig. 2).
2. To ensure optimal growth conditions use a shaking water bath incubator.
3. Filter sterilize using 0.2 μm mixed cellulose ester or polyethersulfone filters.
4. Use polysine adhesion slides as alternative to poly-L-lysine-coated coverslips.
5. For best growth use a shaking water bath incubator.

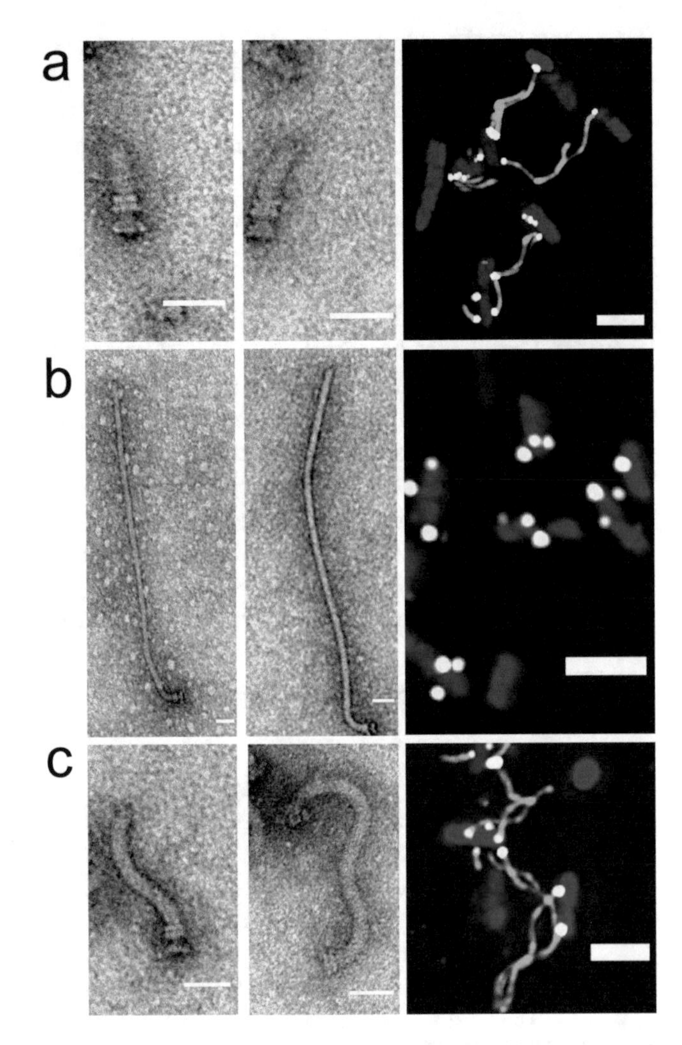

Fig. 2 Example images illustrating immunostaining of hook and filament structures of the bacterial flagellum (**a**) Strain TH16941 (Δ*araBAD*7606::*fliK*+ *flgE*7742::3 × HA Δ*fliK*6140 P*flhDC*5451::Tn*10*dTc[del-25]) was grown under conditions of simultaneous induction of flagellar gene expression and the molecular ruler *fliK*, thereby resulting in the production of hooks of physiological length and filament formation as a hallmark for the substrate specificity switch. Left panel, representative electron micrograph images of hooks isolated from strain TH16941. Scale bar represents 50 nm. *Right panel*, representative fluorescent microscopy image of anti-hemagglutinin (staining the hook) and anti-flagellin co-immunostaining. DNA (*blue*), hooks (*white*), and flagellar filaments (*green*). Scale bar represents 2 μm. (**b**) Flagellar gene expression, but not *fliK* was induced in strain TH16941. The absence of the molecular ruler resulted in uncontrolled hook growth (polyhook phenotype) and failure to induce the substrate specificity switch. *Left panel*, representative electron micrograph images of hooks isolated from strain TH16941. *Scale bar* represents 50 nm. *Right panel*, representative fluorescent microscopy image of anti-hemagglutinin (staining the hook) and anti-flagellin co-immunostaining. DNA (*blue*) and hooks (*white*). Scale bar represents 2 μm. (**c**) Flagellar gene expression was induced in strain TH16941, followed by late induction of the molecular ruler, FliK. Delayed secretion of FliK retains the ability to terminate hook growth, thereby demonstrating that FliK acts as a secreted molecular ruler that measures hook length in infrequent intervals. *Left panel*, representative electron micrograph images of hooks isolated from strain TH16941. Scale bar represents 50 nm. *Right panel*, representative fluorescent microscopy image of anti-hemagglutinin (staining the hook) and anti-flagellin co-immunostaining. DNA (*blue*), hooks (*white*), and flagellar filaments (*green*). Scale bar represents 2 μm. Adapted with permission from [10]

6. Grow bacteria in TB at 30 °C to limit the number of flagella. A reduced number of flagella per cell facilitates analyses of the microscopy images.
7. To form a flow-cell using parafilm as spacer, put one or two layers of parafilm on the top of a microscope slide (alternatively, use a polysine adhesion slide, *see* **Note 4**) and seal with a coverslip. Put the assembly on a heating block set to 100 °C for several seconds until the parafilm starts melting, thereby gluing the coverslip to the microscope slide.
8. Prepare a humid incubation chamber using a 15 cm diameter petri dish, which contains wet paper towels around the inside perimeter to prevent the chamber from drying out. Incubate in the dark or wrap the chamber with aluminum foil. Place flow cells with the poly-L-lysine-side down onto wooden sticks in the humid incubation chamber during incubation steps.
9. Membrane staining with FM® 4–64 dye is optional and cannot be combined with hook immunostaining using anti-hemagglutinin antibodies conjugated to Alexa Fluor®594 due to overlap in excitation and emission wavelengths.

Acknowledgments

This work was supported by the Helmholtz Association young investigator grant VH-NG-932, and the People Programme (Marie Curie Actions) of the European Union Seventh Framework Programme (grant 334030).

Reference

1. Chevance FFV, Hughes KT (2008) Coordinating assembly of a bacterial macromolecular machine. Nat Rev Microbiol 6:455–465
2. Chevance FFV, Takahashi N, Karlinsey JE, Gnerer J, Hirano T, Samudrala R, Aizawa S, Hughes KT (2007) The mechanism of outer membrane penetration by the eubacterial flagellum and implications for spirochete evolution. Genes Dev 21:2326–2335
3. Cohen EJ, Hughes KT (2014) Rod-to-hook transition for extracellular flagellum assembly is catalyzed by the L-ring-dependent rod scaffold removal. J Bacteriol 196:2387–2395
4. Samatey FA, Matsunami H, Imada K, Nagashima S, Shaikh TR, Thomas DR, Chen JZ, DeRosier DJ, Kitao A, Namba K (2004) Structure of the bacterial flagellar hook and implication for the molecular universal joint mechanism. Nature 431:1062–1068
5. Takahashi N, Mizuno S, Hirano T, Chevance FFV, Hughes KT, Aizawa S (2009) Autonomous and FliK-dependent length control of the flagellar rod in *Salmonella enterica*. J Bacteriol 191:6469–6472
6. Hirano T, Yamaguchi S, Oosawa K, Aizawa S (1994) Roles of FliK and FlhB in determination of flagellar hook length in *Salmonella typhimurium*. J Bacteriol 176:5439–5449
7. Williams AW, Yamaguchi S, Togashi F, Aizawa SI, Kawagishi I, Macnab RM (1996) Mutations in *fliK* and *flhB* affecting flagellar hook and filament assembly in *Salmonella typhimurium*. J Bacteriol 178:2960–2970
8. Patterson-Delafield J, Martinez RJ, Stocker BA, Yamaguchi S (1973) A new *fla* gene in

Salmonella typhimurium--flaR--and its mutant phenotype-superhooks. Arch Mikrobiol 90:107–120

9. Erhardt M, Hirano T, Su Y, Paul K, Wee DH, Mizuno S, Aizawa S-I, Hughes KT (2010) The role of the FliK molecular ruler in hook-length control in *Salmonella enterica*. Mol Microbiol 75:1272–1284

10. Erhardt M, Singer HM, Wee DH, Keener JP, Hughes KT (2011) An infrequent molecular ruler controls flagellar hook length in *Salmonella enterica*. EMBO J 30:2948–2961

11. Keener JP (2010) A molecular ruler mechanism for length control of extended protein structures in bacteria. J Theor Biol 263: 481–489

Chapter 4

Coupling of Flagellar Gene Expression with Assembly in *Salmonella enterica*

Fabienne F.V. Chevance and Kelly T. Hughes

Abstract

There are more than 70 genes in the flagellar and chemosensory regulon of *Salmonella enterica*. These genes are organized into a transcriptional hierarchy of three promoter classes. At the top of the transcriptional hierarchy is the *flhDC* operon, also called the flagellar master operon, which is transcribed from the flagellar class 1 promoter region. The protein products of the *flhDC* operon form a hetero-multimeric complex, $FlhD_4C_2$, which directs σ^{70} RNA polymerase to transcribe from class 2 flagellar promoters. Products of flagellar class 2 transcription are required for the structure and assembly of the hook-basal body (HBB) complex. One of the class 2 flagellar genes, *fliA*, encodes an alternative sigma transcription factor, σ^{28}, which directs transcription from flagellar class 3 promoters. The class 3 promoters direct transcription of gene products needed after HBB completion including the motor force generators, the filament, and the chemosensory genes. Flagellar gene transcription is coupled to assembly at the level of hook-basal body completion. Two key proteins, σ^{28} and FliT, play assembly roles prior to HBB completion and upon HBB completion act as positive and negative regulators, respectively. HBB completion signals a secretion-specificity switch in the flagellar type III secretion system, which results in the secretion of σ^{28} and FliT antigonists allowing these proteins to perform their roles in transcriptional regulation of flagellar genes. Genetic methods have provided the principle driving forces in our understanding of how flagellar gene expression is controlled and coupled to the assembly process.

Key words Flagellar gene expression, Gene regulation, Transposable elements, Genetic manipulations, λ-Red, Real-time PCR, β-Galactosidase assays

1 Introduction

1.1 Background

There are more than 70 genes in the flagellar-chemosensory regulon of *Salmonella enterica*. These genes are organized into a transcriptional hierarchy of three promoter classes that are coupled to the assembly process, as shown in Fig. 1. The flagellar transcriptional hierarchy of *Salmonella* was determined by generating a 38 × 38 matrix of double mutant strains (1,444 strains!) having Tn*10* and Mu*d*I insertions in all known flagellar genes [1]. Strains defective in class 1 flagellar genes failed to transcribe class 2 and class 3 genes. Strains defective in class 2 flagellar genes failed to

Tohru Minamino and Keiichi Namba (eds.), *The Bacterial Flagellum: Methods and Protocols*, Methods in Molecular Biology, vol. 1593, DOI 10.1007/978-1-4939-6927-2_4, © Springer Science+Business Media LLC 2017

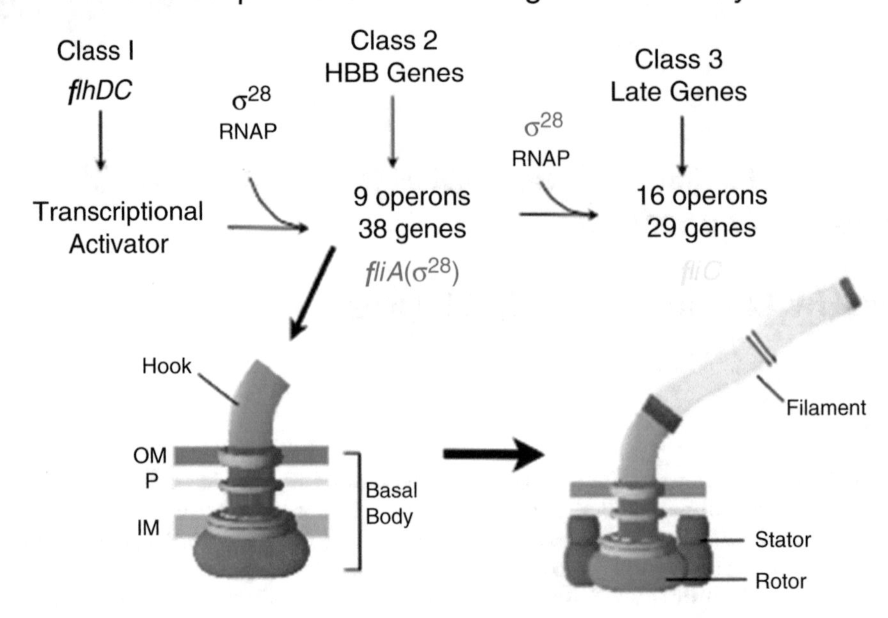

Fig. 1 The flagellar transcriptional hierarchy. (abbreviations: outer membrane (*OM*) periplasm (*P*) and inner membrane (*IM*)). The class 1 genes include the *flhDC* "flagellar master operon"

transcribe class 3 genes, but remained active in class 1 transcription. Insertions in class 3 flagellar genes were proficient in transcription for all flagellar genes. The mechanism behind the coupling of flagellar gene regulation to assembly is significantly (though not completely) understood as described below.

1.2 Regulation of $FlhD_4C_2$ Expression and Class 2 Transcription

At the top of the flagellar transcriptional hierarchy is the *flhDC* operon. Our current understanding of how *flhDC* expression is regulated is diagrammed in Fig. 2. The activators of *flhDC* transcription (cyclic AMP and catabolite activator protein), *flhDC* mRNA translation (CsrA), and $FlhD_4C_2$ active complex assembly (DnaK) were identified as mutants that failed to express flagella [2–4]. The activation of *flhDC* transcription by HilD and inhibitors of FlhDC activity were found using a T-POP transposon derivative of Tn*10d*Tc, described earlier [5]. The transcripts of the divergent P*tetA* and P*tetR* promoters of Tn*10d*Tc terminate at sequences near the outer ends of the transposon. Three T-POP derivatives were isolated and deleted for either the P*tetA* or P*tetR* terminator sequences or both [6]. Since transcription from P*tetA* and P*tetR* is induced by the addition of tetracycline (Tc) or its non-antibiotic analog anhydrotetracycline (ATc), addition of Tc or ATc to cells containing a T-POP element allows for transcription from the P*tetA* or P*tetR* promoters into adjacent DNA sequences. The T-POP transposon allows one to perform the genetic equivalent of a

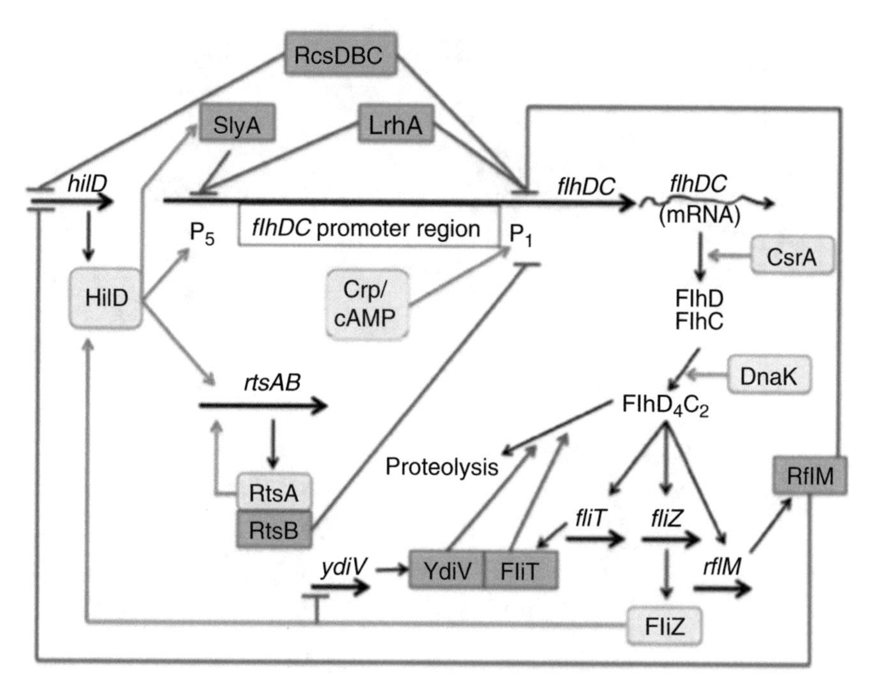

Fig. 2 Regulation of transcription, translation, assembly, and degradation of the $FlhD_4C_2$ complex of *Salmonella enterica*. Positive regulatory elements are boxed in *green* and negative regulators are boxed in *red*

microarray experiment to search for regulators, positive or negative, for any gene that can be assay, such as any gene with a Mud-*lac* insertion. In our lab we primarily use the T-POP derivative missing the P*tetA* terminator. When T-POP was inserted upstream of any of the *rflM*(*ecnR*), *fliT*, *rtsB*, *rcsDCB*, *lrhA*, or *ydiV* operons, addition of Tc or ATc resulted in the inhibition of flagellar class 2 and class 3 transcription [7]. Null T-POP insertions were isolated in the coding sequences of *rflM*, *slyA*, *rcsB*, *lrhA*, *ydiV*, or the *fliT* promoter region causing that resulted in increased *flhDC* activity [8]. Insertions in *clpP* also resulted in increased $FlhD_4C_2$ activity [8]. It was later shown that YdiV and FliT targets $FlhD_4C_2$ to ClpXP protease for degradation [9, 10].

1.3 Regulation of Class 3 Transcription and the Secretion-Specificity Switch

As mentioned above, any null allele affecting hook-basal body (HBB) completion resulted in the inhibition of class 3 transcription including filament genes, *fliC* and *fljB*, and genes of the chemosensory signal transduction pathway. Each flagellar filament comprises ~1% of the total cell protein. The mechanism of the secretion-specificity switch ensures that filament protein, which is secreted and assembled outside the cell, and the chemosensory genes are not transcribed until there is at least one HBB completed for the filament to assemble onto [11, 12]. The secretion-specificity switch also ensures the cessation of hook-rod secretion and results in the inhibition of $FlhD_4C_2$-dependent class 2 transcription [13].

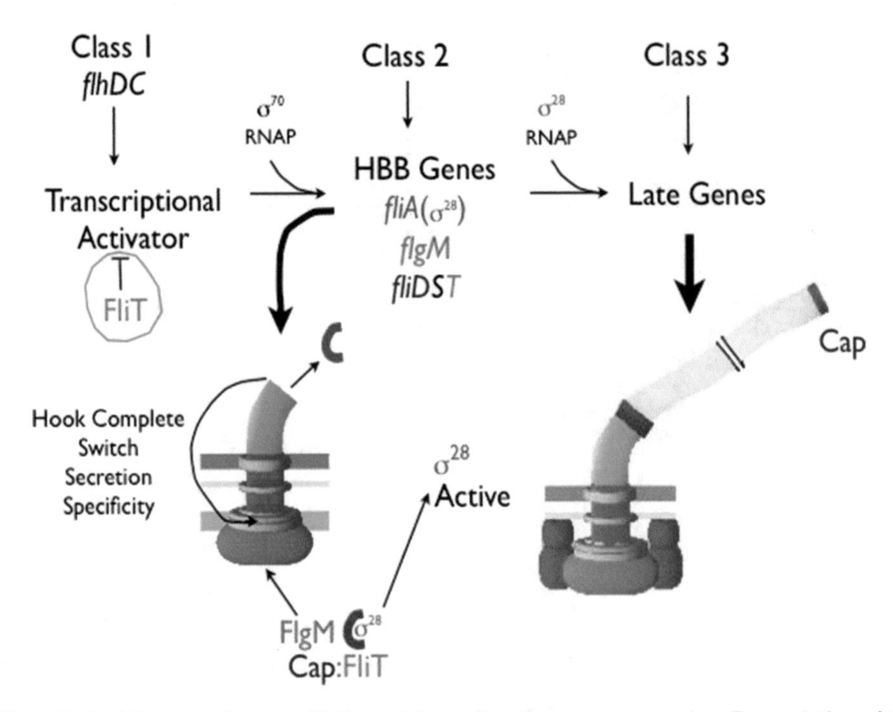

Fig. 3 The effect of the secretion-specificity switch on flagellar gene expression. Transcription of flagellar genes is coupled to HBB completion and the secretion-specificity switch by the secretion of the anti-σ^{28} factor, FlgM, and the anti-FliT factor, FliD(Cap). The σ^{28} protein is expressed from a class 2 promoter along with HBB genes. However, σ^{28} is kept inactive by a direct interaction with the anti-σ^{28} factor FlgM. Thus, FlgM prevents σ^{28}-dependent class 3 transcription prior to HBB completion. The filament cap gene, *fliD*, is also transcribed from a class 2 promoter along with its secretion chaperone gene *fliT*. FliT is an anti-$FlhD_4C_2$ factor that targets $FlhD_4C_2$ to ClpXP protease for degradation. However, binding of FliD to FliT prior to HBB completion inhibits its anti-$FlhD_4C_2$ activity. Upon HBB completion and the FliK-catalyzed secretion-specificity switch at FlhB, FlgM and FliD are secreted as late secretion substrates. Secretion of FlgM releases σ^{28} to transcribe flagellar class 3 promoters, whereas secretion of FliD releases FliT to target $FlhD_4C_2$ to ClpXP protease for degradation and prevent further class 2 transcription

The coupling of HBB completion to class 2 inhibition and class 3 activation is shown in Fig. 3. HBB completion signals the flagellar type III secretion apparatus to switch from rod-hook substrate to late substrate secretion. Class 2 transcription results in expression of the genes needed for HBB completion and genes needed after HBB completion. The genes expressed from class 2 promoters needed after HBB completion include the σ^{28} transcription factor gene, *fliA*. The σ^{28} factor-bound RNA polymerase holoenzyme specifically recognizes flagellar class 3 promoters [14]. Also expressed with HBB gene products is the anti-σ^{28} factor FlgM that inhibits σ^{28}-dependent class 3 promoter transcription prior to HBB completion [15, 16], the filament cap protein, FliD (17), and the FliD secretion-chaperone FliT [17], which prevents FliD degradation in the cytoplasm prior to HBB completion. Upon HBB completion the secretion-specificity switch is catalyzed, FlgM and FliD are secreted from the cell, releasing σ^{28} and FliT to perform their second functions as regulators to initiate class 3 transcription and to inhibit further class 2 transcription, respectively. In the absence of FliD, FliT targets $FlhD_4C_2$ to the ClpXP protease for degradation [10]. Thus, HBB completion coincides with cessation of class 2 transcription and initiation of class 3 transcription.

The FliK protein can be thought of as a ruler or tape measure that catalyzes the secretion specificity switch upon HBB completion [18]. FliK is secreted as a rod-hook substrate and takes temporal measurements of rod-hook length during HBB construction [19]. The mechanism of FliK action to catalyze the flagellar secretion specificity switch is diagrammed in Fig. 4. Once the rod-hook reaches a minimal length of ~65 nm, the C-terminus of a FliK molecule ($FliK_C$), in the process of being secreted, interacts with the FlhB component of the flagellar type III secretion system [20]. FlhB is the gatekeeper for secretion-substrate selectivity and is presumed to undergo a conformation change upon interaction with $FliK_C$ [21].

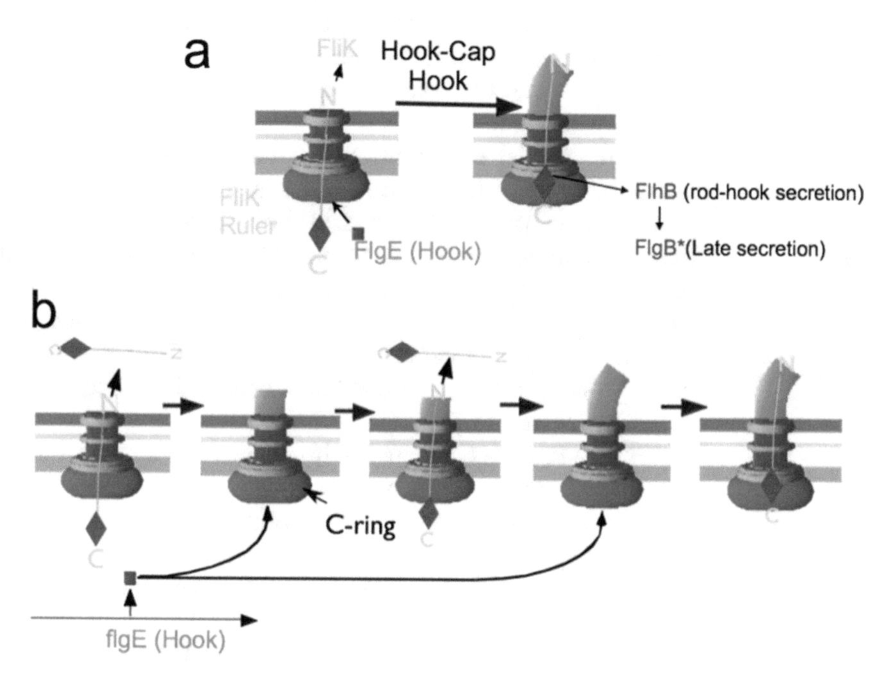

Fig. 4 Mechanism of the flagellar secretion-specificity switch from rod-hook substrate secretion to filament (late substrate). (**a**) The FliK protein is secreted during rod-hook assembly, but is not incorporated into the flagellar structure [25]. If the C-terminus of FliK can interact with the FlhB gatekeeper of the flagellar type III secretion apparatus, a conformational change in FlhB to FlhB* results in the change from rod-hook substrate selection to late or filament type substrate specificity. (**b**) The FliK acts as a molecular ruler (or tape measure) taking temporal measurements during rod-hook assembly. Secretion of FliK is relatively slow. However, once the N-terminus of FliK begins to exit the central rod or rod-hook channel, FliK is secreted at a rate that is too fast to allow the C-terminus of FliK to interact with FlhB and catalyze the secretion-specificity switch. It is only when the rod-hook reaches a minimal length, whereby the N-terminus of FliK remains in the secretion channel when the C-terminus of Flik is in physical proximity to FlhB that the conformational change in FlhB to late secretion mode occurs

1.4 Overview of Methods

1.4.1 Phage Methods

Standard genetic manipulations in *Salmonella enterica* are carried out using the generalized transducing phage P22. Norton Zinder isolated P22 from a lysogenic strain of *Salmonella* by UV induction. The utility of P22 as a transducing phage was greatly improved by Horst Smieger. Dr. Schmieger isolated a P22 variant, P22 *HT*, which was defective in DNA packaging (HT). The HT alleles were then combined with an integrase-defective allele (*int-201*) to create the generalized transducing phage P22 *HT105/1 int-201*. The *HT* mutations are in the DNA packaging enzyme, the gene 9 protein. Gene 9 recognizes specific DNA sequences of P22 DNA to initiate packaging into empty phage heads. The HT (high transducing) alleles of gene 9 are defective in DNA sequence specificity so that 50% of the DNA packaged is chromosomal DNA. The *int* allele of P22 *HT int* prevents integration into the host chromosome. This allows one to infect *Salmonella* at a high number of phage particles relative to bacterial cells, the multiplicity of infection or moi. P22 *HT int* will lysogenize, existing as a free circular DNA particle in the lysogenic state, but when a cell, lysogenic for P22 *HT int* divides, only one daughter receives the P22 genome. This allows the researcher to easily isolate phage-free cells after P22 *HT int* infection so that they can be used as recipients for further transduction experiments.

The flagellar transcriptional hierarchy was characterized using a combination of transposons and bacteriophage χ (Chi) selection. Chi has been extremely useful in the isolation of spontaneous and insertion mutants in flagellar genes. Chi is a lytic phage that attaches to *Salmonella* flagella. Flagella rotation allows flagella-attached Chi to migrate to the cell surface and inject its DNA. Cells unable to produce functional flagella are resistant to Chi infection and killing, which provides a powerful selection for strains defective in flagellar formation.

Both P22 *HT int* and Chi phages are grown by a simple method in our lab. First, a single-phage plaque is isolated. Phage is serially diluted into a sterile saline solution mixed with cells and top agar and spread onto an L-plate. Overnight incubation at 37 °C produces isolate plaques. We use isolated plaques to create my initial parental phage lysate. I typically add 0.1 ml diluted phage to 0.1 ml of cells from an overnight culture. We then add 1 ml of L broth followed by 3 ml of molten (ca. 55 °C) top agar before plating. The extra L broth added results in a larger plaque size, which is useful for isolating Chi plaques. Using a Pasteur pipette we poke a plaque so that the agar and phage in the plaque enter the pipette tip. We then blow the agar plug into 1 ml of an overnight cell culture, add 4 ml of L broth containing E-salts and 0.2% D-glucose (LE-dex). I then incubate the tube with shaking for 6–9 h, pellet the cell debris by table-top centrifugation, and pour the phage lysate into a sterile tube. We add a few drops of $CHCl_3$ and vortex to sterilize the culture and store the phage lysate at 4 °C. These parental lysates can then be used to make P22 or Chi phage broth by diluting the phage to ~5×10^7 plaque-forming units into LE-dex. The P22 broth is used to grow phage lysates on any cell culture for transduction purposes. The Chi broth is used to make large volume of Chi lysates for Chi phage-resistant mutant selections.

1.4.2 Transposon Methods

The transposons used routinely in the lab are derivatives of either transposon Tn*10* or bacteriophage Mu. The three Tn*10* derivatives that are routinely used are Tn*10d*Tc, Tn*10d*Cm, and T-POP. All three Tn*10* derivatives are deleted for Tn*10* transposase, which is provided from plasmid-expression vectors pNK972 and pNK2880. Plasmid pNK972 encodes ampicillin resistance and constitutively expresses wild-type Tn*10* transposase from the *tac* promoter. Plasmid pNK2880 encodes ampicillin resistance and constitutively expresses a mutant Tn*10* transposase from the *tac* promoter with altered target specificity alleles *ats-1* and *ats-2*. The *ats* alleles of Tn*10* transposase greatly reduce the frequency of transposition into so-called hot-spot DNA target sites, thus providing more random target mutagenesis than the wild-type Tn*10* transposase. However, the frequency of transposition is reduced by the *ats* alleles by a factor of about 100. Historically, we use derivatives of bacteriophage Mu for the construction of *lac* transcriptional and translational fusions (the Mu*d-lac* vectors) to flagellar genes of interest. New Mu*d* vectors have been recently developed for making translation fusions of flagellar genes to either β-lactamase (Bla) or green fluorescent protein (GFP).

The transposons Tn*10d*Tc and Tn*10d*Cm retain the Tn*10* transposase recognition sequences, which flank tetracycline (Tc) or chloramphenicol (Cm) drug resistance genes. The Tc resistance of Tn*10d*Tc is from the parent Tn*10* transposon and includes the *tetR* and *tetA* genes. The *tetA* gene product is an efflux pump that couples Tc export from the cytoplasm to the proton motive force. The *tetR* gene product is a repressor that inhibits both *tetA* and *tetR* transcription in the absence of Tc. The *tetR* and *tetA* genes are divergently transcribed and the TetR site of binding and repression is located in the DNA region between the *tetR* and *tetA* promoter sites. We refer to the Tc-resistance cassette from Tn*10*, which includes the inducible *tetR* and *tetA* genes as the *tetRA* cassette. The Tn*10d*Cm transposon derivative was constructed by replacing the *tetRA* cassette of Tn*10d*Tc with a constitutively expressed chloramphenicol acetyl transferase (*cat*) gene.

The donor strains for making insertions of Tn*10d*Tc and Tn*10d*Cm carry these transposons on the F plasmid originally from *E. coli* that was moved into *Salmonella*. Using recipient strains that do not contain the F plasmid, inheritance of Tn*10d*Tc and Tn*10d*Cm must occur by transposition. P22 grown on either the Tn*10d*Tc or Tn*10d*Cm donor strain is used to infect a recipient *Salmonella* strain carrying a Tn*10* transposase expression plasmid (pNK972 or pNK2880). For Tn*10d*Tc transposition experiments, the donor phage and recipient cells are plated directly onto L plates containing 15 μg/ml Tc. For Tn10dCm experiments, the donor phage and recipient cells are allowed to incubate for 30 min at 37 °C prior to plating onto L-Cm plates to allow for phenotypic expression of Cm resistance. Because Tc is bacteriostatic, not bactericidal, phenotypic expression is not required, but we find that our numbers of Tc-resistant (Tc^R) colonies are as much as tenfold higher if the phage + cell mixture is incubated at 37 °C prior to plating on L-Tc plates. When using Chi phage to select for Tn*10d*Tc or Tn*10d*Cm insertions in flagellar genes, 0.5 ml donor phage and 0.5 ml recipient cells are mixed in a sterile tube and allowed to sit at 37 °C. After 30 min, 4.5 ml of L broth is added and the mixture is allowed to incubate for another 30 min at 37 °C. This allows for expression of the drug resistance gene from the donor and loss of flagellar gene expression for insertions in flagellar genes. Phage Chi is then added at a multiplicity of infection (moi) of 5. To 1 ml of the Chi-infected mixture, 3 ml of molten (ca. 55 °C) top agar is added and the mixture is plated onto L-Tc or L-Cm selection plates. Following overnight incubation at 37 °C isolated Tc^R or Cm^R colonies are picked and made phage-free on green indicator plates. The purified insertion mutants are then tested on motility plates for insertions that have resulted in the loss of motility.

The Mu*d-lac* derivatives have proven to be powerful tools analyzing the regulation of flagellar gene expression. The original derivatives Mu*d*I and Mu*d*II were designed by Malcolm Casadaban and collaborators to use *lac* transcriptional (operon) and translational (gene) fusions, respectively, to assay gene expression for any gene of interest [22, 23]. Transposition-defective derivatives of Mu*d*I (Mu*d*A and Mu*d*J) and Mu*d*II (Mu*d*B and Mu*d*K) were later isolated to allow for the isolation of regulatory mutants affecting Lac expression [24–27]. Our lab routinely uses the Mu*d*J and Mu*d*K vectors in gene regulation studies of the flagellar system of *Salmonella*. The Mu*d*J and Mu*d*K elements are lab-engineered transposons that contain a constitutively expressed kanamycin-resistance gene and are used for making transcriptional and translational fusions to lactose genes. Mu*d*J and Mu*d*K are used to make *lac* operon fusions and *lacZ* gene fusions to genes of interest, respectively. When Mu*d*J is inserted in a gene in the correct orientation, the *lac* genes are expressed from the promoter of the inserted gene. The Mu*d*J element contains a ribosome-binding site (RBS) upstream of the *lac* operon resulting in translation of *lacZ* independent of translation of the inserted gene. Because the *lac* operon is now expressed from the promoter of the disrupted gene, any regulatory signal that affects the promoter is reflected as changes in the levels of *lac* operon transcription. Since expression of the *lacZ* gene can be easily detected on indicator plates and quantified by assaying β-galactosidase activity, the expression of the promoter can be easily studied in vivo. The Mu*d*K element has a truncated *lacZ* gene that lacks a RBS. Mu*d*K must be inserted in a gene in both the correct orientation and the correct reading frame in order for *lacZ* to be expressed. Thus, the *lacZ* gene is fused to the N-terminal portion of the inserted gene resulting in a gene fusion whose product is a protein fusion of an N-terminal portion of the inserted gene to LacZ. The site of insertion within a gene determines how much N-terminal sequence is fused to *lacZ*. Expression of *lacZ* is dependent on both the promoter of the inserted gene for transcription and on the RBS of the inserted gene. With Mu*d*K insertions, the *lac* operon expressed from the promoter of the disrupted gene, and translation of the *lacZ* fusion is dependent on the 5′-UTR translation signals of the inserted gene. Any regulatory signal that affects either transcription or translation of the inserted gene will affect *lac* expression. Using a combination of Mu*d*J and Mu*d*K elements in a given experiment, one can assay if regulatory signals affect the gene at the transcriptional or post-transcriptional level. A regulatory affecting the gene's transcription will affect lac expression for both Mu*d*J and Mu*d*K insertions. A post-transcriptional regulator will show a greater effect on *lac* expression with a Mu*d*K insertion than a Mu*d*J insertion. It is important to note that a translational inhibitor can show an effect on both Mu*d*J and Mu*d*K due to the effect of transcriptional polarity. For example, the FljA proteins bind in the 5′-UTR of *fliC* mRNA and inhibit translation. Expression of *fljA* results in a 200-fold reduction in expression of a *fliC*::Mu*d*K insertion due to translational inhibition and a fivefold reduction in *lac* expression with a *fliC*::Mu*d*J insertion due to transcriptional polarity resulting from translational inhibition.

Another example of the use of Mu*d-lac* derivative has been for the isolation of mutants that couple HBB completion to the secretion-specificity switch. The characterization of the mechanism of FlgM inhibition of σ^{28}-dependent class 3 transcription used simple genetic methodology; the results of this methodology are shown in Table 1. A Mu*d-lac* transcriptional fusion element (Mu*d*J) is used to assay flagellar class 3 promoter transcription and produces a red colony phenotype on MacConkey-lactose indicator plates and a white colony phenotype on Tz-lactose indicator plates. This strain is Lac^+ because FlgM is secreted through functional HBB structures allowing for σ^{28}-dependent class 3 transcription. Mutagenesis with transposon Tn*10* derivatives into this reporter strain produced Lac^- mutants at a relatively high frequency of ~1% of the total insertions. Given there are ~4500 genes in *Salmonella* with an average gene size of ~1000 base pairs, this suggested that ~45 genes can be disrupted which resulted in the inhibition of flagellar class 3 transcription. DNA sequence analysis revealed the majority of these genes were defective in the HBB assembly, which was later shown to inhibit FlgM secretion [28].

Table 1
Effects of flagellar mutants on Mu*d-lac* expression phenotypes in hook-basal body mutants defective in assembly prior to PL-ring formation (HBB$^-$) and in mutant strains defective in PL-ring formation (Ring$^-$) [40]

Flagellar genotype				Lac phenotype
HBB$^+$		class	3::Mu*d*J	+
HBB$^-$		class	3::Mu*d*J	–
HBB$^-$	flgM	class	3::Mu*d*J	+
HBB$^-$	fliA*	class	3::Mu*d*J	+
Ring$^-$	flgM	class	3::Mu*d*J	+
Ring$^-$	fliA*	class	3::Mu*d*J	+
Ring$^-$	flk	class	3::Mu*d*J	+
Ring$^-$	flhE	class	3::Mu*d*J	+
Ring$^-$	flk	class	3::Mu*d*J	+
Ring$^-$	flhE	class	3::Mu*d*J	+
Ring$^-$	flhA$_C$	class	3::Mu*d*J	+
Ring$^-$	flgG*	class	3::Mu*d*J	+

* dominant gain-of-function allele

1.4.3 Gene Fusion Methods

In many of our selections and screens we utilize gene fusions other than *lacZ* fusions. For the purpose of cell biology visualization studies by fluorescence microscopy, we construct fusions of green fluorescent protein (GFP) derivatives to flagellar proteins of interest [18]. In assays to measure levels of secretion of flagellar proteins into the periplasm, we use fusions of secreted flagellar proteins to β-lactamase (Bla) lacking its N-terminal "Sec" secretion signal. Bla confers resistance to β-lactams, such as penicillin and ampicillin, but only if it is secreted into the periplasm. Wild-type Bla protein possesses a cleavable, N-terminal type II secretion signal that targets Bla to the Sec-dependent secretion into the periplasm and the cells are ampicillin resistant (Ap^R). If the Sec signal is removed, Bla is retained in the cytoplasm and the cells are ampicillin sensitive (Ap^S). If Bla missing its N-terminal secretion signal is fused to secreted flagellar proteins (Fla), the resulting Fla-Bla fusions can be used to select for Fla-Bla secretion into the periplasm and the level of secretion can be determined by measuring the minimal inhibitory concentration (MIC) of resistance to Ap [8, 29].

Alternatively, a pore-forming domain from a colicin protein fused to a secreted flagellar protein (Fla-Col) can be used as a selection against secretion. Colicins will kill *Salmonella* bacteria by entering through the outer membrane and cell wall layers and then forming a pore in the inner membrane resulting in cell lysis. The colicin will only form the pore from the periplasmic side of the inner membrane. Thus, a Fla-Col fusion is only lethal when it is secreted into the periplasm. Thus, expression and secretion of a Fla-Col fusion into the periplasm provides a strong positive selection for mutants that are defective in the flagellar secretion apparatus of the flagellar secretion signal of the specific Fla-Col construct.

1.4.4 Classic Trail Complementation Method

The process of P22 abortive transduction was initially used to separate flagellar alleles into distinct complementation groups [30–32] and later shown to work for other genetic systems [33]. This was the first example of complementation for bacterial genes. P22 readily transduces chromosomal genes. However, for every productive inheritance of a donor allele by P22-mediated transduction, there are ten events where that same marker is inherited by an abortive transduction event, which requires the products of P22 genes *7*, *16*, and *20* [34]. This method is well described by Benson and Roth [34].

1.4.5 Plasmid Complementation

In any genetic study it is often important to know whether isolated mutants result from the loss or gain of function. A mutant that is recessive to the wild-type allele is almost always due to a loss-of-function mutation. A mutation that is dominant to the wild-type is often caused by a gain-of-function mutation or a loss-of-function allele that produces a defective gene product that can interfere and inhibit the activity or function of the wild-type protein in multimeric interactions of protein complexes. In order to test if a given allele is dominant or recessive to the wild-type gene, one must set up a complementation test where two copies of a given gene, one gene copy carrying the wild-type allele and the other gene copying the mutant allele to be tested. One method developed by the lab or Robert Macnab has been to clone the gene into a pTrc99A plasmid vector where the gene is placed under the control of the *tac* promoter. It has been determined that most flagellar genes cloned into pTrc99A will complement a chromosomal loss-of-function allele for that gene in the absence of IPTG inducer. Thus, the basal level of expression in the high-copy number pTrc99A plasmid vector is enough to complement mutant alleles.

1.4.6 Complementation by Chromosomal Duplication

Other complementation methods involve having a second copy of a given gene duplicated in the chromosome. We have frequently used expression from the arabinose-inducible *araBAD* promoter. In these cases, the gene to be duplicated replaces the *araBAD* structural operon. Since *araBAD* is deleted, added arabinose inducer is not degraded. The concern when complementation is performed using alternative expression systems is the expression levels from the different promoters vary to the point where complementation test cannot be performed. When the *flgM* gene was expressed from the *araBAD* locus, we observed that an excess of FlgM was expressed resulting in over-repression of flagellar class 3 transcription. Mutating the *araBAD* promoter and isolating promoter-down alleles solved this problem. One *araBAD* promoter allele resulted in expression of *flgM* to a level comparable to the level of *flgM* expressed from its native promoter.

Another trick for generating strains for complementation/dominance analysis is to make tandem chromosomal duplications of the native locus. The easiest method is the transduction of earlier versions of the Mu*d*J and Mu*d*K elements referred to as Mu*d*A and Mu*d*B, respectively. The Mu*d*A and Mu*d*B are larger versions of the Mu*d*J and Mu*d*K elements, respectively; too large (~37 kbp and ~35 kbp, respectively) to be transduced by a single P22 transduced fragment. P22 will package and inject 43.5 kbp of chromosomal DNA, but degradation of the DNA ends by RecBCD nuclease allows cotransduction of up to about 32 kbp. Inheritance of Mu*d*A and Mu*d*B by P22-mediated transduction requires that two P22 transduced fragments, containing the left and right ends of the Mu*d*A or Mu*d*B elements, infect a given cell where they can recombine with each other to generate a DNA fragment large enough to allow inheritance of Mu*d*A or Mu*d*B by P22-mediated transduction. When one mixes P22 lysates grown on Mu*d*A or Mu*d*B insertions in different chromosomal locations, the Mud recombinant in the recipient can come from the different donor Mu*d* elements and when inherited in the chromosome by recombination of the DNA ends flanking the donor Mu*d* elements will result in duplication or deletion of the region of the chromosome [26]. In this way, tandem duplications between two defined chromosomal loci are generated to create duplicated copies for complementation/dominance analysis of alleles of any gene of interest.

1.4.7 Targeted Mutagenesis and Gene Construction Using tetRA Recombineering Method

We routinely use a selection/counter-selection method called recombineering to perform targeted mutagenesis in the generation of specific mutations, site-directed mutagenesis and in the construction of gene-fusions that are not associated with transposon mutagenesis (i.e., Mu*d* transposons). The recombineering method utilizes the recombination functions of bacteriophage lambda (λ-Red), which will carry out high-frequency recombination utilizing homologous DNA sequences of only ~30–50 bases in length. This means that one can use DNA oligonucleotides to make any changes on the chromosome of interest. We typically use the *tetR-tetA* genes of transposon Tn*10*, which we refer to as a *tetRA* cassette, for selection/counter-selection, but other DNA cassettes, such as *sacB*-Km^R, are readily available and used in many labs for this purpose. The *tetA* gene of Tn*10* encodes an efflux pump that confers tetracycline resistance (Tc^R) by pumping Tc from the cytoplasm using the proton gradient of the proton motive source to drive Tc secretion. The *tetA* and *tetR* genes of *tetRA* are divergently transcribed from a central promoter region. The *tetR* gene encodes a repressor the binds the central promoter region and represses both *tetA* and *tetR* gene transcription. The binding of Tc to TetR results in a conformational change that causes DNA-bound TetR to come off the DNA and allow transcription of *tetA* and *tetR* until Tc is no longer present and TetR in the absence of Tc will bind and inhibit transcription. While the presence of the TetA efflux pump in the cytoplasmic membrane confers resistance to Tc, it also makes the cells sensitive to certain lipophilic compounds that are toxic. Fusaric acid is an example of a toxic lipophilic compound that will kill cells expressing TetA in the membrane at a lower concentration that it will kill cells lacking TetA. Thus, one can select for the insertion of the *tetRA* element by selecting for Tc^R, and one can select for the removal of *tetRA* on plates containing fusaric acid.

For the construction of a deletion mutant, we typically PCR-amplify the *tetRA* cassette with DNA oligonucleotides that will amplify from the two ends of the *tetRA* cassette. One DNA oligonucleotide will have 42 bases of DNA homology to the 5′-end of the target gene and 18 bases of homology to one end of *tetRA*. The other DNA oligonucleotide will have 42 bases of DNA homology to the 3′-end of the target gene and 18 bases of homology to the other end of *tetRA*. We use DNA oligonucleotides that are 60 bases in length simply because the price of a DNA oligonucleotide increases substantially from the company we buy our DNA oligonucleotides from when the length exceeds 60 bases in length. The amplified *tetRA* cassette will have 42 bases of flanking homology on each end to the target DNA sequence to allow for λ-Red recombination to insert the *tetRA* cassette. We then spend the money to purchase a DNA oligonucleotide that is 80 bases long where the first 40 bases are homologous to the 5′-target and the last 40 bases are homologous to the 3′-target sequences. An 18 nucleotide oligonucleotide that is homologous to one end of the 80 base oligonucleotide is often used to generate a double-stranded donor DNA strand by primer extension following hybridization of the two DNA oligonucleotides to each other. Once the donor DNA is ready it is introduced into the *tetRA*-containing recipient that is also expressing the λ-Red genes by electroporation followed by selection on fusaric acid Tc^S-selection plates.

One trick for making the targeted mutagenesis using *tetRA* work well is to be sure that the isolated *tetRA* insertion mutants are cleaned up nonselectively on L medium and screened for the ability to grow in the presence of tetracycline (Tc^R), and screened for tight non-growth on Tc^S selection plates. The PCR amplification of the *tetRA* cassette can result in mutagenesis of the *tetRA* genes resulting in an insertion mutant with a leaky phenotype that allows it to grow on both Tc^R and Tc^S plates. Such insertion mutants cannot be used in the next step of the experiment that requires *tetRA* cassette replacement.

Targeted, in-frame deletions of flagellar genes are made leaving the first and last five amino acid codons of the targeted gene. This is to help ensure that the deletion will not have a polar effect on downstream gene expression when the targeted gene is in an operon with other genes. The deletion mutant is then checked that it is not polar on other genes by complementation analysis. Using λ-Red recombination the *araBAD* genes are deleted and replaced with a *tetRA* cassette. This Δ*araBAD*::*tetRA* construct is used over and over again for the replacement of the *tetRA* cassette with any gene of interest for use in complementation test analyses. For example, replacement of the Δ*araBAD*::*tetRA* tetRA cassette with the *fliS*+ gene results in an Δ*araBAD*::*fliS*+ construct, which results in a wild-type *fliS* gene expressed from the arabinose-inducible *araBAD* promoter (P*araBAD*::*fliS*+). This is moved into a Δ*fliS* strain to test for complementation by first introducing a *leu*::Tn*10* insertion selecting for Tc^R and screening for leucine auxotrophy (inability to grow on minimal salts-glucose medium without added leucine). The leucine biosynthetic operon and the arabinose degradation operons are linked by P22 transduction. When P22 transducing phage grown on the P*araBAD*::*fliS*+ donor strain is used to transduce a *leu*::Tn*10* recipient strain to Leu^+ (growth on minimal salts-glucose medium without added leucine), about 50% of the Leu^+ transductants become unable to grow on arabinose as a sole carbon source (Ara^-) and have inherited the P*araBAD*::*fliS*+ allele by co-transduction. The resulting strain that is Δ*fliS* P*araBAD*::*fliS*+ can be tested for complementation of the Δ*fliS* by the addition of arabinose to express the P*araBAD*::*fliS*+ construct. The strain should also be tested for any effects of the Δ*fliS* allele on expression of the adjacent *fliD* and *fliT* genes that are co-transcribed with *fliS* in the *fliDST* operon by assaying for FliD and FliT activities in the Δ*fliS*-allele mutant background.

2 Materials

2.1 Phages

The generalized transducing phage of S. typhimurium P22 *HT105/1 int-201* is used in all transductional crosses [35].

1. P22 Broth: 200 ml LB; 2 ml 50 × E-salts (36) ; 0.8 ml 50% Glucose; 0.1 ml of P22 stock (5×10^{10} pfu/ml); add the E X 50, glucose, and P22 stock to 200 ml of sterile LB. P22 Broth will last for months at room temperature and decades at 4 °C.
2. Phage Stock: After growing a 1 ml overnight culture in L broth, add 4 ml P22 broth and then incubate at 37 °C with aeration (shaking or rotating) for 6 h or longer. After incubation, pellet the cell debris by centrifuging. Pour the phage containing supernatant into sterile test tube containing a few drops of chloroform. Vortex on high until a frothy foam appears at the surface of the liquid (~10 s). Store at 4 °C.

2.2 Culture Media and Buffers

1. Lysis broth (LB) is composed of 10 g Bacto tryptone 5 g Bacto yeast extract and 5 g NaCl per liter.
2. Solid media (plates) is composed of LB supplemented with 12 g/L Amresco agar or 15 g/L Bacto agar.
3. Phage indicator media (*see* **Note 1**).

We routinely use green plates to "clean up" colonies following transduction. Transduction plates contain mixtures of bacteria and phage. Exposure to P22 phage on transduction plates can result in the isolation of transductants that are P22-resistant. This is bad because these colonies cannot be used for P22 transduction again. Thus, transductions are single colony isolated on green plates and tested for phage-sensitivity so that they can be used again in P22 transductional crosses. Green plates provide a color indicator for bacteria that are phage-infected. When a colony is phage infected, lysing bacteria release acidic compounds; on green indicator plates they produce a blue color phenotype. Cells that are either phage-free or phage-resistant produce a light green/yellow color on green plates. Following transduction, colonies are single-colony isolated on green plates. Potential phage-free colonies that are light in color are picked and cross-streaked against a virulent P22 phage (P22 H5) to distinguish phage-free, phage-sensitive cells from phage-resistant cells, as described by Maloy [36]. Only phage-free, phage-sensitive cells are kept for further experimental analysis.

There are two types of green indicator mediums that are used to purify phage-free, phage-sensitive transductants following a P22-mediated transductional cross: Mint green and Standard green. Standard green plates contain the alizarin yellow dye, which can be toxic to many marked bacteria, especially those defective in cell wall metabolism. The dyes in mint green plates are less toxic and are preferred by many *Salmonella* labs. We prefer the Standard green plates and whenever a toxicity issue arises we use LB plates to purify products of transductional crosses. This is because the standard green plates produce sharper color differences between phage-infected and phage-free cells.

Mint green plates are prepared in two 1 L flasks labeled A and B. Flask A contains 10 g Bacto tryptone, 5 g Bacto yeast extract, 5 g NaCl, 2.5 g D-glucose, 12 g Ameresco agar, and 500 ml H_2O. Flask B contains 0.5 g $K_2HPO_4 \bullet 3H_2O$ and 500 ml H_2O. Flasks A and B are sterilized by autoclave and allowed to cool to ~55 °C. To Flask B is added 1.25 ml of a 1% (w/v) sterile stock solution of Evans Blue dye and 2.5 ml of a 1% (w/v) sterile stock solution of uranine. The contents of flasks A and B are mixed and poured into sterile petri plates. Standard green plates are composed of 6.2 g glucose, 6.6 g Bacto Tryptone, 0.84 g Bacto yeast extract, 4.1 g NaCl, 12 g Ameresco Agar, 0.05 g methyl blue, and 0.5 g of alazarin yellow per liter.

4. Antibiotics are added to LB at the final concentrations: 100 μg/ml sodium ampicillin, 12.5 μg/ml chloramphenicol, 50 μg/ml kanamycin sulfate, 15 μg/ml tetracycline-HCl, or 0.5 μg/ml anhydrotetracycline (ATc).
5. L-arabinose is supplemented to 0.2% (w/v) as needed.
6. ATc plates used to select for tetracycline-sensitive selection are prepared as follows: flask A consists of 12 g Amresco agar, 5 g Bacto tryptone, and 5 g yeast extract in 500 ml H_2O; flask B consists of 10 g NaCl, 10 g $NaH_2PO_4 \bullet H2O$ in 500 ml H_2O. Following autoclave sterilization, the flasks A and B are cooled to ca. 55 °C. To flask B is added 2.5 ml of 0.2 mg/ml anhydroteracycline (ATc) in 50% ethanol (stored at 4 °C), 5 ml of 2.4 mg/ml fusaric acid in dimethylformamide (stored at −20 °C), and 5 ml 20 mM $ZnCl_2$. Flasks A and B are mixed prior to pouring. ATc plates are protected from light and stored at 4 °C prior to use.
7. Lactose Indicator media (*see* **Note 2**).

A 200× X-gal solution is 0.8% of 5-bromo-4-chloro-3-indolyl-β-D-galactopyranoside (X-gal) dissolved in dimethylformamide and stored at 4 °C. Plates containing X-Gal are prepared by adding 5 ml of 0.8% stock solution to 1 L of medium. MacConkey-lactose (Mac-lac) plates are prepared by dissolving 50 g of MacConkey lactose powder (BD sciences) in 1 L H_2O sterilized, cooled, and poured. Tetrazolium-lactose (Tz-lac) plates are prepared.

8. Minimal E Salts.

50 × E salts are prepared as follows: 9.6 g of $MgSO_4 \bullet 7H_2O$; 100 g of Citric Acid; 655 g of $K_2HPO_4 \bullet 3H_2O$; 175 g $NaNH_4HPO_4 \bullet 4H_2O$ are added to 300 ml dH_2O in a 2 L beaker and cover with foil. Heat on high, but do not allow to boil. Add $MgSO_4 \bullet 7H_2O$ and dissolve completely. Add citric acid and allow to dissolve completely. Then add $K_2HPO_4 \bullet 3H_2O$ and allow to dissolve completely. Finally, add $NaNH_4HPO_4 \bullet 4H_2O$ and allow to dissolve completely. Add dH_2O to a final volume of 1 L. Cool at 4 °C and add 25 ml chloroform before placing on the shelf for storage at room temperature.

2.3 β-Galactosidase Assays Described by Miller [37]

1. 37°C and 30°C water bath.
2. Refrigerated centrifuge.
3. Standard spectrophotometer (readings at OD_{650}, 550, and 420 nm).
4. Iced-cold buffered saline: 0.85% NaCl in 100 ml of 1× Saline buffer.
5. 10× Saline buffer: 110 g $Na_2HPO_4 \bullet 7H_2O$, 30 g of KH_2PO_4, 750 ml H_2O, adjust pH to 7, and volume to 1 L.

6. 10% SDS ($NaC_{12}H_{25}SO_4$): wear a mask when weighing SDS, dissolve 100 g of SDS (sodium dodecyl sulfate) in 800 ml of water, heat to about 68°C to assist dissolution, adjust to pH 7.2 with a few drops of HCl, and then adjust volume to 1 L. Do not autoclave.
7. β-Mercaptoethanol.
8. Chloroform.
9. Z-buffer: 4.27 g Na_2HPO_4, 2.75 g $NaH_2PO_4 \bullet H_2O$, 0.375 g KCl, 0.125 g $MgSO_4 \bullet 7\ H_2O$. Adjust pH and volume to 7.0 and 500 ml. Do not autoclave.
10. ONPG solution: 4 mg/ml in Z-buffer. Prepare fresh.
11. 1 M $NaCO_3$ solution in water.

2.4 Real-Time PCR

1. Ice-cold phenol/ethanol mix (5% (v/v) phenol/95% (v/v) ethanol).
2. Tris-EDTA (TE) buffer, pH 7.5: 10 mM Tris–HCl, pH 7.5, 1 mM EDTA.
3. Lysis buffer: 50 mg/ml lysozyme in TE buffer pH 7.5.
4. Refrigerated centrifuge.
5. RNA kit (for example Qiagen kit).
6. Nanodrop spectrophotometer.
7. RNAse-free water.
8. DNAse I.
9. Random primers (random decamers or hexamers according to kit).
10. PCR thermocycler or heating block (thermomixer).
11. Thermoscientific Maxima first strand cDNA synthesis kit for RT-qPCR.
12. Biorad-Evagreen super Mix.
13. Electronic pipettes.
14. PCR plates (biorad).
15. Microseal B film (biorad).

3 Methods

3.1 Measurement of Flagellar Gene Expression Levels by β-Gal Assays (See Note 3)

1. Grow overnight (O.N.) culture of your strain(s) of interest (do not forget a wild-type control!).
2. Grow three independent colonies (independent, biological replicates) if you plan to perform statistical analysis.
3. Dilute O.N. culture 1:100 in 3 ml fresh media and grow for about 2.5 h at 37 °C to mid-log phase (OD_{600} ~ 0.4–0.6 = 2×10^8 cells/ml).

4. Spin cells down in a refrigerated centrifuge and resuspend pellet in 3 ml of ice-cold buffered saline and keep on ice.
5. Mix 25 ml of Z-buffer with 2.25 ml of water, 0.25 ml of 10% SDS, and 68 μl of β-mercaptoethanol.
6. Add 0.55 ml of the above mixture to each tube, plus two extra for blanks!
7. Add 0.1 ml of chloroform.
8. Add 0.5 ml of cells (or saline for blanks). The amount of cells depends on the activity of the gene studied. Dilute the cells if necessary.
9. Vortex shortly and incubate for 5 min with shaking at 30 °C.
10. While shaking, prepare a 4 mg/ml solution of ONPG in Z-buffer.
11. Add 0.2 ml ONPG to each tube periodically starting the clock on adding the ONPG to the first tube.
12. Time 10 min then stop each reaction with 0.5 ml 1 M $NaCO_3$.
13. Continue shaking for 2 min, then measure OD at wavelengths 420 and 550.
14. Warm up the cell cultures and take their OD at wavelength 650.
15. Calculate enzyme activity as follows:

$$\text{Activity} = \frac{OD_{420} - (1.75 \times OD_{550})}{(OD_{650} \times \text{time} \times \text{vol})} \times \frac{1\text{nmol}}{0.0045\text{ml.cm}} \times 1.7\text{ml}$$

where:
time = min of reaction.
vol = volume cell suspension (0.5 ml).
ε420 = o-nitrophenol =0.0045 OD420/nmol.
Activity is expressed in nmol/min/OD_{650}.

3.2 Real-Time Quantitative PCR with Bio-Rad CFX96 (Reverse Transcription and Quantitative PCR) (See Note 4)

3.2.1 Isolation of Total RNA

1. Grow overnight (O.N.) culture of your strain(s) of interest (do not forget a wild-type control!).
2. Grow three independent colonies (independent, biological replicates) if you plan to perform statistical analysis.
3. Dilute O.N. culture 1:100 in 2 ml fresh media and grow for about 2.5 h at 37 °C to mid-log phase (OD_{600} ~ 0.4–0.6 = 2 × 10^8 cells/ml).
4. Mix 1 ml log-phase culture with 0.8 ml ice-cold phenol/ethanol mix (5% (v/v) phenol/95% (v/v) ethanol) to stabilize RNA [38].
5. Spin 10 min at 4000 rpm in eppindorf microfuge, discard supernatant.
6. Resuspend pellet in 100 μl lysis buffer (TE pH 7.5, 50 mg/ml lysozyme).
7. Incubate 5 min at room temperature.

8. Further purify RNA using the manufacturer kit RNA instructions (e.g., Qiagen).
9. Quantify RNA using NanoDrop (be sure to switch to RNA measuring mode).

3.2.2 DNase I Treatment of RNA

1. Treat 4 μg of RNA with 2.5 units DNase I for 30 min at 37 °C.
2. Perform RNA purification as outlined in the manufacturers' protocol.
3. Elute RNA in two times 10 μl RNase-free water (total 20 μl).
4. Quantify RNA using NanoDrop (be sure to switch to RNA measuring mode).
5. Expect yield of 200 ng/μl and $A_{260/280}$ ~ 2.1–2.2 and $A_{260/230}$ ~ 2.3–2.5.

3.2.3 Reverse Transcription of RNA

1. Use 1.5 μg RNA in 20 μl reaction for reverse transcription (it is important to use equal amounts of input RNA in each RT sample).
2. Use random primers (random decamers or hexamers according to kit).
3. Heat RNA and random primers to 80 °C for 3 min.
4. Add MMLV reverse transcriptase, buffer, and dNTPs.
5. Perform reverse transcription for 1 h at 42°C, afterward heat up to 92°C for 10 min to inactivate the reverse transcriptase.
6. Quantify cDNA using NanoDrop (be sure to switch to DNA measuring mode).
7. Expect yield of 600–800 ng/μl and $A_{260/280}$ ~ 2.1–2.2 and $A_{260/230}$ ~ 2.0.

3.2.4 Quantitative Real-Time PCR Using SYBR Green

1. Adjust cDNA concentration for all samples to 10 ng/μl (about 80-fold) (high expressed gene like *fliC*). For medium to low expressed gene, adjust cDNA concentration to 30 ng/μl.
2. As a control, also dilute RNA (no reverse transcription control) about 80-fold and check for genomic DNA contamination.
3. Set up 10 μl qPCR reactions using Bio-Rad EvaGreen supermix.
4. Set up mastermix (be sure to set up 5–10% excess to account for pipetting errors):

   5 μl of 2× EvaGreen supermix.

   1.5 μl of 3 μM primer 1.

   1.5 μl of 3 μM primer 2.
5. Pipette 8 μl of mastermix into each well using electronic pipettes.
6. Pipette 2 μl sample (20 ng) into the appropriate wells.

7. Seal plate with Microseal B film.
8. Vortex briefly and spin down samples.
9. Run fast qPCR with subsequent melt curve:
   (a) 98°C for 30 s.
   (b) 98°C for 5 s.
   (c) 63°C for 10 s.
   (d) plate read.
   (e) go to **step 2** for 39 more times.
   (f) 98°C for 5 s.
   (g) 63°C for 10 s.
   (h) melt curve 65–95°C with 0.5°C increments for 5 s.
   (i) plate read.

3.3 Targeted Mutagenesis Using λ-Red

3.3.1 tetRA Element Chromosomal Insertion Using λ-Red Recombination

Preparation of Donor DNA Fragment

1. PCR the *tetRA* cassette from genomic DNA from any strain containing a Tn*10d*Tc insertion with primers that have additional sequence for recombination into your gene (*see* **Note 5**).
2. PCR program: 95 °C, 3 min: 1 cycle

   95°C, 30 s, 49 °C, 30 s, 72 °C, 2 min: 30 cycles.

   72°C, 10 min: 1 cycle.
3. Prepare clean ~2 kb of DNA product using Qiagen kit; elute in 20 μl EB buffer.

The λ-Red Recombination Reaction

1. Grow strain pKD_{46}(P*araBAD*-λ-red)/LT2 in LB Amp_{100} at 30°C O.N.
2. Subculture 1:100 in LB Amp_{100} + 0.2% arabinose (final conc.) at 30 °C.
3. Harvest cells when OD_{600} is ~0.6–0.8 (25 ml).
4. Wash cells twice with cold MilliQ (pure) water and resuspend in MQ water (~200 μl).
5. Electroporate PCR *tetRA* donor DNA (~100 ng up to 300 ng) into 50 μl of prepared cells; add 1 ml LB and incubate at 37 °C for 30 min.
6. Plate 300 μl onto LB plates with added Tc (15 μg/ml)—incubate O.N. at 37 °C.
7. Streak Tc^R colonies for isolation on LB plates at 42 °C and screen for Tc^R Ap^S colonies.
8. Streak *tetRA* insertion colonies on zinc-fusaric acid (Tc-sensitive) plates (*see below*) and choose the most Tc^S colonies (no growth on Tc^S plates). This is very important if you will use the *tetRA* construct for subsequent replacement (below).
9. Confirm strain by PCR.

3.3.2 tetRA Cassette Replacement via λ-Red Recombination

1. Streak a strain containing a *tetRA* cassette that you want to replace (made above).
2. Electroporate plasmid pKD46 (Ap^R, 30 °C) into the *tetRA*-containing strain.
3. Grow an O.N. culture of this strain in LB Ap_{100} at 30°C.
4. Subculture 1:100 in 25 ml LB Ap_{100} + 0.2% arabinose; incubate at 30 °C.
5. Harvest cells when OD_{600} reaches ~0.6–0.8.
6. Wash cells twice with 25 ml cold MQ water and resuspend in ~200 μl water.
7. Electroporate your PCR fragment (designed to replace the *tetRA* cassette) (~100–300 ng DNA) into 50 μl of prepared cells, add 1 ml LB and incubate at 37°C for 1 h.
8. Make dilutions into buffered saline and plate 0.1 ml of 10^0, 10^{-1}, and 10^{-2} dilutions onto fresh Tc^S plates (*see* recipe below). It is VERY IMPORTANT to make a cell only control!!
9. Incubate overnight at 42 °C if using chlorotetracycline Tc^S plates; if using andhydro-Tc^S plates incubation can be at 30 °C, 37 °C, or 42 °C.
10. Restreak potential isolates on Tc^S plates and incubate at 42 °C and then LB plates at 37 °C.
11. Pick and patch on LB-Tc and LB plates at 37 °C, and confirm the colonies have lost the *tetRA* cassette.
12. Check putative isolates by PCR.
13. Confirm construct by sequencing analysis.

4 Notes

1. Indicator plates are routinely used to isolate phage-free, phage-sensitive strains following P22 transduction. *Salmonella* labs use two types of plates for "cleaning up" colonies following transduction: green and mint green plates. Both types use glucose fermentation to indicate phage infection or sensitivity. In poorly buffered medium with excess glucose (1%) cell cytoplasm becomes acidic. On green plates colonies with any amount of cell lysis are blue in color and light green for colonies that are phage-free or phage-resistant. The light green colonies are then cross-streaked on another green plate against virulent phage to determine which colonies are phage sensitive and can be used in further transduction experiments. The standard green plates contain acid blue and a slightly toxic dye, alizarin yellow. Mint green plates use Evans Blue and uranine as pH indicator dyes for cell lysis. The advantage of Mint green plates is reduced toxicity. The advantage of standard green

plates is that the blue cell lysis color is more pronounced. Also, different types of mutations can produce colonies varying in color from wild type making them easy to identify. Also, many mutants defective in periplasmic or outer membrane proteins, such as the Tol proteins, will not grow on standard green plates, which may be a useful screen/selection for the study of such proteins or transport systems. Finally, when tittering phage that produce small plaques such as Chi, green plaques facilitate the visualization of these plaques due to the blue color on the green background.

2. A second important class of indicator media is used for determining relative levels of *lac* expression with Mu*d*J and Mu*d*K fusions. Three indicator plate types are used to estimate levels of *lac* expression: X-Gal, MacConkey-lactose (Mac-lac), and tetrazolium-lactose (Tz-lac). X-Gal is an indicator of β-galactosidase (β-gal, LacZ) levels only as it does not require the LacY permease to enter the cytoplasm. The Lac activity assayed on Mac-lac and Tz-lac indicator plates require lactose fermentation and therefore require expression of both LacZ and LacY. Mac-lac and Tz-lac are pH indicator media. Lactose fermentation results in the acidification of colonies. Lac activities produce opposite color changes on Mac-lac and Tz-lac plates; Lac^+ colonies are red on Mac-lac and white on Tz-lac while Lac^- colonies are white on Mac-lac and red on Tz-lac. It takes higher expression of the *lac* operon to produce a Lac^+ phenotype on Mac-lac than on X-gal containing media and even higher expression to be Lac^+ on Tz-lac plates. Using a combination of X-gal, Mac-lac, and Tz-lac plates, one can estimate expression levels. On X-gal, colonies of cells producing less than 5 units of β-gal activity are white turning light blue as cells approach levels of 10 β-gal units and turn blue as cells approach β-gal levels around 20 units and higher. Thus, it takes very low β-gal activity to produce blue colonies on X-gal. One caveat I have observed is that the presence of tetracycline in the medium interferes with detection of Lac activity with X-gal. I have not looked into the reason for this. Figure 5 shows the range of color phenotypes on Mac-lac and Tz-lac indicator plates used in estimating *lac* expression levels. This range works well using Mu*d*J transcriptional fusions. In the construction of the Mu*d*J reporter, a cloning error occurred resulting in the duplication-inversion of one end of Mu. This cloning aberration results in a large transcriptional terminator the end of the Mu*d*J element upstream of the *lac* operon reporter comprising a 48-base perfect palindrome with 48 bases on each side flanking an 8-base loop. This transcriptional terminator has provided an advantage when setting up selections for regulatory mutants. For example, in a hook-basal body mutant that is

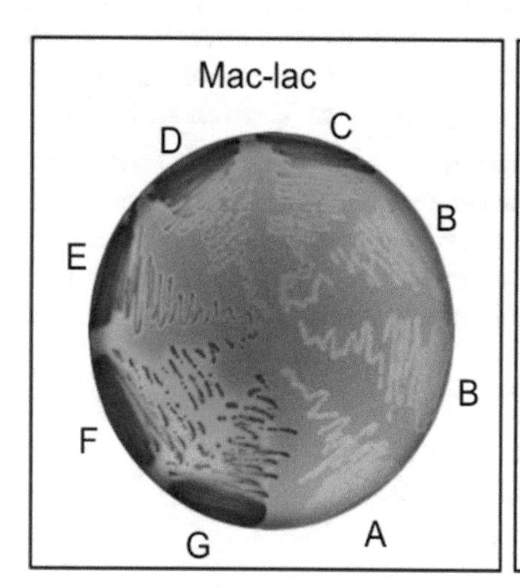

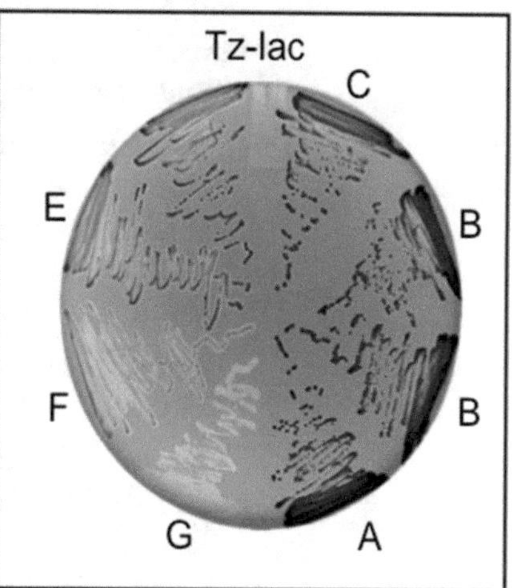

Fig. 5 Effects of *lac* expression on colony phenotypes on Mac-lac and Tz-lac indicator plates. The measures β-gal activities of the cells shown above in Miller units are as follows: *A* = 10, *B* = 15, *C* = 30, *D* = 45, *E* = 60, *F* = 80, *G* = 220

unable to secrete FlgM, inhibition of *fliC*::Mu*d*J expression is sufficient to prevent growth on minimal lactose plates, whereas inhibition of *fliC*::Mu*d*K expression is not enough to prevent growth on lactose. Thus, without this stem-loop sequence in Mu*d*J we would have not been able to carry out many of the selections that allowed us to determine processes of flagellum assembly and coupled gene regulatory mechanisms.

3. It is good to run a pilot experiment or at least check your strains in indicator plates. If it only shows activity on X-gal, then you need to add 0.5 ml of cells where if it is white on TTC-lac plates, you will need sometimes as little as 0.01 ml. What you want is an OD_{420} around 0.5.
4. Always wear gloves when working with RNA!
   Check the quality of RNA on denaturing or native agarose gel before going further.
   In order to just check the quality of RNA, a native agarose gel is fine as long as you denature the RNA using at least 60% formamide:
   Mix about 1 μg RNA with 1× loading dye and 60% formamide and heat for 5 min at 65 °C.
   Run 1% agarose gel at 5 V/cm.
   Salmonella enterica does not possess an intact 23S rRNA! You will see three major bands if you have isolated nondegraded total RNA (2× former 23S rRNA bands and the 16S rRNA band) [39].
   Use electronic pipettes for pipetting samples, etc. for qPCR. Precision is the key!

For gene expression analysis it is important to normalize mRNA input levels to at least one reference gene. During log-phase growth, the following genes are suitable as control genes: *flk*, *gyrB*, *rpoA*, *rpoB*, *rpoD*, *gapA*, and *hisG*.

5. The *tetRA* primers should be designed to integrate into the chromosome in an orientation-dependent manner such that with the addition of tetracycline, you know the direction of the individual *tetA* and *tetR* transcripts. The *tetA* promoter transcribes at levels about fivefold higher than P*tetR* transcript levels.

   Primer TetR: 5′ 40 bp (target DNA sequence)-TTA-AGACCCACTTTCACATT-3′ stop codon of *tetR* (Tm = 54°C).

   Primer TetA: 5′-42 bp(target DNA sequence c′)-CTAAGCACTTGTCTCCTG-3′ c′ stop codon of *tetA* (Tm = 54 °C).

Acknowledgments

This work was supported by PHS grant GM056141 from the National Institutes of Health.

References

1. Kutsukake K, Ohya Y, Iino T (1990) Transcriptional analysis of the flagellar regulon of *Salmonella typhimurium*. J Bacteriol 172:741–747
2. Shi W, Zhou Y, Wild J, Adler J, Gross CA (1992) DnaK, DnaJ, and GrpE are required for flagellum synthesis in *Escherichia coli*. J Bacteriol 174:6256–6263
3. Wei BL, Brun-Zinkernagel AM, Simecka JW, Pruss BM, Babitzke P, Romeo T (2001) Positive regulation of motility and *flhDC* expression by the RNA-binding protein CsrA of *Escherichia coli*. Mol Microbiol 40:245–256
4. Yokota T, Gots JS (1970) Requirement of adenosine 3′, 5′-cyclic phosphate for flagellar formation in *Escherichia coli* and *Salmonella typhimurium*. J Bacteriol 103:513–516
5. Singer HM, Kuhne C, Deditius JA, Hughes KT, Erhardt M (2014) The *Salmonella* Spil virulence regulatory protein HilD directly activates transcription of the flagellar master operon *flhDC*. J Bacteriol 196:1448–1457
6. Rappleye CA, Roth JR (1997) A Tn*10* derivative (T-POP) for isolation of insertions with conditional (tetracycline-dependent) phenotypes. J Bacteriol 179:5827–5834
7. Wozniak CE, Lee C, Hughes KT (2009) T-POP array identifies EcnR and PefI-SrgD as novel regulators of flagellar gene expression. J Bacteriol 191:1498–1508
8. Erhardt M, Hughes KT (2010) C-ring requirement in flagellar type III secretion is bypassed by FlhDC upregulation. Mol Microbiol 75:376–393
9. Takaya A, Erhardt M, Karata K, Winterberg K, Yamamoto T, Hughes KT (2012) YdiV: a dual function protein that targets FlhDC for ClpXP-dependent degradation by promoting release of DNA-bound FlhDC complex. Mol Microbiol 83:1268–1284
10. Sato Y, Takaya A, Mouslim C, Hughes KT, Yamamoto T (2014) FliT selectively enhances proteolysis of FlhC subunit in $FlhD_4C_2$ complex by an ATP-dependent protease ClpXP. J Biol Chem 289:33001–33011
11. Hughes KT, Gillen KL, Semon MJ, Karlinsey JE (1993) Sensing structural intermediates in bacterial flagellar assembly by export of a negative regulator. Science 262:1277–1280
12. Kutsukake K (1994) Excretion of the anti-sigma factor through a flagellar substructure couples flagellar gene expression with flagellar assembly in *Salmonella typhimurium*. Mol Gen Genet 243:605–612
13. Yamamoto S, Kutsukake K (2006) FliT acts as an anti-$FlhD_2C_2$ factor in the transcriptional

control of the flagellar regulon in *Salmonella enterica* serovar Typhimurium. J Bacteriol 188:6703–6708

14. Ohnishi K, Kutsukake K, Suzuki H, Iino T (1990) Gene *fliA* encodes an alternative sigma factor specific for flagellar operons in *Salmonella typhimurium*. Mol Gen Genet 221:139–147
15. Gillen KL, Hughes KT (1993) Transcription from two promoters and autoregulation contribute to the control of expression of the *Salmonella typhimurium* flagellar regulatory gene *flgM*. J Bacteriol 175:7006–7015
16. Ohnishi K, Kutsukake K, Suzuki H, Iino T (1992) A novel transcriptional regulation mechanism in the flagellar regulon of *Salmonella typhimurium*: an antisigma factor inhibits the activity of the flagellum-specific sigma factor, sigma F. Mol Microbiol 6:3149–3157
17. Kutsukake K, Ide N (1995) Transcriptional analysis of the *flgK* and *fliD* operons of *Salmonella typhimurium* which encode flagellar hook-associated proteins. Mol Gen Genet 247:275–281
18. Erhardt M, Singer HM, Wee DH, Keener JP, Hughes KT (2011) An infrequent molecular ruler controls flagellar hook length in *Salmonella enterica*. EMBO J 30:2948–2961
19. Minamino T, Gonzalez-Pedrajo B, Yamaguchi K, Aizawa S, Macnab RM (1999) FliK, the protein responsible for flagellar hook length control in *Salmonella*, is exported during hook assembly. Mol Microbiol 34:295–304
20. Minamino T, Ferris HU, Moriya N, Kihara M, Namba K (2006) Two parts of the T3S4 domain of the hook-length control protein FliK are essential for the substrate specificity switching of the flagellar type III export apparatus. J Mol Biol 362:1148–1158
21. Fraser GM, Hirano T, Ferris HU, Devgan LL, Kihara M, Macnab RM (2003) Substrate specificity of type III flagellar protein export in *Salmonella* is controlled by subdomain interactions in FlhB. Mol Microbiol 48:1043–1057
22. Casadaban MJ, Chou J (1984) In vivo formation of gene fusions encoding hybrid beta-galactosidase proteins in one step with a transposable Mu-*lac* transducing phage. Proc Natl Acad Sci U S A 81:535–539
23. Casadaban MJ, Cohen SN (1979) Lactose genes fused to exogenous promoters in one step using a Mu-*lac* bacteriophage: *in vivo* probe for transcriptional control sequences. Proc Natl Acad Sci U S A 76:4530–4533
24. Groisman EA (1991) In vivo genetic engineering with bacteriophage Mu. Methods Enzymol 204:180–212
25. Hughes KT, Roth JR (1984) Conditionally transposition-defective derivative of Mu d1(Amp Lac). J Bacteriol 159:130–137
26. Hughes KT, Roth JR (1985) Directed formation of deletions and duplications using Mud(Ap, lac). Genetics 109:263–282
27. Hughes KT, Roth JR (1988) Transitory *cis* complementation: a method for providing transposition functions to defective transposons. Genetics 119:9–12
28. Gillen KL, Hughes KT (1991) Negative regulatory loci coupling flagellin synthesis to flagellar assembly in *Salmonella typhimurium*. J Bacteriol 173:2301–2310
29. Lee HJ, Hughes KT (2006) Posttranscriptional control of the *Salmonella enterica* flagellar hook protein FlgE. J Bacteriol 188:3308–3316
30. Lederberg J (1956) Linear inheritance in transductional clones. Genetics 41:845–871
31. Lederberg J, Iino T (1956) Phase variation in *Salmonella*. Genetics 41:743–757
32. Stocker BAD (1956) Abortive transduction of motility in *Salmonella*; a non-replicated gene transmitted through many generations to a single descendant. J Gen Microbiol 15:575–598
33. Ozeki H (1956) Abortive transduction in purine-requiring mutants of Salmonella typhimurium. Carnegie Institute of Washington, Genetic Studies of Bacteria, Publication 612:97–106
34. Benson NR, Roth JR (1997) A *Salmonella* phage-P22 mutant defective in abortive transduction. Genetics 145:17–27
35. Davis RW, Botstein D, Roth JR (1980) Advanced bacterial genetics. Cold Spring Harbor Laboratory, Cold Spring Harbor, NY
36. Maloy SR (1990) Experimental techniques in bacterial genetics. Jones and Bartlett, Boston, MA
37. Miller J (1972) Experiments in molecular genetics. Cold Spring Harbor Laboratory, Cold Spring Harbor, NY
38. Simm R, Remminghorst U, Ahmad I, Zakikhany K, Romling U (2009) A role for the EAL-like protein STM1344 in regulation of CsgD expression and motility in *Salmonella enterica* serovar Typhimurium. J Bacteriol 191:3928–3937
39. Winkler ME (1979) Ribosomal ribonucleic acid isolated from *Salmonella typhimurium*: absence of the intact 23S species. J Bacteriol 139:842–849
40. Aubee JI, Olu M, Thompson KM (2016) The i6A37 tRNA modification is essential for proper decoding of UUX-Leucine codons during RpoS and IraP translation. RNA 22:729–742

Chapter 5

Dynamic Measures of Flagellar Gene Expression

Santosh Koirala and Christopher V. Rao

Abstract

Many genes are required to assemble flagella. These genes encode not only the structural elements of the flagellum but also a number of regulators that control how the flagellar genes are temporally expressed during the assembly process. These regulators also specify the likelihood that a given cell will express the flagellar genes. In particular, not all cells express the flagellar genes, resulting in mixed populations of motile and non-motile cells. Nutrients provide one signal that specifies the motile fraction. In this chapter, we describe two methods for measuring flagellar gene expression dynamics using fluorescent proteins in *Salmonella enterica*. Both the methods can be used to investigate the mechanisms governing flagellar gene expression dynamics.

Key words *Salmonella*, Flagella, Gene expression, Fluorescence, Flow cytometry

1 Introduction

1.1 Background

Salmonella enterica (hereafter referred to simply as *Salmonella*) uses flagella to swim in liquids and swarm over surfaces. Over 50 genes divided among at least 17 operons are involved in flagellar assembly. These genes are not constitutively expressed but rather expressed in a temporal hierarchy that mirrors the assembly process itself. As a brief background (see the references [1, 2] for further details), the promoter for these genes can be divided into three classes based on how they are temporally expressed. A single class 1 promoter region controls the expression of the *flhDC* operon, which encodes the $FlhD_4C_2$ master regulator. The activity of the *flhDC* promoter is controlled by a large number of global regulators that respond to different intracellular signals. $FlhD_4C_2$, in turn, activates the class 2 promoters, which control the expression of the genes encoding the hook-basal-body. $FlhD_4C_2$ also activates the expression of the flagellar-specific sigma factor σ^{28}, which then activates the class 3 promoters. These promoters control the expression of the genes encoding the flagellar filament, motor proteins, and the chemotaxis pathway.

Tohru Minamino and Keiichi Namba (eds.), *The Bacterial Flagellum: Methods and Protocols*, Methods in Molecular Biology, vol. 1593, DOI 10.1007/978-1-4939-6927-2_5, © Springer Science+Business Media LLC 2017

Flagellar gene expression occurs in a sequential manner, where first the single class 1 promoter is activated, then the class 2 promoters are activated, and lastly the class 3 promoters are activated [3, 4]. By activated, we mean the time when expression from these promoters can be detected using transcriptional fusions to reporter genes. A number of studies have used transcriptional fusions to investigate the mechanisms that govern flagellar gene expression dynamics. These experiments were initially performed primarily using microplate readers, where gene expression was measured using transcriptional fusions to either fluorescent proteins or the luciferase operon from *Photorhabdus luminescens* [3, 5–9]. Both of these reporters enable in vivo measurements of promoter activity without having to lyse the cells. In these experiments, the cells are grown in 96-well microplates within the readers. Typically, fluorescence or luminescence measurements are taken every 30 min. In addition, the absorbance (OD_{600}) is measured to account for cell growth.

More recently, a number of studies have employed flow cytometry to measure gene expression dynamic at single-cell resolution [10–14]. These studies have shown that flagellar gene expression is bistable in *Salmonella*, where only a fraction of the cells express the flagellar genes. For example, we have recently found that nutrients can be used to tune the fraction of cells that express the flagellar genes [10]. Specifically, the amount of yeast extract added to the growth medium was found to specify the fraction of cells expressing the flagellar genes (Fig. 1). No single nutrient within yeast extract appears to control this response—rather, the response appears to be controlled by nutritional/energetic state of the cells,

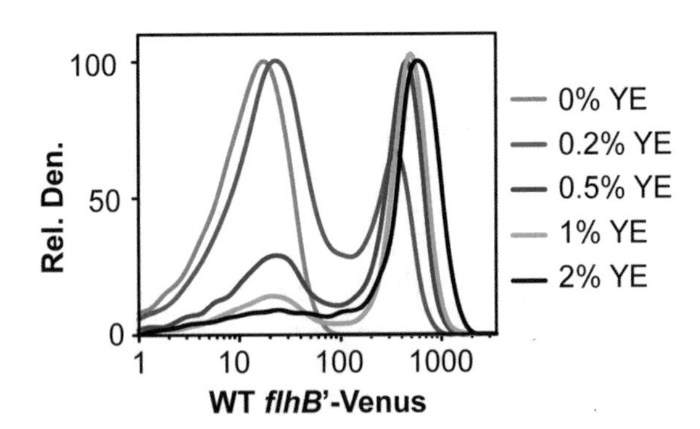

Fig. 1 Flagellar gene expression exhibits a bistable response to nutrients. The histograms show the distribution of cells expressing the flagellar genes from the class 2 *flhB* promoter, as determined using transcriptional fusions to the fluorescent protein Venus. In these experiments, the cells were grown in VBE medium with 0.2% glucose and the specified concentration of yeast extract for 5 h prior to analysis by flow cytometry. This histogram for cells not expressing Venus is identical to the histogram of cells expressing Venus and growing in 0% yeast extract

as many individual nutrients are capable of inducing gene expression (though not as well as yeast extract). Two proteins, YdiV and FliZ, were found to govern the bistable response of the class 2 promoters to nutrients (a separate mechanism separately controls the response of the class 3 promoters [11, 12]). Briefly, YdiV represses flagellar gene expression by binding to FlhD subunit of the $FlhD_4C_2$ complex and preventing it from activating the class 2 promoters [15, 16]. FliZ, which is expressed from a hybrid class 2/3 promoter, activates flagellar gene expression by repressing YdiV expression [17]. These two proteins form a double negative-feedback loop, where YdiV represses the expression of the flagellar genes, including FliZ, and FliZ represses the expression of YdiV. Bistability results from the mutual repression of these two proteins. When either gene is deleted, gene expression is homogeneous within the population, where all cells exhibit similar levels of promoter activity (Fig. 2). Furthermore, nutrients repress the expression of YdiV. In particular, when nutrients concentrations are high, YdiV expression is low and more cells express the flagellar genes. However, when nutrient concentrations are low, YdiV expression is high and fewer cells express the flagellar genes.

1.2 Overview of the Methods

This chapter describes our protocols for measuring the dynamics of flagellar gene expression in *Salmonella*. The first describes our protocol for using fluorescence microplate readers to measure gene expression for cells grown in rich medium. The second describes our protocol for using flow cytometry to measure gene expression at single-cell resolution for cells grown in minimal media supplemented with varying concentrations of yeast extract. The choice of growth medium does not affect the protocols in anyway. It simply reflects how we commonly perform these experiments in our

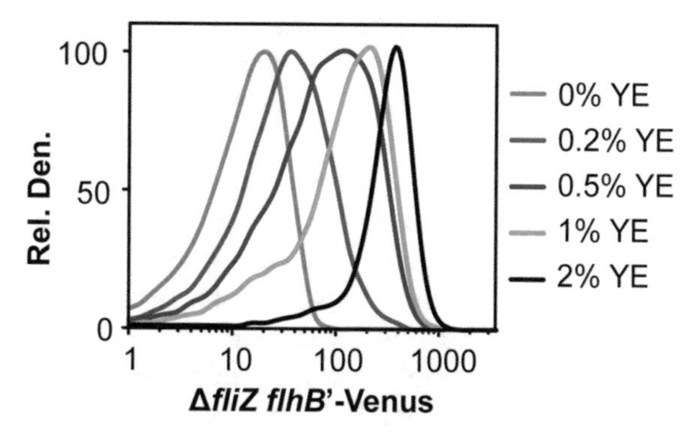

Fig. 2 Flagellar gene expression exhibits a homogeneous response to nutrients in a Δ*fliZ* mutant. The histograms show the distribution of cells expressing the flagellar genes from the class 2 *flhB* promoter, as determined using transcriptional fusions to the fluorescent protein Venus. In these experiments, the cells were grown in VBE medium with 0.2% glucose and specified concentration of yeast extract for 5 h prior to analysis by flow cytometry

laboratory, where the former is commonly used to screen the behavior of different regulatory mutants and the latter to perform more detailed analysis at single-cell resolution. Both protocols assume that promoter activity is measured using single-copy transcriptional fusion to the fluorescent protein Venus [18]. These protocols can easily be adjusted for other fluorescent proteins.

In the first section (Subheading 3.1), we provide our protocol for measuring gene expression using fluorescence microplate readers. In these experiments, the cells are grown directly within the microplate reader using 96-well microplates with clear bottoms to allow fluorescence and absorbance readings. This enables repeated fluorescence and absorbance measurements during cell growth without having to transfer samples between test tubes and the reader. Measurements are typically made every 30 min. Absorbance (OD_{600}) is used to track cell growth and to normalize the fluorescence readings to account for cell density. Multiple strains and promoters can potentially be tested in a single experiment. However, each plate should contain multiple technical replicates to account for variability across the plate. We have found that at least six replicates are necessary to provide reproducible data. In particular, the replicates are averaged to provide a single fluorescence and absorbance reading at a given time point. The experiments are then performed on at least three separate days to provide a measure of biological variability. The samples should ideally be placed in wells spatially distributed across the plate rather than clustering them in a single row or column to account for potential spatial variability within the reader. To prevent evaporation, the plates are sealed with an oxygen-permeable membrane. These experiments are commonly performed at 30 °C. The reason is that we observe better dynamics resolution at this temperature (Fig. 3). We generally find it difficult to see a separation between the class 2 and class 3 promoters at 37 °C. The same trick can also be employed when using flow cytometry.

The microplate reader experiments provide only a bulk measure of gene expression, in the sense that they measure expression from the entire population of cells within a well. In these regards, they measure the average level of gene expression within the population as determined using transcriptional fusions. As a consequence, they do not capture heterogeneity within the population. The assay is still useful in our opinion as it is relatively easy to perform and enables many samples to be tested in a single experiment. As noted above, the microplate assay is useful to rapidly screen different regulatory mutants. In particular, this assay can be used to determine how different mutants affect the dynamics of gene expression, at least when averaged over the entire population of cells.

In the second section (Subheading 3.2), we describe our protocol for measuring gene-expression dynamics using flow cytometry. In these experiments, the cells are first grown in test tubes at

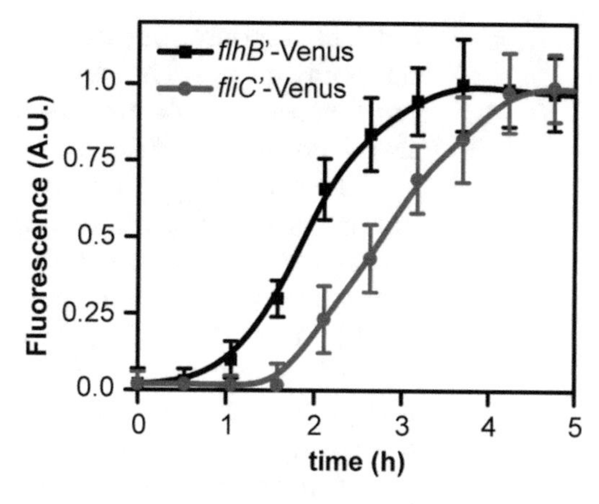

Fig. 3 The class 2 (*flhB*) and class 3 (*fliC*) promoters are expressed in a temporal hierarchy. Expression from these promoters was determined using single-copy transcriptional fusions to the fluorescent protein Venus. In the experiments, the cells were grown in LB in 96-well microplates within a fluorescent microplate reader. The fluorescence value, given in arbitrary units (A. U.), is normalized by the absorbance (OD_{600}) account for cell density. The error bars denote the standard deviation from three biological replicates. Note that class 3 gene expression lags class 2 gene expression by approximately 1 h

37 °C. They are then resuspended in a buffer containing the fluorescent DNA stain, 4′-6-diamidino-2-phenylindole (DAPI), and chloramphenicol. DAPI is used to fluorescently stain the cells. This provides a simple method for distinguishing cells from other debris within the growth medium in the flow-cytometry data. Alternatively, cells can be identified from forward and side-scattered light. Chloramphenicol is used to arrest protein synthesis and effectively lock the cells in their current state. We often collect samples at only a single time point, typically after 5 h during the late exponential phase of growth. However, multiple time points can also be sampled. However, the cells should be suspended first in phosphate buffered saline containing chloramphenicol and then stored on ice. The cells should not be stored in the DAPI staining buffer as it contains the detergent IGEPAL CA-630.

Any flow cytometer can be used provided it can measure DAPI and Venus fluorescence. Fluorescence should be measured from approximately 100,000 cells in order to generate clear histograms. We use the commercial software package FCS Express (De Novo Software) to process and analyze the data. We do not have any experience using other software packages though suspect they are all quite similar. We do not discuss flow cytometry per se as many protocols have already been published on the topic (*see* [19]). The only points we wish to emphasize are the following. The same gates should be used for all samples. In addition, you should run both positive and negative controls. When examining the bistable nutrient response, our negative control is a Δ*flhDC* mutant and our positive control is a Δ*ydiV* mutant. In addition, a Δ*fliZ* mutant can be used to generate a homogenous response to nutrients for class 2 promoters (Fig. 2).

The protocol can be extended to multiple promoters. While not discussed in this chapter, we previously measured simultaneous expression from the class 2 *flhB* promoter and *ydiV* promoter using transcriptional fusions to Venus and the cyan fluorescent protein (CFP), respectively [20]. We have also measured simultaneous expression from the class *flhB* promoter and class 3 *fliC* promoter using transcriptional fusions to mCherry [21] and Venus, respectively (unpublished). The only requirements are fluorescent proteins will minimal spectral overlap and a flow cytometer with the requisite lasers for excitation.

In both protocols, exponential growth following dilution in fresh medium is used to initiate flagellar gene expression. Alternatively, gene expression can be artificially induced using mutants where the native *flhDC* promoter is replaced with the tetracycline-inducible *tetRA* promoter from transposon *Tn*10 (P*flhDC*::*tetRA*) [7, 8]. In these experiments, flagellar gene expression is induced using 10 ng/mL anyhdrotetracycline (aTc) at a fixed yeast extract concentration (0.2%). Note that yeast extract inhibits expression from the *tetRA* promoter so it cannot be used as an orthogonal inducer of flagellar gene expression. Rather, aTc is used to tune the fraction of motile cells. This mutant (P*flhDC*::*tetRA*) can be used to demonstrate that flagellar gene expression exhibits hysteresis (see [10] for details).

For both protocols, we have found that single-copy transcriptional fusions integrated into the chromosome provide better temporal resolution than plasmid-based transcriptional fusions. The reason is presumably due to copy number and unequal plasmid partitioning between cells. We routinely use the so-called CRIM method developed by Haldimann and Wanner to generate these single-copy transcriptional fusions [22].

One limitation of using fluorescent proteins is that they are stable with relatively long half-lives. In these regards, they do not provide an instantaneous measure of promoter activity but instead measure the time-integrated activity of the promoter. As an example, the data in Fig. 3 only show when the promoters are being activated. More detailed analysis is required in order to determine when the promoters are being deactivated [23, 24]. This is less of an issue when considering steady-state measures of gene expression using flow cytometry as only a single time-point is considered (Fig. 1). While more instantaneous measures can be obtained using destabilized variants containing protease-degradation tags [20], these variants yield much weaker signals.

We conclude this section by noting that both protocols are short and do not involve many steps. However, these experiments are extremely sensitive to pipetting errors and the initial conditions of the experiments (e.g., starting OD_{600}). In addition, pre-warmed media and plates should always be used to minimize variability. Often it takes multiple repetitions before these protocols can be mastered. A useful benchmark in our experience is to see whether you can observe a large delay (30–60 min) in the induction of the class 2 and class 3 promoters using the time-course fluorescence assay (Fig. 3).

2 Materials

2.1 Time-Course Gene Expression Assay

1. *Salmonella* cells containing a single-copy promoter fusion to fluorescent protein Venus.
2. Pre-warmed (30 °C) Lysongeny broth (LB): 10 g/L tryptone, 5 g/L yeast extract, 10 g/L NaCl.
3. Pyrex 9820 test tubes with caps.
4. Corning™ 96-Well Clear-Bottom Black Polystyrene Microplate.
5. Breath-Easy gas permeably membrane.
6. 30 °C incubator with test-tube shaker for overnight cell growth.
7. Tecan Infinite M1000 Pro microplate reader. We have also used the same protocol with a Tecan Safire 2 microplate reader.

2.2 Single-Cell Gene Expression Assay

1. *Salmonella* cells containing a single-copy promoter fusion to fluorescent protein Venus.
2. Vogel-Bonner minimal E (VBE) medium: 200 mg/L $MgSO_4.7H_2O$, 2 g/L citric acid monohydrate, 10 g/L anhydrous K_2HPO_4, 3.5 g/L $NaNH_4PO_4$ supplemented with 0.2% glucose and yeast extract at the specified concentrations.
3. Pyrex 9820 test tubes with caps.
4. Falcon™ 5 mL round-bottom polystyrene tubes.
5. DAPI staining media: 100 mM Tris, 150 mM NaCl, 1 mM $CaCl_2$, 0.5 mM $MgCl_2$, 0.1% IGEPAL CA-630, 14.3 μM 4′-6-diamidino-2-phenylindole (DAPI), and 50 μg/mL chloramphenicol.
6. Flow cytometer with violet (403 nm) and blue (488 nm) laser and a standard filter set. We use a BD LSR II Flow Cytometry Analyzer (BD Biosciences). The pacific Blue channel (excitation, 405 nm; emission, 450/50 nm) is used to measure DAPI fluorescence and the fluorescein isothiocyanate (FITC) channel (excitation, 488 nm; emission, 530/30 nm) to measure Venus expression.
7. Software to analyze data. We use FCS Express (De Novo Software).

3 Methods

3.1 Time-Course Gene Expression Assay

1. Grow cells overnight in test tubes at 30 °C with constant shaking (200 rpm). The cells are grown in 2 mL of medium. Do not let overnight cultures grow for more than 16 h.
2. Subculture the cells to an OD_{600} = 0.02 using pre-warmed (30 °C) LB (*see* **Note 1**).
3. Grow the cells to an OD_{600} = 0.15 in test tubes at 30 °C with constant shaking (200 rpm).
4. Transfer 150 μL of the culture to an individual well on a pre-warmed (30 °C) 96-well microplate. For each sample, fill at least six wells on the plate.
5. Seal the plate with a BreathEasy membrane. Alternatively, you can overlay the wells with mineral oil if oxygen-limited growth is desired.
6. Transfer the plate to a microplate reader.
7. Grow the cells within the microplate reader at 30 °C with constant orbital shaking (168 rpm).
8. Measure the fluorescence (Ex: 515 nm/Em: 528 nm with 5 nm bandwidth) and absorbance (600 nm) every 30 min (*see* **Note 2**).
9. Normalize the fluorescence values by the optical density (OD_{600}) to account for changes in cell density. The integrated activity of the promoter at a specific time point is given by fluorescence divided by the optical density. These calculations should be done on a well-by-well basis and the average value reported for the plate. Finally, normalize the activities to a maximum value of 1.0. Normalization facilitates the comparison between promoters of different strength (*see* Fig. 2).

3.2 Single-Cell Gene Expression Assay

1. Grow cells overnight at 37 °C in 2 mL of VBE medium supplemented with 0.2% glucose and 0.2% yeast. Do not let overnight cultures grow for more than 16 h.
2. Subculture the cells in 5 mL of fresh, pre-warmed (37 °C) VBE medium, supplemented with desired concentration of yeast extract, to a starting OD_{600} = 0.05 (1:50–1:100 dilution). Make sure to always check the starting OD_{600} of the fresh culture.
3. Grow cells at 37 °C for 5 h. Alternatively, cells may be grown for shorter or longer periods of time for time-course experiments (*see* **Note 3**). We find that it takes approximately 5 h for the steady-state distributions to develop. This corresponds to the late exponential phase of growth.
4. Transfer 5 μL of culture into 1 mL DAPI buffer and incubate at 30 min in the 5 mL polystyrene tubes. Do not store the cells in DAPI buffer as it contains the detergent IGEPAL CA-630.

If time-course experiments are performed, then resuspend 1 mL of culture into 1 mL of PBS containing 50 μg/mL chloramphenicol. Transfer to DAPI staining buffer only when growth experiments are complete.

5. Use a flow cytometer to measure cells. We typically analyze 100,000 cells. Do not rush the experiments. Keep the cell count below 5000 cells/s.
6. Analyze the data using flow cytometry software (*see* **Note 4**).

4 Notes

1. The initial cell densities need to be fixed at the same value for all cells. These should be measured as subtle differences can yield significantly different response. Use fresh pre-warmed media.
2. The gain should be set so that you obtain the strongest signal possible while simultaneously minimizing background fluorescence. The fluorescence gain should be fixed for all experiments. Bandwidths can also be increased to increase signal. However, larger bandwidths lower the signal-to-noise ratio. A negative-control strain can be used to determine the degree of autofluorescence. In general, the flagellar promoters are strong so that the signal-to-noise ratio is large.
3. When performing time-course experiments, a separate culture in a separate tube should be grown for each time point so not to disturb growth for later time points.
4. An example of the raw and processed data is provided in Fig. 4. As discussed in the text, the DAPI signal is used to identify cells from debris. We use 2-D plots where cells are plotted according to the side scatter (SSC-A) and DAPI signal (Pacific Blue). We use a rectangular DAPI gate to select the region of interest on the 2D plots (Fig. 4b). The same gate must be applied to generate histograms of FITC signal for all samples. Low DAPI signals indicate cell debris, whereas very high signal indicates cells clumped together. We also smooth the histograms to remove local fluctuations in the raw histograms. The default setting is to use 1024 bins which gives a lot of local fluctuation in cell count (Fig. 4c). We typically use 64 bins, which makes the data smoother (Fig. 4d). FCS express has a smoothing function that can be used to further smooth the data. Also, data can be normalized to a peak value. We chose 100.

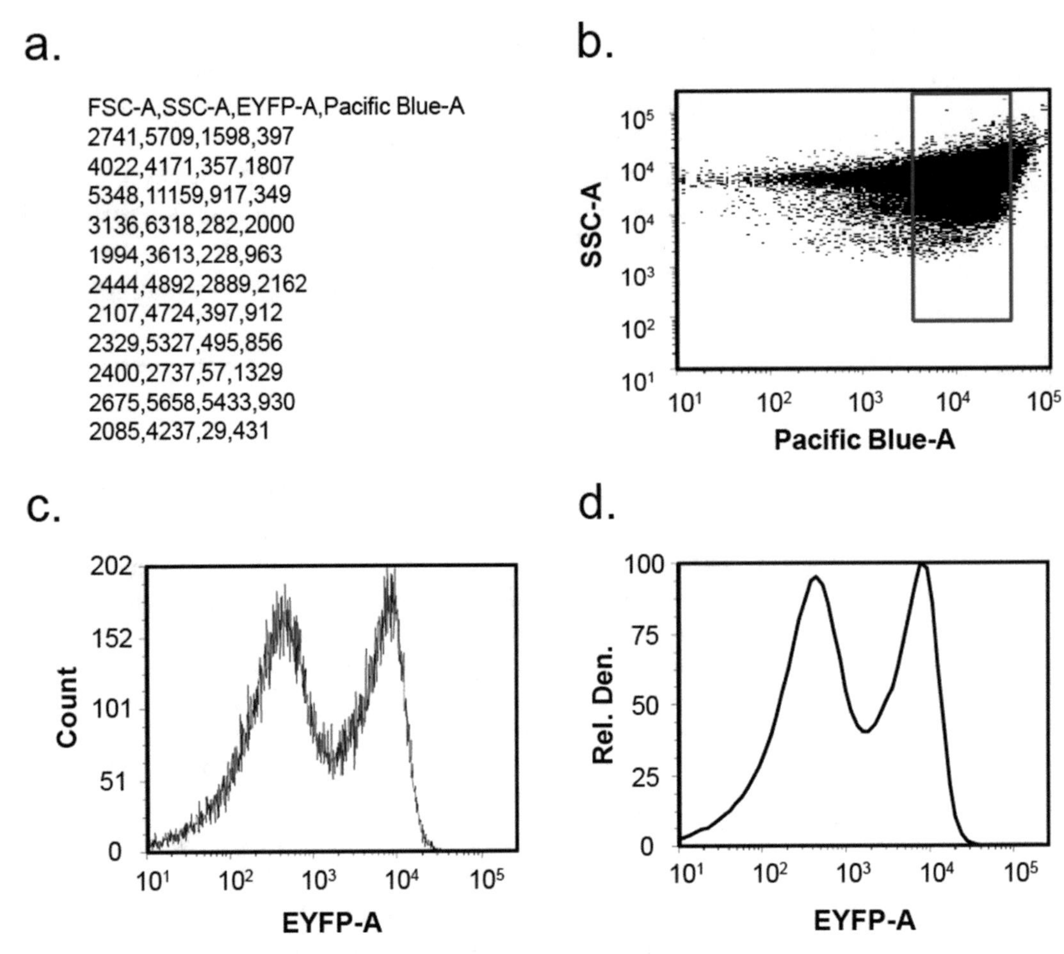

Fig. 4 The steps involved in analyzing flow cytometry data. (**a**) The raw data contains the measured intensity of forward scatter, side-scatter, Pacific Blue/DAPI, and E-YFP/FITC signals for each cell. (**b**) Cells are identified by gating the data based on the DAPI fluorescence signal. (**c**) A univariate histogram with 1024 bins is then generated based on measured Venus fluorescence using the E-YFP/FITC channel. (**d**) The number of bins is reduced to 64 and the histograms are smoothed to remove local fluctuations in the raw histograms

References

1. Chilcott GS, Hughes KT (2000) Coupling of flagellar gene expression to flagellar assembly in Salmonella enterica serovar typhimurium and *Escherichia coli*. Microbiol Mol Biol Rev 64(4):694–708
2. Chevance FF, Hughes KT (2008) Coordinating assembly of a bacterial macromolecular machine. Nat Rev Microbiol 6(6):455–465. doi:10.1038/nrmicro1887
3. Kalir S, McClure J, Pabbaraju K, Southward C, Ronen M, Leibler S, Surette MG, Alon U (2001) Ordering genes in a flagella pathway by analysis of expression kinetics from living bacteria. Science 292(5524):2080–2083. doi:10.1126/science.1058758
4. Karlinsey JE, Tanaka S, Bettenworth V, Yamaguchi S, Boos W, Aizawa SI, Hughes KT (2000) Completion of the hook-basal body complex of the Salmonella typhimurium flagellum is coupled to FlgM secretion and fliC transcription. Mol Microbiol 37(5):1220–1231
5. Kalir S, Alon U (2004) Using a quantitative blueprint to reprogram the dynamics of the flagella gene network. Cell 117(6):713–720. doi:10.1016/j.cell.2004.05.010
6. Kalir S, Mangan S, Alon U (2005) A coherent feed-forward loop with a SUM input function prolongs flagella expression in *Escherichia coli*. Mol Syst Biol 1(2005):0006. doi:10.1038/msb4100010

7. Saini S, Floess E, Aldridge C, Brown J, Aldridge PD, Rao CV (2011) Continuous control of flagellar gene expression by the sigma28-FlgM regulatory circuit in Salmonella enterica. Mol Microbiol 79(1):264–278. doi:10.1111/j.1365-2958.2010.07444.x
8. Brown JD, Saini S, Aldridge C, Herbert J, Rao CV, Aldridge PD (2008) The rate of protein secretion dictates the temporal dynamics of flagellar gene expression. Mol Microbiol 70(4):924–937.doi:10.1111/j.1365-2958.2008.06455.x
9. Saini S, Brown JD, Aldridge PD, Rao CV (2008) FliZ Is a posttranslational activator of FlhD4C2-dependent flagellar gene expression. J Bacteriol 190(14):4979–4988. doi:10.1128/JB.01996-07
10. Koirala S, Mears P, Sim M, Golding I, Chemla YR, Aldridge PD, Rao CV (2014) A nutrient-tunable bistable switch controls motility in Salmonella enterica serovar Typhimurium. MBio 5(5):e01611–e01614. doi:10.1128/mBio.01611-14
11. Saini S, Koirala S, Floess E, Mears PJ, Chemla YR, Golding I, Aldridge C, Aldridge PD, Rao CV (2010) FliZ induces a kinetic switch in flagellar gene expression. J Bacteriol 192(24):6477–6481. doi:10.1128/JB.00751-10
12. Stewart MK, Cummings LA, Johnson ML, Berezow AB, Cookson BT (2011) Regulation of phenotypic heterogeneity permits Salmonella evasion of the host caspase-1 inflammatory response. Proc Natl Acad Sci U S A 108(51):20742–20747. doi:10.1073/pnas.1108963108
13. Stewart MK, Cookson BT (2014) Mutually repressing repressor functions and multi-layered cellular heterogeneity regulate the bistable Salmonella fliC census. Mol Microbiol 94(6):1272–1284. doi:10.1111/mmi.12828
14. Cummings LA, Wilkerson WD, Bergsbaken T, Cookson BT (2006) In vivo, fliC expression by Salmonella enterica serovar Typhimurium is heterogeneous, regulated by ClpX, and anatomically restricted. Mol Microbiol 61(3):795–809. doi:10.1111/j.1365-2958.2006.05271.x
15. Wada T, Morizane T, Abo T, Tominaga A, Inoue-Tanaka K, Kutsukake K (2011) EAL domain protein YdiV acts as an anti-FlhD4C2 factor responsible for nutritional control of the flagellar regulon in Salmonella enterica serovar typhimurium. J Bacteriol 193(7):1600–1611. doi:10.1128/JB.01494-10
16. Takaya A, Erhardt M, Karata K, Winterberg K, Yamamoto T, Hughes KT (2012) YdiV: a dual function protein that targets FlhDC for ClpXP-dependent degradation by promoting release of DNA-bound FlhDC complex. Mol Microbiol 83(6):1268–1284. doi:10.1111/j.1365-2958.2012.08007.x
17. Wada T, Tanabe Y, Kutsukake K (2011) FliZ acts as a repressor of the ydiV gene, which encodes an anti-FlhD4C2 factor of the flagellar regulon in Salmonella enterica serovar typhimurium. J Bacteriol 193(19):5191–5198. doi:10.1128/JB.05441-11
18. Nagai T, Ibata K, Park ES, Kubota M, Mikoshiba K, Miyawaki A (2002) A variant of yellow fluorescent protein with fast and efficient maturation for cell-biological applications. Nat Biotechnol 20(1):87–90. doi:10.1038/nbt0102-87
19. Hawley TS, Hawley RG (2011) Flow cytometry protocols, Methods in molecular biology, vol 699. Humana, New York
20. Andersen JB, Sternberg C, Poulsen LK, Bjorn SP, Givskov M, Molin S (1998) New unstable variants of green fluorescent protein for studies of transient gene expression in bacteria. Appl Environ Microbiol 64(6):2240–2246
21. Shaner NC, Campbell RE, Steinbach PA, Giepmans BN, Palmer AE, Tsien RY (2004) Improved monomeric red, orange and yellow fluorescent proteins derived from Discosoma sp. red fluorescent protein. Nat Biotechnol 22(12):1567–1572. doi:10.1038/nbt1037
22. Haldimann A, Wanner BL (2001) Conditional-replication, integration, excision, and retrieval plasmid-host systems for gene structure-function studies of bacteria. J Bacteriol 183(21):6384–6393. doi:10.1128/JB.183.21.6384-6393.2001
23. Leveau JH, Lindow SE (2001) Predictive and interpretive simulation of green fluorescent protein expression in reporter bacteria. J Bacteriol 183(23):6752–6762. doi:10.1128/JB.183.23.6752-6762.2001
24. Ronen M, Rosenberg R, Shraiman BI, Alon U (2002) Assigning numbers to the arrows: parameterizing a gene regulation network by using accurate expression kinetics. Proc Natl Acad Sci U S A 99(16):10555–10560. doi:10.1073/pnas.152046799

Part II

Structure of the Bacterial Flagellum

Chapter 6

Purification and Characterization of the Bacterial Flagellar Basal Body from *Salmonella enterica*

Shin-Ichi Aizawa

Abstract

The bacterial flagellum is a motility organelle. The flagellum is composed of three main structures: the basal body as a rotary engine embedded in the cellular membranes and cell wall, the long external filament that acts as a propeller, and the hook acting as a universal joint that connects them. I describe protocols for the purification of the filament and hook-basal body from *Salmonella enterica* serovar Typhimurium.

Key words Basal body, Cytoplasmic membrane, Electron microscopy, Filament, Hook, Lipopolysaccharide, Peptidoglycan layer

1 Introduction

1.1 Background

Flagella are anchored in the membrane layers of the cell surface. There are two membrane layers in Gram-negative bacteria: the outer and inner membranes. Between the two membranes, there is a peptidoglycan layer. The main body of the flagellar base is a complex of proteins and called the basal body. The basal body has a unique structure, composed of a four-ring structure and a rod penetrating the rings. A ring-shaped protein complex interacts with each layer and called by the name of the layer: M ring for the inner membrane, P ring for the peptidoglycan layer, and the L ring for the lipopolysaccharide (LPS) in the outer membrane. There is one more ring just above the M ring, which seems to interact with no layer and thus called supramembrane (S) ring [1–3]. The S ring is made of a protein that is the component protein (FliF) of the M ring. Therefore, FliF alone forms the MS ring complex [4]. The function of the S ring is not clear. However, since all flagellar basal bodies so far studied have the S ring, it may be essential for the formation of the stable structure as a torque generator.

The physico-chemical properties of the membranes are all different from each other. The inner membrane can be easily

Tohru Minamino and Keiichi Namba (eds.), *The Bacterial Flagellum: Methods and Protocols*, Methods in Molecular Biology, vol. 1593, DOI 10.1007/978-1-4939-6927-2_6, © Springer Science+Business Media LLC 2017

solubilized with mild detergents such as triton X-100. In contrast, the outer membrane is physically strong and not easy to solubilize due to several factors; (a) the peptidoglycan layer that supports the outer membrane and gives physical strength to it; (b) lipopolysaccharide, which prevents chemicals from freely passing through; (c) pili or fimbriae that stem from the outer membrane; and (d) capsule, which is another membranous layer outside the outer membrane. Here, I describe a method for purification of intact flagella from *Salmonella enterica* serovar Typhimurium, a well-studied Gram-negative species [5]. The method can be applied to almost any bacterial species with minor modifications.

1.2 Overview of the Methods

Before describing the method for purification of the flagellar basal body, I am going to describe the method for purification of flagellar filaments alone (Subheading 3.1). Although the basal body occupies only 1% in size of the flagellum, it can be destroyed in the middle of the rod by shear force inevitably applied to the cell during purification steps [6]. Learning the purification method of filaments will be helpful to avoid losing the basal body from the filaments.

Isolation of filaments is carried out by shearing force. The filament, the major part of the flagellum, is exposed to outside the cell and, thus, can be detached from the cell body by physical forces externally applied such as vortexing the culture media or passing through a pipette with a narrow mouth. Filaments thus isolated retain the hook attached at the proximal end [6]. For the purpose to analyze flagellin, this method is quick and simple to achieve.

Intact flagellum is the filament that retains the hook-basal body attached at the proximal end. In the 1970s, there were two methods for the isolation of the intact flagella from *E. coli* by DePamphilis and Adler [1–3] and from *Bacillus subtilis* by Demitt and Simon [7]. In their methods, cells are lysed using a combination of a lytic enzyme such as lysozyme and a detergent such as Triton X-100. They succeeded in isolation of the intact flagella but not in purification. I further developed and simplified their methods and established a method for purification of intact flagella (Subheading 3.2). The purification method is mostly based on methods for purifying flagellar filaments using a combination of low- and high-speed spins. For the efficient isolation of intact flagella, mild treatment through the procedure is required to avoid filaments falling off the cell body as mentioned above.

2 Materials

Prepare all solutions using Milli-Q water and analytical grade reagents and then autoclave them at 121 °C for 20 min.

2.1 Strains

1. *Salmonella enterica* serovar Typhimurium SJW1103 (wild type for motility and chemotaxis).

2.2 Culture Media

1. Luria broth (LB): 1% (w/v) Bacto tryptone, 0.5% (w/v) Yeast extract, 1.0% (w/v) NaCl.
2. LB agar plate (LA): 1% (w/v) Bacto tryptone, 0.5% (w/v) Yeast extract, 1.0% (w/v) NaCl, 1.5% (w/v) Bacto agar.

2.3 Cell Growth and Harvest

1. 37 °C shaker.
2. Spectrophotometer.
3. Low-speed Centrifuge.
4. Ice.

2.4 Preparation of Flagellar Filaments

1. Phosphate buffered saline (PBS): 10 mM phosphate buffer, 1.0% (w/v) NaCl, pH 7.0.
2. Pasteur pipettes.
3. High-speed centrifuge.

2.5 Preparation of Flagella with the Basal Body Attached

1. Sucrose solution: 10% (w/v) sucrose, 0.1 M Trizma base (*see* **Note 1**).
2. Spatulas.
3. Pasteur pipettes
4. Stirrer, stirrer bars
5. 2 mg/mL lysozyme in DW (freshly prepared).
6. 100 mM EDTA, pH 7.0.
7. 10% (w/v) Triton X-100.
8. 100 mM $MgSO_4$.
9. Cesium chloride (CsCl).
10. TET buffer: 10 mM Tris–HCl, pH 8.0, 5 mM EDTA, 0.1% (w/v) Triton X-100.
11. Saturated ammonium sulfate.
12. Polyethylene glycol (PEG 20,000) powder.
13. Acidic solution: 50 mM glycine–HCl, pH 3.0, 0.1% (w/v) Triton X-100.
14. Beckman SW27.1 swing rotor and the buckets.

2.6 SDS-PAGE

1. SDS-loading buffer (2×): 125 mM Tris–HCl, pH 6.8, 4% (w/v) sodium dodecyl sulfate (SDS), 20% (w/v) glycerol, 0.002% (w/v) bromophenol blue.
2. 1 M Tris.
3. 12.5% SDS-polyacrylamide gel and gel apparatus for SDS-PAGE.

4. Running buffer for SDS-PAGE: 30 g Tris, 144 g Glycine, 10 g SDS per liter.
5. 95 °C heating block.

3 Methods

3.1 Purification of Flagellar Filaments

1. Inoculate 5 mL of overnight culture of *Salmonella* cells into 100 mL of a LB medium and incubate at 37 °C with shaking until the cells grow into the stationary phase or overnight.
2. Harvest cells by low-speed spin (8000 × *g*, 15 min, 4 °C).
3. Discard the supernatants.
4. Dissolve the cell pellets in 10 mL of PBS (*see* **Note 2**).
5. Repeat pipetting cell suspension with a pipette with a narrow mouth for 10–20 times.
6. Observe the cells by microscopy, and if there are still motile cells, repeat the pipetting in **step 5**.
7. Dilute the suspension with 20 mL of PBS.
8. Remove cells by low-speed spin. Repeat spin to remove cells completely.
9. Transfer supernatants into a 50 mL centrifuge tube.
10. Collect filaments by high-speed spin (15,000 × *g*, 30 min, 4 °C).
11. Add PBS to cover the pellets and leave them standing in cold overnight.
12. Collect softened pellets in a tube or a bottle.
13. Examine the purity of filaments by SDS-PAGE and electron microscopy. Purified flagellar filaments will give a single band of flagellin in SDS gel. In some species such as *Bdellovibrio*, a filament is composed of several flagellins and gives several bands (Fig. 1) [8]. By electron microscopy, samples are dominated by filaments. Among them, pili and fimbriae are often contaminated as minor components (Fig. 2).

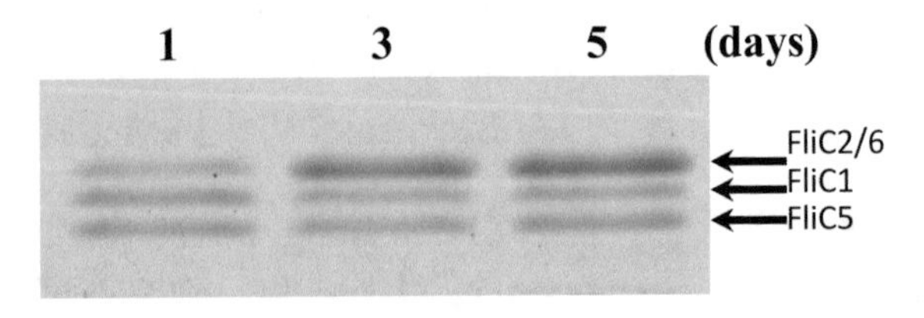

Fig. 1 The band pattern of multi-flagellin filament of *Bdellovibrio* in SDS gel

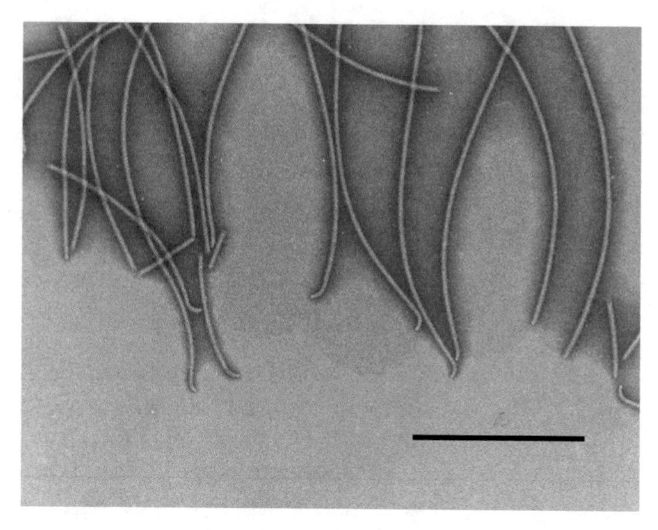

Fig. 2 Electron micrograph of purified filaments from *Salmonella* Typhimurium. One end of each filament attaches the hook and the distal rod, which look different from the filament by PTA negative staining. *Bar* indicates 500 nm

3.2 Isolation and Purification of Intact Flagella

1. Inoculate 5 mL of overnight culture of *Salmonella* cells into 100 mL of fresh LB and incubate at 37 °C with shaking for 4 h or until the cells grow into the late-log phase.
2. Harvest cells by low-speed spin (3000 × *g*, 15 min, 4 °C).
3. Discard the supernatants.
4. Add 10 mL of the sucrose solution to the cell pellets. Cells are gently dissolved using a spatula and a wide-bore pipette.
5. Add lysozyme solution at a final concentration of 0.1 mg/mL. Place the pipette tip at the bottom of the cell suspension and add the solution very slowly (*see* **Note 3**).
6. Add EDTA solution at a final concentration of 10 mM). Place the pipette tip at the bottom of the cell suspension and add the solution very slowly (*see* **Note 3**).
7. Take the beaker out of ice to room temperature for lysozyme to work well. Check sphaeroplast formation by microscopy; keep stirring the cell suspension until cells become round. It usually takes 30 min to complete.
8. Add a nonionic detergent Triton X-100 solution at a final concentration of 1% (w/v). The cloudy solution instantly becomes translucent and the center of the solution rises because of the chromosomal DNA.
9. Keep stirring suspension at room temperature until the viscosity of the suspension decreased due to DNA degradation by endogenous DNases (*see* **Note 4**).

10. Add $MgCl_2$ solution at a final concentration of 10 mM into the lysate. For *Salmonella* species, endogenous DNase digests DNA within 30 min without Mg^{2+} addition. However, most Gram-negative species do require Mg^{2+} for endogenous DNase to work.
11. Add EDTA solution at a final concentration of 1 mM after the lysate becomes thin to prevent re-aggregation of cell membranes and walls in the presence of Mg^{2+}.
12. Keep the cell lysate on ice until use.

Up to this step, intact flagella are completely isolated from the cell body. For further purification of intact flagella, there are two methods: Treat with alkaline pH or apply onto a CsCl density gradient. Both methods are used to remove or dissolve the outer membrane vesicles (OMV), which was most difficult to remove in the original method.

3.2.1 Alkaline Treatment

OMVs are solubilized by alkaline treatment in the presence of EDTA. This is a harsh condition for proteins, but the flagella are strong enough to survive this treatment.

1. Adjust pH of the cell lysate to ten by adding drops of 1 N NaOH solution to suspension constantly stirred. Frequently check pH of the solution with a slip of a pH indicator paper.
2. Remove unlysed cells by low-speed spin (5000 × *g*, 15 min, 4 °C). Repeat twice.
3. Transfer the supernatants into polyallomer centrifuge tubes. The lysate is alkaline, and thus polycarbonate centrifuge tubes cannot be used.
4. Collect intact flagella by high-speed spin (30,000 × *g*, 60 min, 4 °C).
5. Discard the supernatants and wipe the tube wall with Kimwipe.
6. Add TET buffer to cover the pellets and leave them standing in cold overnight.
7. Collect flagella softened in TET buffer into a tube and store in cold until use.

3.2.2 Purification of Intact Flagella by CsCl Density Gradient Centrifugation

This method can separate intact flagella from OMV due to their different buoyant density in a CsCl density gradient.

1. Add saturated ammonium sulfate at a final concentration of 33% (w/v) or polyethylene glycol (PEG) powder at a final concentration of 10% (w/v) to the cell lysate to make aggregates of flagella.
2. Collect aggregated flagella by low-speed spin (5000 × *g*, 15 min, 4 °C).

3. Discard the supernatants carefully not to disturb the flagella paste on the wall of the centrifuge tube.
4. Add 1 ml of TET buffer and wash the flagella paste on the wall.
5. Put the flagella suspension into dialysis bags and dialyze against TET buffer. Change the dialysis solution after 1 h and keep dialyzing overnight.
6. Dilute dialyzed solution to the volume of centrifuge tubes with TET buffer, add CsCl powder directly into dialyzed solution at a final concentration of 33% (w/v), and pour the mixture into centrifuge tubes.
7. Spin the solution in Beckman SW27.1 buckets at 55,000 × *g* for 16 h at 20 °C.
8. Collect the flagellar band in the gradient with a Pasteur pipette. Flagella form a thick band 3/4 of the way down the tube, membrane fragments form a diffuse band above, and the cell debris collects at the bottom.
9. Dialyze the collected flagella sample against TET buffer.
10. Store the sample in cold until use

3.2.3 Purification of the Hook-Basal Body Complex from Intact Flagella

Flagellar filament occupies 99% of the flagellum, judging from its size (10 μm filament vs, 100 nm HBB). To analyze HBB, the abundant filaments have to remove. There are two ways to remove filament: acidic pH or heat treatments. I describe acidic pH treatment method.

1. Add the acidic solution into the flagellar filaments with stirring and leave it for 1 h at room temperature (*see* **Note 5**).
2. Spin the suspension by ultracentrifugation (100,000 × *g*, 60 min, 4 °C).
3. Discard the supernatants, and wipe the tube wall with Kimwipe to remove the residual flagellin.
4. Add TET buffer into the pellets.
5. Store in cold until use.
6. The purity of HBB sample is examined by electron microscopy (EM) and SDS polyacrylamide gel electrophoresis (PAGE).
7. Purified HBB will give several bands of protein components of the basal body in SDS gel (Fig. 3). The hook protein (42 kDa) is the major component, followed by residual flagellin (52 kDa) in various amounts from preparation to preparation. FliF (65 kDa), the component protein of the MS ring complex, often gives a second band at 63 kDa due to unknown degradation. Flagellar component proteins usually do not contain cysteine. An exception is FlgI (38 kDa in the presence of β-mercaptoethanol), the component protein of the P ring, which gives a band at 37 kDa in the absence of BME due to two cysteine residues contained in the protein.

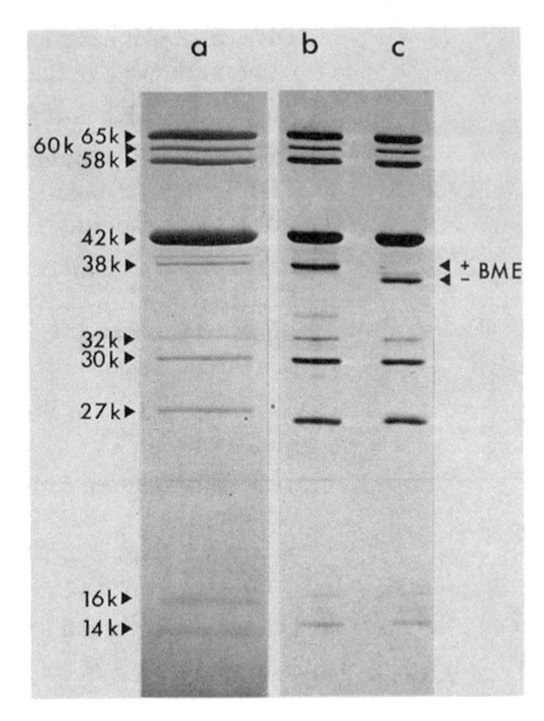

Fig. 3 SDS-gel band pattern of purified HBB from *Salmonella* Typhimurium

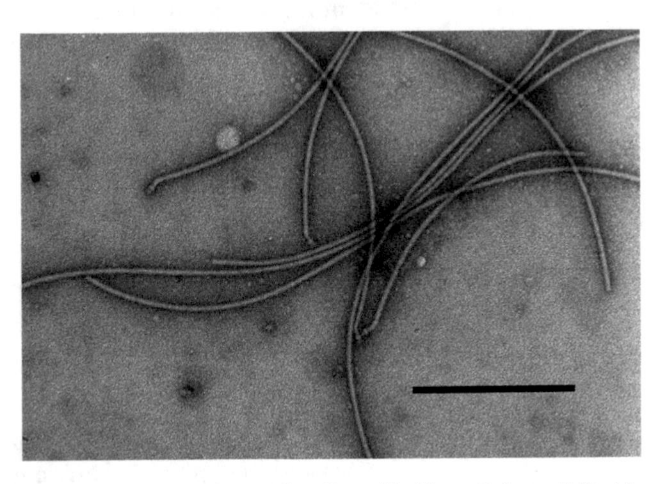

Fig. 4 Electron micrograph of intact flagella purified from *Salmonella* Typhimurium. *Bar* indicates 100 nm

By EM, intact flagella retain basal bodies connected to the filament through the hook, which appears uniquely curved in a hook shape (Fig. 4). Isolated HBB is composed of the hook, four rings, and the rod (Fig. 5).

The method for the purification of intact flagella from *Salmonella* Typhimurium has been applied to other species with minor modifications [9]. Some examples are shown in references [10–15].

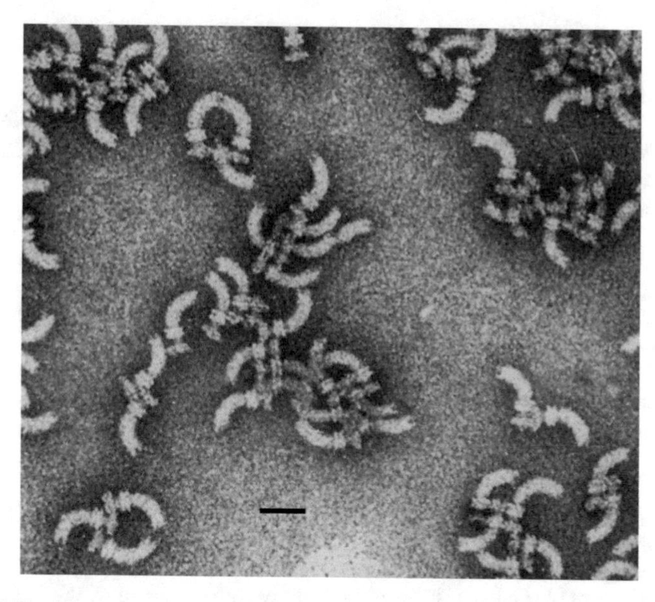

Fig. 5 Electron micrograph of HBB purified from *Salmonella* Typhimurium. *Bar* indicates 100 nm

4 Notes

1. pH is not adjusted. This is important, because EDTA works better at alkaline pH.
2. Cells pelleted by centrifugation should be thoroughly homogenized using a spatula and a wide-bore pipette. Treat them gently to avoid shearing off flagella from cells. Pelleted cells make aggregates. Take time to disperse aggregates into single cells.
3. When adding lysozyme and EDTA solutions to the cell suspension, place the pipette tip at the bottom of the beaker and add solution very slowly. Dropping those solutions from above is not recommended, because those solutions will stay on the top of cell suspension, which contains 20% sucrose.
4. By adding Triton X-100 solution, the cloudy cell suspension instantly becomes translucent and the center of the solution rises because of chromosomal DNA. In *Salmonella*, endogenous DNase digests the DNA within 30 min without Mg addition. However, most Gram-negative species do require Mg for DNase to work. For some species, addition of Mg is not sufficient. Then add a tiny amount of powdered DNase. Confirm the complete digestion of DNA using a pipette; if DNA remains, drops of lysate tend to stay at the tip.
5. To disassemble filaments of the intact flagella, add more than ten times volume of the acid solution into the sample solution. After ultracentrifugation to collect HBB in the pellet, remove residual flagellin remained on the wall of the centrifuge tube with Kimwipe.

References

1. DePamphilis ML, Adler J (1971) Purification of intact flagella from *Escherichia coli* and *Bacillus subtilis*. J Bacteriol 105:376–383
2. DePamphilis ML, Adler J (1971) Fine structure and isolation of the hook-basal body complex of flagella from *Escherichia coli* and *Bacillus subtilis*. J Bacteriol 105:384–395
3. DePamphilis ML, Adler J (1971) Attachment of flagellar basal bodies to the cell envelope: specific attachment to the outer, lipopolysaccharide membrane and the cytoplasmic membrane. J Bacteriol 105:396–407
4. Ueno T, Oosawa K, Aizawa SI (1994) Domain structures of the MS ring component protein (FliF) of the flagellar basal body of *Salmonella typhimurium*. J Mol Biol 236:546–555
5. Aizawa SI, Dean GE, Jones CJ, Macnab RM, Yamaguchi S (1985) Purification and characterization of the flagellar hook basal body complex of *Salmonella typhimurium*. J Bacteriol 161:836–849
6. Okino H, Isomura M, Yamaguchi S, Magariyama Y, Kudo S, Aizawa SI (1989) Release of flagellar filament hook rod complex by a *Salmonella typhimurium* mutant defective in the m ring of the basal body. J Bacteriol 171:2075–2082
7. Dimmitt K, Simon M (1971) Purification and thermal stabilization of intact *Bacillus subtilis* flagella. J Bacteriol 105:369–375
8. Iida Y, Hobley L, Lambert C, Sockett E, Aizawa SI (2009) Roles of multiple flagellins in flagellar formation and flagellar growth post bdelloplast- lysis, in *Bdellovibrio bacteriovorus*. J Mol Biol 394:1011–1021
9. Aizawa SI (2014) The Flagellar World. Elsevir, Oxford
10. Kubori T, Okumura M, Kobayashi N, Nakamura D, Iwakura M, Aizawa SI (1997) Purification and characterization of the flagellar hook-basal body complex of *Bacillus subtilis*. Mol Microbiol 24:399–410
11. Kobayashi K, Saitoh T, Shah DSH, Ohnishi K, Goodfellow IG, Sockett RE, Aizawa SI (2003) Purification and characterization of the flagellar basal body of *Rhodobacter sphaeroides*. J Bacteriol 185:5295–5300
12. Kanbe M, Shibata S, Jenal U, Aizawa S-I (2005) Protease susceptibility of the *Caulobacter crescentus* flagellar hook-basal-body; a possible mechanism of flagellar ejection during cell differentiation. Microbiology 151:433–438
13. Kanbe M, Yagasaki J, Zehner S, Göttfert M, Aizawa SI (2007) Characterization of two sets of sub-polar flagella in *Bradyrhizobium japonicum*. J Bacteriol 189:1083–1089
14. Haya S, Tokumaru Y, Abe N, Kaneko J, Aizawa SI (2011) Characterization of lateral flagella in *Selenomonas ruminantium*. Appl Environ Microbiol 77:2799–2802
15. Uchida K, Jang MS, Ohnishi Y, Horinouchi S, Hayakawa M, Fujita N, Aizawa SI (2011) Characterization of spore flagella in *Actinoplanes missouriensis*. Appl Environ Microbiol 77:2559–2562

Chapter 7

Design and Preparation of the Fragment Proteins of the Flagellar Components Suitable for X-Ray Crystal Structure Analysis

Katsumi Imada

Abstract

Terminal disordering in a monomeric state is a common structural property among bacterial flagellar axial proteins. The conformational flexibility of disordered regions of a protein often disturbs its crystallization. Moreover, disordered regions sometimes cause the aggregation problem. Therefore, trimming disordered regions is essential for crystallization of this type of proteins. In this chapter, we describe a simple but powerful method to determine the stable core and metastable fragments of target proteins for crystallization. This method including limited proteolysis in combination with SDS-PAGE and MALDI-TOF mass spectrometry can be applied to almost any proteins containing disordered regions.

Key words Limited proteolysis, MALDI-TOF, Crystallization, Core fragment, Disordered region

1 Introduction

The bacterial flagellum is a huge molecular assembly composed of more than 20,000 protein subunits of about 50 different types of proteins. The structural components of the flagellum are divided into two classes, the basal body rings and the tubular axial structure [1, 2]. Each of the component proteins has specific regions responsible for tight connection with its neighboring subunits to form the assembly. These regions are usually unfolded in the monomeric state and tend to cause undesired aggregates when a protein is overexpressed. The flagellar axial proteins share disordered N and C terminal regions that fold into an α-helical bundle structure in the flagellum [3–8]. The disordered regions tend to form filamentous aggregates at high protein concentrations required for crystallization [9–12]. Therefore, determination and removal of the unfolded regions are essential for crystallization as well as the characterization of functionally important folded regions of the protein subunit. Limited proteolysis in combination with

Tohru Minamino and Keiichi Namba (eds.), *The Bacterial Flagellum: Methods and Protocols*, Methods in Molecular Biology, vol. 1593, DOI 10.1007/978-1-4939-6927-2_7,

SDS-PAGE and MALDI-TOF mass spectrometry is a simple but powerful method to determine the stable core region of the protein [8–10]. Here, we describe a systematic method to determine the core region of the protein or the protein complex for crystallization. It should be noted that the smallest core fragment is not always crystallized with a high enough quality for structural analysis. Metastable fragments that show partially resistant to proteolysis often grow into crystals suitable for X-ray analysis [8, 9, 13–16]. Therefore, monitoring the time-course of the degradation pattern is important to detect the metastable fragments.

The method is in two parts: (1) the preliminary proteolytic experiment; and (2) the time-course monitoring of degradation. Initially, an appropriate ratio of the protease to the sample protein is determined in a small-scale proteolytic experiment. In the next experiment, the digestion process is monitored by sampling the degradation products. The proteolytic fragments are analyzed by SDS-PAGE and MALDI-TOF mass spectrometry to determine the core and metastable fragments. Several experiments are usually carried out with proteases of different specificities. Trypsin, GluC, carboxypeptidase, and aminopeptidase I are often used. Typical examples of the time-course digestion profiles are shown in Fig. 1. Once the stable core and the metastable regions are determined, the region will be overexpressed and purified for crystallization. It would be possible to obtain a large amount of the core or metastable fragments from the protein by scaling up and adjusting the proteolytic condition.

This technique can be widely applied to any other proteins that contain disordered regions and is useful for the preparation of the core fragment that may be used for the characterization of the protein [17].

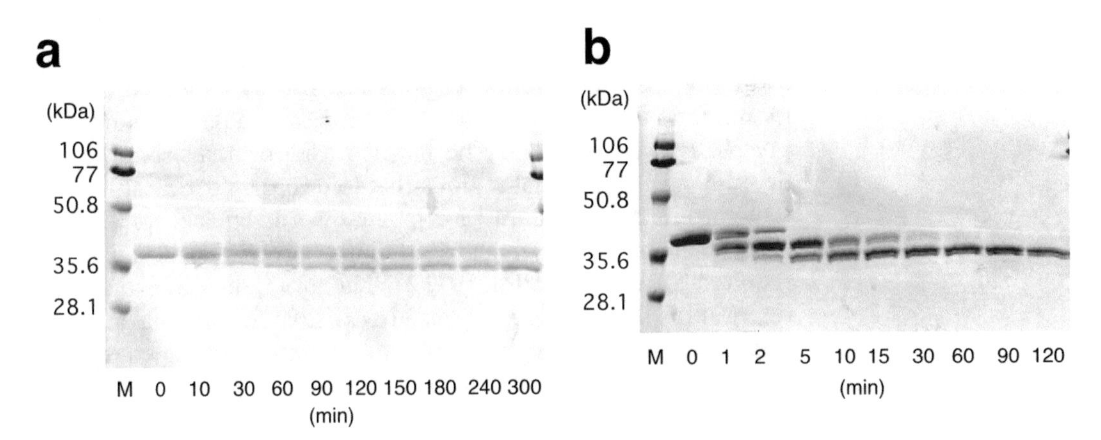

Fig. 1 Typical profiles of limited proteolysis. The digestion profiles of FlgL by (**a**) carboxypeptidase Y and (**b**) trypsin. The *left-end lane* is the molecular weight marker

2 Materials

Prepare all solution using Mill-Q water.

2.1 For Limited Proteolysis (Pre liminary Experiment) and Limited proteolysis (Time-Course Experiment)

1. SDS sample buffer.
2. Purified protein solution: 0.5 mg/mL purified protein in 20 mM HEPES-NaOH, pH8.0 (*see* **Note 1**).
3. Eppendorf tubes.
4. Trypsin solution: 0.05 mg/mL trypsin in 20 mM HEPES–NaOH, pH 8.0 (*see* **Note 1**).
5. GluC solution: 0.05 mg/mL endoprotease Glu-C (V8 protease) in 20 mM HEPES–NaOH, pH 8.0 (*see* **Notes 1** and **2**).
6. Carboxypeptidase Y (CPY) solution: 0.1 mg/mL CPY in 20 mM HEPES–NaOH, pH 8.0 (*see* **Note 1**).
7. Aminopeptidase I (API) solution: 0.25 mg/mL API in 20 mM HEPES–NaOH, pH 8.0 (*see* **Notes 1** and **3**).
8. Electric digital timer.
9. Two heating blocks (one for proteolytic reaction (*see* **Note 4**), and the other for SDS sample preparation).
10. 12% poly acrylamide gel and gel apparatus for SDS-PAGE.

2.2 For Limited Proteolysis (The Time-Course Monitoring of Degradation)

1. Sinapinic acid.
2. Acetonitrile.
3. 1% trifluoroacetic acid (TFA) solution (*see* **Note 5**).
4. MALDI-TOF mass spectrometer.

3 Methods

3.1 Limited Proteolysis (Preliminary Experiment)

1. Prepare two 1.5 mL micro tubes with 10 μL of SDS sample buffer. Label the tubes with the sampling time (0 and 90 min).
2. Take 50 μL of protein solution (0.5 mg/mL) in a 1.5 mL micro tube.
3. Incubate the protein solution at 27 °C for 10 min on a heating block (*see* **Note 4**).
4. Take 5 μL of the protein solution and add to the tube with 10 μL SDS sample buffer labeled with 0 min.
5. Heat the SDS sample at 95 °C for 2 min on the heating block, and then cool down to room temperature.
6. Add 5 μL of protease solution to the tube with the protein solution and keep at 27 °C on the heating block (*see* **Note 4**) for 90 min.

7. Take 5 μL of the reaction mixture and add it to the tube with 10 μL SDS sample buffer labeled with 90 min.
8. Heat the SDS sample at 95 °C for 2 min on a heating block, and then cool down to room temperature.
9. Repeat **steps 1–8** for other three proteases, or do simultaneously.
10. Analyze the sample before and after digestion (0 and 90 min, respectively) by SDS-PAGE.
11. If the protein sample is not degraded at all, increase the amount of protease (*see* **Notes 3** and **6**).
12. If the protein sample is completely degraded, decrease the amount of protease (Add 1 μL of protease solution instead of 5 μL at **step 6**).

3.2 Limited Proteolysis (The Time-Course Monitoring of Degradation)

1. Add 1 mg of sinapinic acid in the solution containing 30 μL of acetonitrile, 60 μL of Mill-Q water, and 10 μL of 1% TFA (*see* **Note 5**). Centrifuge the solution to remove the insoluble matter.
2. Prepare nine 1.5 mL micro tubes with 10 μL SDS sample buffer. Label the tubes with the sampling time (0, 1, 3, 5, 10, 15, 30, 60, and 90 min). Place the tubes properly in the rack to avoid confusion.
3. Take 100 μL of protein solution (0.5 mg/mL) in a 1.5 μL micro tube.
4. Incubate the protein solution at 27 °C for 10 min on a heating block (*see* **Note 4**).
5. Take 5 μL of the protein solution and add it to the tube with 10 μL SDS sample buffer labeled with 0 min.
6. Heat the mixture at 95 °C for 2 min on the heating block, and then cool down to room temperature.
7. Take 0.5 μL of the protein solution and apply it on the MALDI plate.
8. Add 0.5 μL of the matrix solution immediately on the top of the protein drop.
9. Add 10 μL of protease solution to the tube with protein solution and keep at 27 °C on the heat block (*see* **Note 4**).
10. Take 5 μL of the reaction mixture and mix it with 10 μL of the SDS sample buffer at the sampling time (1, 3, 5, 10, 15, 30, 60, 90 min after the addition of the protease).
11. Heat the mixture at 95 °C for 2 min on the heat block, and then cool down to room temperature.
12. Immediately after setting the SDS sample on the heat block, take 0.5 μL of the protein solution from the reaction mixture, and apply it on the MALDI plate.

13. Add 0.5 μL of the matrix solution immediately on the top of the protein drop.
14. Repeat **steps 9–12** until the last sample.
15. Repeat **steps 1–13** for other three proteases, or do simultaneously.
16. Analyze the samples (0, 1, 3, 5, 10, 15, 30, 60, 90 min) by SDS-PAGE.
17. Analyze the samples on the MALDI plate (0, 1, 3, 5, 10, 15, 30, 60, 90 min) with the MALDI-TOF mass spectrometer (*see* **Note 7**).
18. Assign each proteolytic fragment by comparing the molecular mass obtained by mass spectrometry with the calculated mass from the amino acid sequence of possible proteolytic fragments (*see* **Note 8**).
19. Compare the degradation pattern by each protease analyzed by SDS-PAGE with that by mass spectrometry, and assign the regions resistant to the protease (Fig. 2, *see* **Note 9**).
20. Together with the results of four proteases, determine the core and (if any) metastable regions.

4 Notes

1. HEPES-NaOH, pH 7.0–8.0, and Tris–HCl, pH 7.0–8.0 are both useable, but use the same buffer for the preparation of the protein and protease solution. Do not include EDTA. If the original protein solution contains EDTA, remove it with dialysis or using desalting column.
2. GluC also cleaves the C-terminal side of aspartic acid in the phosphate buffer. No cleavage occurs between Glu (Asp) and Pro.
3. The protease activity of API is relatively low. If the protease activity is not high enough, add $CaCl_2$ to the reaction mixture at a final concentration of 1 mM. API cannot cleave the N-terminal side of proline. If the N-terminus of the digested fragment is proline, the fragment may still have a disordered N-terminal region.
4. A water bath can be used instead of a heating block.
5. If the protein fragments are difficult to detect by MALDI-TOF mass spectrometry, use 3% TFA solution.
6. For example, increase the protease concentration by five times.
7. Make sure all the sample drops are completely dried up.
8. PeptideMass (http://web.expasy.org/peptide_mass/) [18, 19] and ProtParam (http://web.expasy.org/protparam/) [19]

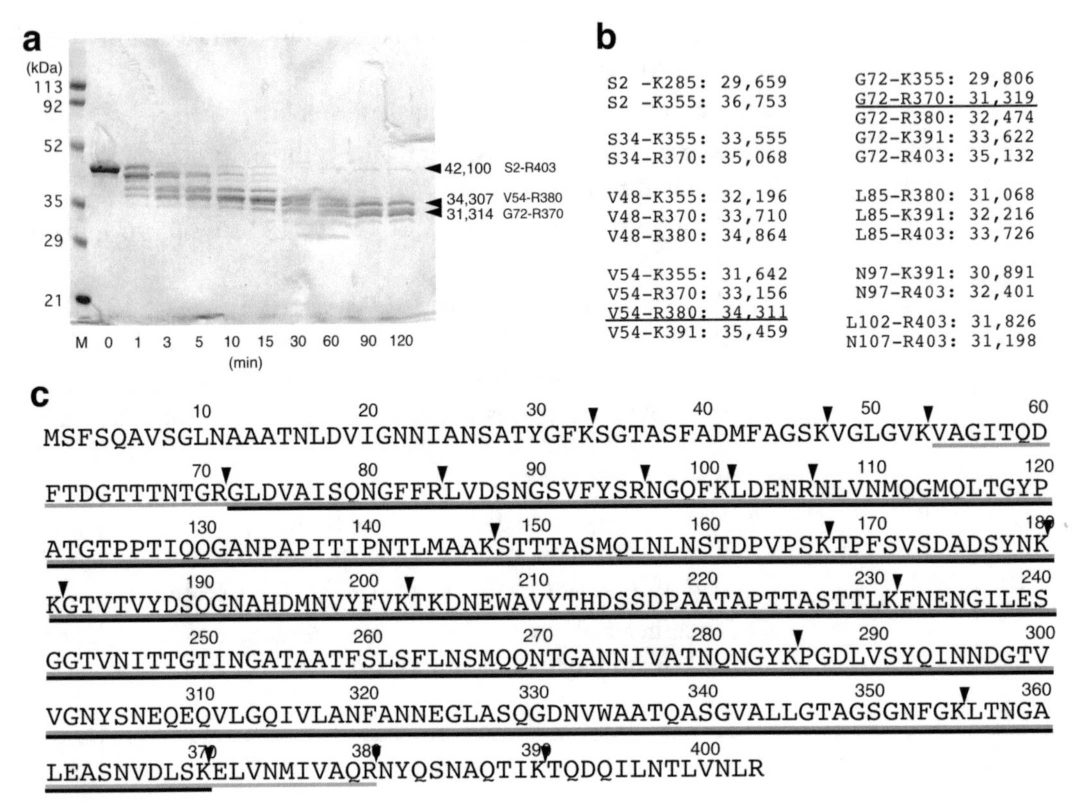

Fig. 2 An example of the determination of the core fragment. (**a**) The digestion profile of FlgE by trypsin. Two major fragments were obtained after 120 min digestion. The left-end lane is the molecular weight marker. The molecular mass of the major fragments and the purified FlgE used in this experiment measured by MALDI-TOF mass spectrometry are shown in the right of the figure. The measured molecular mass of FlgE was 42,100, indicating that the N-terminal methionin of this sample is missing. (**b**) The molecular mass of the tryptic fragments of FlgE calculated from the amino acid sequence. The fragments whose molecular mass is close to the measured values are listed. V54-R380 and G72-R370 show the best match to the MALDI-TOF data. (**c**) Amino acid sequence of FlgE. The trypsin cleavage sites are indicated by *arrow heads*. The regions corresponding to the two major fragments, V54-R380 and G72-R370, are shown in *gray* and *black bars*, respectively. In this case, G72-R370 is a core fragment and V54-R380 is a metastable fragment

in the proteomics tools on the ExPASy server are useful for the calculation of the mass of the fragment from the amino acid sequence.

9. The peak heights of the MALDI-TOF mass spectra do not reflect the amount of the fragments.

Acknowledgment

This work was supported in part by JSPS KAKENHI Grant Numbers JP15H02386 and MEXT KAKENHI Grant Numbers JP23115008.

References

1. Macnab RM (2003) How bacteria assemble flagella. Annu Rev Microbiol 57:77–100
2. Minamino T, Imada K, Namba K (2008) Mechanisms of type III protein export for bacterial flagellar assembly. Mol Biosyst 4:1105–1115
3. Vonderviszt F, Aizawa S, Namba K (1991) Role of the disordered terminal regions of flagellin in filament formation and stability. J Mol Biol 221:1461–1474
4. Vonderviszt F, Ishima R, Akasaka K, Aizawa S (1992) Terminal disorder: a common structural feature of the axial proteins of bacterial flagellum? J Mol Biol 226:575–579
5. Mimori-Kiyosue Y, Vonderviszt F, Namba K (1997) Locations of terminal segments of flagellin in the filament structure and their roles in polymerization and polymorphism. J Mol Biol 270:222–237
6. Yonekura K, Maki-Yonekura S, Namba K (2003) Complete atomic model of the bacterial flagellar filament by electron cryomicroscopy. Nature 424:643–650
7. Fujii T, Kato T, Namba K (2009) Specific arrangement of alpha-helical coiled coils in the core domain of the bacterial flagellar hook for the universal joint function. Structure 17:1485–1493
8. Saijo-Hamano Y, Uchida N, Namba K, Oosawa K (2004) In vitro characterization of FlgB, FlgC, FlgF, FlgG, and FliE, flagellar basal body proteins of *Salmonella*. J Mol Biol 339:423–435
9. Samatey FA, Imada K, Vonderviszt F, Shirakihara Y, Namba K (2000) Crystallization of the F41 fragment of flagellin and data collection from extremely thin crystals. J Struct Biol 132:106–111
10. Samatey FA, Matsunami H, Imada K, Nagashima S, Namba K (2004) Crystallization of a core fragment of the flagellar hook protein FlgE. Acta Crystallogr D 60:2078–2080
11. Samatey FA, Imada K, Nagashima S, Vonderviszt F, Kumasaka T, Yamamoto M, Namba K (2001) Structure of the bacterial flagellar hook and implication for the molecular universal joint mechanism. Nature 410:331–337
12. Samatey FA, Matsunami H, Imada K, Nagashima S, Shaikh TR, Thomas DR, Chen JZ, DeRosier DJ, Kitao A, Namba K (2004) Structure of the bacterial flagellar hook and implication for the molecular universal joint mechanism. Nature 431:1062–1068
13. Saijo-Hamano Y, Matsunami H, Namba K, Imada K (2013) Expression, purification, crystallization and preliminary X-ray diffraction analysis of a core fragment of FlgG, a bacterial flagellar rod protein. Acta Crystallogr Sect F Struct Biol Cryst Commun 69:547–550
14. Minamino T, González-Pedrajo B, Oosawa K, Namba K, Macnab RM (2002) Structural properties of FliH, an ATPase regulatory component of the Salmonella type III flagellar export apparatus. J Mol Biol 322: 281–290
15. Uchida Y, Minamino T, Namba K, Imada K (2012) Crystallization and preliminary X-ray analysis of the FliH-FliI complex responsible for bacterial flagellar type III protein export. Acta Crystallogr Sect F Struct Biol Cryst Commun 68:1311–1134
16. Imada K, Minamino T, Uchida Y, Kinoshita M, Namba K (2016) Insight into the flagella type III export revealed by the complex structure of the type III ATPase and its regulator. Proc Natl Acad Sci USA 113:3633–3638
17. Nakano N, Kubori T, Kinoshita M, Imada K, Nagai H (2010) Crystal structure of Legionella DotD: insights into the relationship between type IVB and type II/III secretion systems. PLoS Pathog 6:e1001129
18. Wilkins MR, Lindskog I, Gasteiger E, Bairoch A, Sanchez J-C, Hochstrasser DF, Appel RD (1997) Detailed peptide characterisation using PEPTIDEMASS—a World-Wide Web accessible tool. Electrophoresis 18:403–408
19. Gasteiger E, Hoogland C, Gattiker A, Duvaud S, Wilkins MR, Appel RD, Bairoch A (2005) Protein identification and analysis tools on the ExPASy server. In: Walker JM (ed) The proteomics protocols handbook. Humana Press, New York, pp 571–607

Chapter 8

Structural Analysis of the Flagellar Component Proteins in Solution by Small Angle X-Ray Scattering

Lawrence K. Lee

Abstract

Small angle X-ray scattering is an increasingly utilized method for characterizing the shape and structural properties of proteins in solution. The technique is amenable to very large protein complexes and to dynamic particles with different conformational states. It is therefore ideally suited to the analysis of some flagellar motor components. Indeed, we recently used the method to analyze the solution structure of the flagellar motor protein FliG, which when combined with high-resolution snapshots of conformational states from crystal structures, led to insights into conformational transitions that are important in mediating the self-assembly of the bacterial flagellar motor. Here, we describe procedures for X-ray scattering data collection of flagellar motor components, data analysis, and interpretation.

Key words Small angle X-ray scattering, Protein structure, Protein dynamics, Bacterial flagellar motor, Motility, Chemotaxis, Synchrotron radiation

1 Introduction

1.1 SAXS Analysis of Flagellar Motor Components

Small angle X-ray scattering (SAXS) can be used to determine structural properties of proteins in solution. Combined with other structural characterization methods, in particular X-ray crystallography, SAXS can be a powerful method to dissect molecular mechanisms. This is especially true of flagellar motor components, such as switch complex proteins in the rotor [1], stator complexes [2, 3], hook [4], and filament proteins [5], which are necessarily dynamic, adopting distinct conformational states to facilitate different modes of function. SAXS analysis provides information on the shape of these components in solution including precision measurements on the radius of gyration (Rg), molecular weight (and hence stoichiometry), estimates of absolute dimensions, and average shape. Furthermore, where multiple conformational states exist, SAXS can provide an estimate of the equilibrium proportion of molecules in each state and hence the free energy associated with the transition from one state to the other [6], and be used to

Tohru Minamino and Keiichi Namba (eds.), *The Bacterial Flagellum: Methods and Protocols*, Methods in Molecular Biology, vol. 1593, DOI 10.1007/978-1-4939-6927-2_8,

assess the flexibility of the protein construct [7]. High-resolution structural models can be compared directly to SAXS data to produce atomic models of the structure and dynamics of flagellar motor components in solution [6]. Analysis of proteins by SAXS is also in principle not limited by size. Indeed, scattering intensity scales as a proportion of the number of electrons in the molecule squared [8], so analysis of larger particles can require less material. Thus, SAXS also provides an interesting opportunity to structurally characterize large flagellar motor assemblies in solution, provided that sufficient pure and monodispersed sample can be obtained.

Since flagellar motor subunits have the capacity to self-assemble, data collection on these proteins may have a higher chance of success using a simultaneous purification and data collection method. Thus, all of our SAXS data collections on flagellar motor components have been performed in this way. Two such techniques have been established. The first entails coupling the eluate from size exclusion chromatography directly into the X-ray beam path. This technique, known as size-exclusion chromatography SAXS (SEC-SAXS), has become the gold standard for obtaining high-quality SAXS data not only because it ensures that samples are as pure and monodispersed as possible but also can provide an ideal buffer sample on either side of the protein elution peak [9]. The second was more recently invented and allows coupling of the eluate from differential ultracentrifugation (DU) to SAXS data collection (DU-SAXS) [10]. The method may become particularly important for obtaining SAXS data on large multi-subunit assemblies such as the BFM, which can often be more readily purified with DU. The disadvantage of simultaneous purification with SAXS data collection is that samples can be diluted substantially, data collection requires more time, and synchrotron radiation source is required to obtain sufficient photon flux to collect useful data in the time frame of the purification procedure. Moreover, detectors with a sufficient sensitivity, resolution, and fast read-out rate will generally only be accessible at synchrotron SAXS beamlines.

1.2 Basics of SAXS

X-ray scattering occurs as a result of coherent secondary wavelets that are scattered from atoms within a molecule after interacting with an incident X-ray beam. Interference from wavelets scattered from all atoms within a single molecule results in a diffraction pattern and SAXS experiments record the rotationally averaged diffraction from all molecules within the X-ray beam path. Because of the reciprocal relationship between diffraction patterns and the distance between scattering atoms, real space information about the particle's 3D structure such as absolute dimensions, Rg and interatomic distance distribution can be obtained by calculating an indirect Fourier Transform from scattering data [11, 12]. Conversely, 3D shapes can be generated *ab initio* with simulated annealing methods, which are consistent with the scattering data and hence represent the shape of the scattering molecule in solution [13].

In this book chapter, we describe the methods used for SAXS data collection and processing of the flagellar motor protein FliG but couched in general terms with references to other techniques that may be particularly useful for large, dynamic, or flexible flagellar motor components.

2 Materials

2.1 Protein Sample

1. 10 mg/mL purified FliG (*see* **Note 1**).
2. Size exclusion chromatography (SEC) column (e.g., Superdex 200) (*see* **Notes 2** and **3**).
3. Multi-angle laser light scattering (MALLS) instrument (*see* **Note 2**).

2.2 Data Collection

1. SEC-SAXS experiments require an appropriate size exclusion chromatography column and a liquid chromatography (LC) machine. In SAXS experiments with FliG, we used a 23 mL Superdex 200 SEC column (GE Healthcare) (*see* **Note 3**).
2. If molecular weight calculations are required, an inline MALLS machine is ideal to simultaneously measure protein concentration during data collection.
3. For differential ultracentrifugation coupled with SAXS (DU-SAXS) [10] experiments, a Gradient Station (Biocomp, Fredericton, NB, Canada) was used to fractionate samples after ultracentrifugation.
4. Buffers: For flagellar motor components, we used 0.1 M Tris–HCl, pH 7.5 and NaCl. For FliG SAXS, the concentration of NaCl was varied to stabilize different protein conformations however, for initial SAXS characterization we generally use 250 mM NaCl to reduce the risk of interparticle interactions as a result of extreme low or high ionic strengths. (*see* **Note 4** for the preparation of buffers for DU-SAXS).

2.3 Data Processing

1. Data reduction was performed with beamline-specific software package at the Australian Synchrotron known as *ScatterBrain* (Australian Synchrotron, Clayton Australia, https://www.synchrotron.org.au/ aussyncbeamlines/saxswaxs/software-saxswaxs) (*see* **Note 5**).
2. Unless otherwise specified, all programs for data analysis were from the ATSAS suite of programs developed by Dimitri

   Svergun and coworkers for the analysis of biological SAXS samples [14].

3. Custom scripts and software were written in Unix C-shell and C++ respectively.

3 Methods

3.1 SAXS Data Collection

1. Install SEC column onto FPLC and equilibrate system (Fig. 1) (*see* **Note 6**).
2. We typically load 100 μL of samples onto a 23 mL SEC column and set flow rate to 0.5 mL/min (*see* **Note 7**).
3. Simultaneously start SAXS data collection. We typically use 5 s exposure of 11 KeV X-rays per frame (41.7 μL per frame) with a flux of 4×10^{12} photons per second (*see* **Note 8**).

3.1.1 DU-SAXS

1. After DU, eluate is obtained by depressing a hollow piston upon the ultracentrifuge tube containing the desired separated sample. This forces the eluate through tubing, which is piped directly to a quartz capillary in the X-ray beam path [10] (Fig. 1).
2. The piston can be compressed at a uniform speed resulting in an eluate flow rate of 0.83 mL/min, while SAXS data is simultaneously collected with 5 s exposure at 11 KeV with a flux of 4×10^{12} photons per second (*see* **Note 8**).

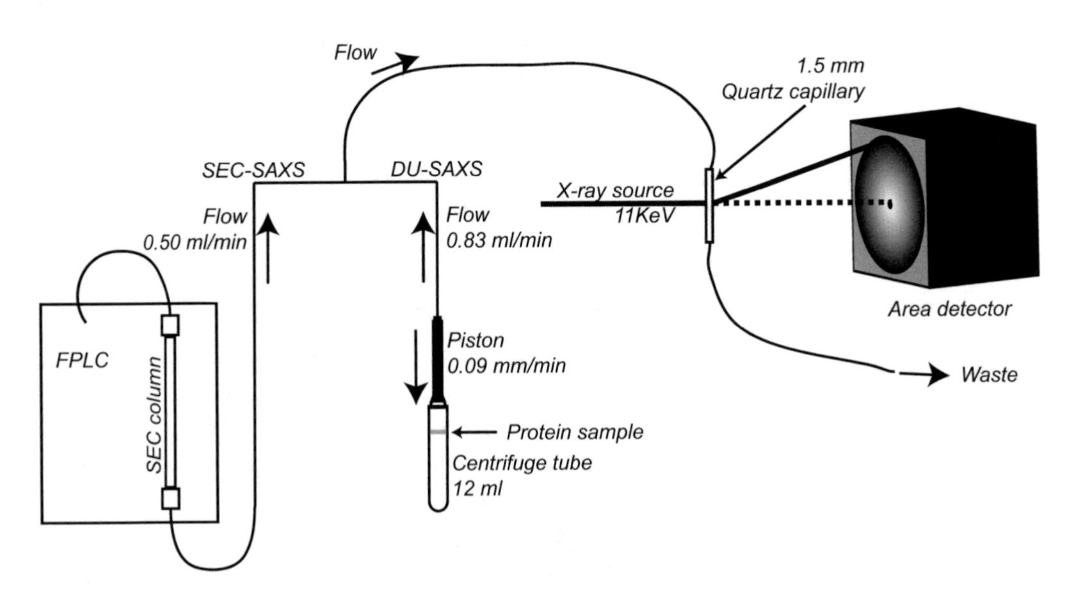

Fig. 1 Experimental setup. Eluate from either a size exclusion chromatography column or from a differential ultracentrifugation experiment is piped directly into a quartz capillary for simultaneous fractionation and SAXS data collection

3. Since the density of the buffer is changing over the duration of the experiment, to obtain an appropriate buffer that is equivalent to the sample, data is collected from a second "blank" ultracentrifuge tube that is identical except with vehicle buffer rather than sample added before DU.

3.2 SAXS Data Analysis

3.2.1 Data Reduction and Visual Inspection of Elution Profiles

1. Data are reduced from X-ray diffraction images to two-dimensional scattering profiles and variance by radially averaging pixel intensity. Thus, SAXS data are often stored as a 3-column ASCII file consisting of wave vector magnitude (q), intensity, and error, which are trivial to manipulate with custom-written software or scripts (*see* **Note 5**).
2. It is important to visualize elution profiles to determine which data frames consist of data from sample and those consisting of appropriate buffer for subtraction. In addition, visualization provides an initial qualitative assessment of potential problems with sample or experimental conditions during data collection, such as baseline drift, changes in beam flux, asymmetric, or overlapping elution peaks. It is useful to generate elution profiles over a range of scattering angles, which provides additional information about the scattering properties of the sample. For example, trace contaminants of higher molecular weight species such as oligomers or aggregates will be more obvious in low-q data and potentially absent in high-q data. Conversely, buffer drift will only be apparent in high q but not low q data. In addition, plotting elution profiles at different q-ranges provides an indication of whether the scattering from the protein decayed to background at high-scattering angles, which can be useful for performing the buffer subtraction described below. Elution profiles are therefore generated by averaging data over bins spanning the entire q-range of the experiment and plotted vs frame using custom-written software and scripts Unix shell scripts (Fig. 2).
3. Elution profiles are used to identify data frames consisting of protein sample as well as data frames that consist of buffer only. The latter consist of ~20 data frames that are as close as possible to the sample elution peak but where there is no detectable scattering (Fig. 2). Data from these buffer frames are averaged and used for all subsequent buffer subtractions (*see* **Note 9**).
4. The averaged buffer is then subtracted from each data frame of the experiment using the DATOP module of the ATSAS suite of programs [14] (*see* **Note 10**).

3.2.2 Assessing the Monodispersity of Protein Sample Across Elution Peak

1. Sample monodispersity is assessed by calculating the radius of gyration (Rg) of the scattering particle across the elution peak. This can be automated after subtracting buffer using the program AUTORG from the ATSAS suite of programs [14] (Fig. 2).

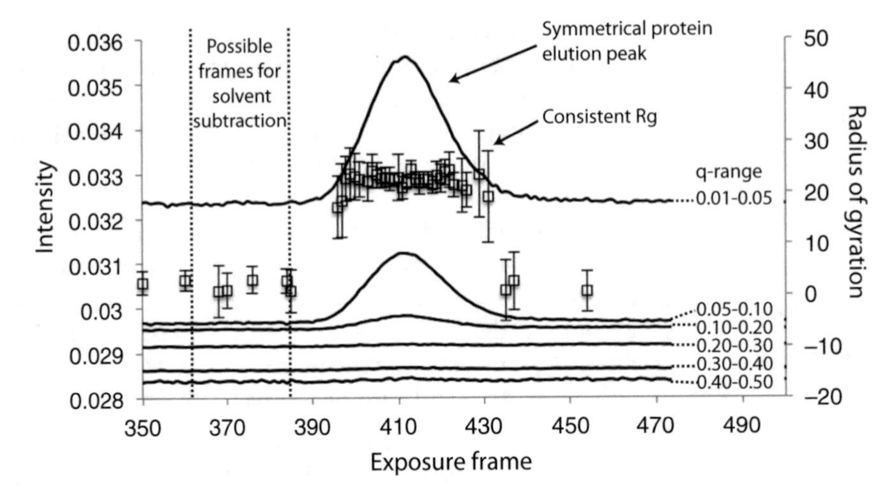

Fig. 2 SAXS elution profile from flagellar motor protein FliG. X-ray scattering intensity for different averaged q scattering angles is plotted over the exposure frame number. Calculated radius of gyration for each frame is also plotted as squares with errors depicting one standard deviation. A region close to the protein peak where an appropriate solvent sample may be extracted is highlighted

2. Scattering data from the sample of interest is then obtained by averaging subtracted data with a consistent *Rg* within the same elution peak. A consistent *Rg* is required for SAXS data to be useful for structural analysis.

3.2.3 Interatomic Distance Distribution Plots

1. To visualize scattering data in real space an indirect Fourier Transform is applied resulting in a $P(r)$ profile, which describes the probable frequency of interatomic vector lengths (r) scaled by the number of electrons in the atom pairs, using the software GNOM [12]. Calculating the $P(r)$ requires the assumption that $P(r) = 0$ when $r = 0$ and at the maximal linear dimension of the molecule (Dmax). Thus, unlike the *Rg*, which is precisely measured with SAXS, the D_{max} is a model parameter rather than a definitive result and needs to be defined carefully and systematically.
2. The D_{max} can be estimated by releasing the restraints on $P(r)$ and systematically generating $P(r)$ profiles with a range of specified maximum dimensions. Because of the intrinsic flexibility of proteins, the $P(r)$ profile should asymptotically approach zero at the maximum distance detected in the sample. If the D_{max} is restrained to a distance substantially more or less than this, the $P(r)$ plateaus and hovers around zero or fails to reach zero at all respectively. A D_{max} can therefore be nominally defined as the shortest vector length (r) above which the $P(r)$ profile hovers around zero (Fig. 3).
3. Next the goodness of fit of a $P(r)$ profile to experimental data is assessed as well as a perceptual measure of whether the quality of the $P(r)$ plot based on *a priori* knowledge [12]. These are automatically quantified by GNOM with several statistical

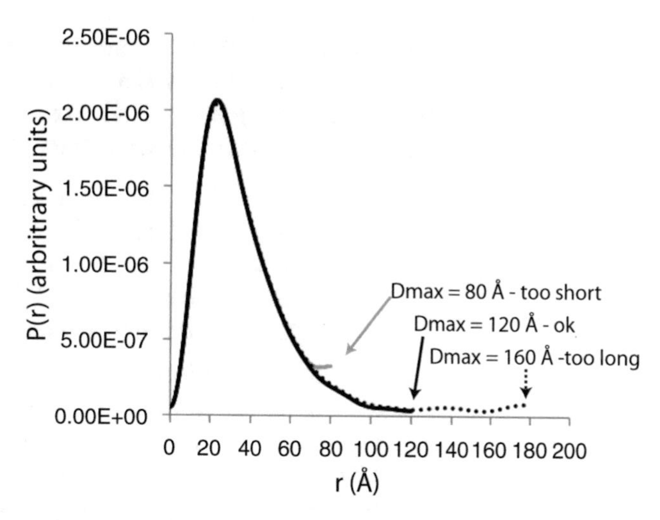

Fig. 3 P(r) profile from construct of flagellar motor protein FliG. 3 P(r) plots are shown with different specified maximum dimensions (80 Å in gray, 120 Å in black, and 160 Å in dotted black). No restraints for P(0) and P(D_{max}) to be equal to zero have been applied

methods that provide an overall absolute assessment of the fit quality (*see* **Note 11**).

4. Agreement between $P(r)$ profiles and scattering data can be optimized by altering two other parameters that are required to calculate the $P(r)$ profile. The first is the range of data points to be used in the analysis. Data points in the lowest scattering angles should be inspected for any influence from the beam-stop, which prevents the direct X-ray beam from reaching the detector. These scattering artifacts are mostly masked but can influence the data points at the lowest scattering angles, resulting in sharp nonsystematic changes in scattering intensity. Such anomalous data points should be ignored in further analysis. In addition, scattering intensity decreases at higher q angles often degrading into noise at a certain q angle above which data points should also be omitted from further analysis. A second "smoothing" parameter is required to avoid fitting fluctuations in scattering profiles from experimental variance or noise, where data is weak. This is akin to the matrix weight in X-ray crystallographic refinement protocols and is automatically calculated with GNOM based on the quality of fit to experimental data and various perceptual criteria [12] but may require manual alteration to be increased if the function begins to model noise, or decreased if obvious scattering features are ignored.

3.2.4 Subtraction of Solvent Blank

Subtraction of a well-matched solvent blank is vital for meaningful scattering data analysis. Particular care is required for obtaining well-matched solvents when a buffer blank is prepared separately; however, poorly matched solvent blanks can be problematic even when performing SEC-SAXS. Here, we outline strategies to ensure that buffer subtractions have been performed appropriately.

1. A good test of a well-matched solvent blank is to release restraints on the $P(r)$ profile and see whether the $P(r)$ profile approaches zero at $r = 0$. In principle, a positive or negative $P(0)$ value can be indicative of the solvent being under or over subtracted respectively (*see* **Note 11**).
2. Another indication of a well-matched solvent blank is that subtracted data at high-q normally approaches and hovers around zero. If data decays to a stable baseline above or below zero than data has been under or over subtracted. However, this assumes that scattering intensity from the protein has decayed to noise at high-q, which may not necessarily be the case. The elution profile (Fig. 2) can be used to assess this: scattering intensity from the protein of interest has likely decayed to noise if there is no detectable scattering at high q at the location of the elution peak.
3. Subtle under or over subtractions can be accounted for by applying a scale factor to the buffer solution before subtraction (*see* **Note 12**).
4. Buffer subtraction is more involved when performing DU SAXS because each data frame will have a separate equivalent solvent blank. The appropriate procedure for buffer subtraction with DU-SAXS is outlined in detail in Hynson *et al.* [10].

3.3 Interpretation of SAXS Data

3.3.1 Assessing Data Quality

1. Sample monodispersity can be quickly assessed by generating a "Guinier plot" where scattering data is plotted as $\log(\mathit{Intensity})\ vs\ q^2$. At low scattering angles ($q \times R_g < 1.0$ *or* 1.3 depending on the shape of the scattering particle), a Guinier plot should approximate to a straight line where the Rg is proportional to the square root of the slope m, $(R_g = \sqrt{-3m})$. It is also useful for assessing data quality. A systematically upturned or downturned Guinier plot is indicative of aggregation and interparticle interference respectively. It is difficult to recover useful data from an aggregated sample. In contrast, interparticle interference should be less pronounced in more dilute samples and hence useful data may be obtained from more dilute regions of an elution peak.
2. A more accurate estimate of R_g can be calculated from the $P(r)$ profile, which accounts for the entire dataset rather than just low q data. Nonetheless, the R_g calculated from Guinier and $P(r)$ should be similar and drastic differences can be indicative of poor sample quality.

3. The intensity at zero scattering angle ($I(0)$) is proportional to the concentration of the scattering particle and the molecular weight. $I(0)$ can be precisely determined by extrapolating the Guinier plot to a zero scattering angle and hence if the concentration of the scattering sample is known the molecular weight can also be estimated with a high level of precision. With SEC-SAXS sample concentration at each frame can be determined from the refractive index of the sample or ideally with parallel inline MALLS. A sample with the same molecular weight and Rg will have the same shape; hence, SAXS can provide a very stringent test of sample monodispersity and stoichiometry (Fig. 2).
4. A summary table quantifying SAXS data quality and the solution structure properties of the scattering particle should be prepared according to the guidelines set out by Jacques *et al.* [15].

3.3.2 Ab nitio Shape Restorations

1. The shape of the scattering particle in the solution can be qualitatively estimated by performing *ab initio* shape restorations using the DAMMIN program in the ATSAS suite, which generates shapes represented by dummy atoms that are consistent with scattering data via a simulated annealing approach. The consistency between models and experimental data is quantified statistically in log files generated for each simulation.
2. It is useful to perform multiple shape restorations independently of the same dataset because several models may fit the scattering data equally well. The variance between different shape restorations can therefore provide an indication of the confidence that a particular shape is representative of the scattering protein in solution. This variance can be quantified with the program DAMAVER, which calculates a normalized spatial distribution (NSD) statistic [16] and generates an average model. If all or most simulated models have a similar shape, then we can have greater confidence that the averaged shape reflects the shape of the protein in solution.

3.3.3 Comparison with High-Resolution Models and Between Different Datasets

1. A theoretical scattering profile can be generated from atomic coordinates using the program CRYSOL, providing an effective means to compare high-resolution structural models to experimental scattering data. The goodness of fit is automatically quantified statistically with CRYSOL and should also be assessed qualitatively by overlaying theoretical and experimental scattering plots. The latter can provide information on how the model may differ. For example, a more elongated structure should result in relatively higher scattering intensity at low scattering angles because of the inverse relationship between distance in real space and scattering angle.

2. Interdomain motions such as those found in the flagellar motor protein FliG [6] can also be modeled using an ensemble optimization method EOM [17], which is implemented in the ATSAS suite [14] or a similar implementation known as BIBLIOMOD [18].

3.3.4 Assessment of Flexibility

The dynamic nature of flagellar motor components, such as the stator protein MotB and rotor protein FliG, has led to proposed models for assembly and function that incorporate intrinsic flexibility [2, 6, 19, 20]. Scattering from disordered proteins decays more slowly at increasing scattering angles and hence flexibility within scattering particles can be measured with SAXS.

1. A straightforward way to assess protein flexibility is to generate a "Kratky plot" $I \bullet q^2$ *vs* q where scattering from partially folded or highly flexible proteins increases at high q, but plateaus around zero for globular proteins [8]. However, the increase in scattering in Kratky plots for flexible proteins relative to globular proteins is only observable at relatively high q scattering angles where scattering signal often degrades to noise.
2. An alternative approach that utilizes the Porod-Debye Law, can detect protein flexibility at lower q scattering angles. Data is plotted as $I \bullet q^4$ *vs* q^4 where scattering from ordered globular regions plateaus whereas scattering from intrinsically flexible proteins increases at higher scattering angles [7].

4 Notes

1. There are two sample requirements for a successful SAXS experiment. The first is that protein is ultrapure and monodispersed, the second is that protein is of sufficient concentration for an interpretable signal. The latter is dependent on the size of the protein. Protein samples for SAXS experiments performed on FliG (~40 kDa) were generally prepared at ~10 mg/mL, but for some constructs, as little as ~3 mg/mL still provided sufficient scattering signal for further analysis.
2. Before performing SAXS experiments, it is useful to first characterize the monodispersity of the protein sample by visualizing elution profiles from size exclusion chromatography (SEC) experiments, ideally coupled with multi-angle laser light scattering (MALLS). A consistent molecular weight across a symmetrical elution peak is a good predictor of a successful SAXS experiment.
3. We performed all SAXS analysis on flagellar motor proteins in combination with simultaneous purification with SEC [9], which we describe later. However, we note that it is also pos-

sible to collect useful SAXS data directly without SEC. This has the advantage of not requiring synchrotron radiation and that data collection can occur without diluting the sample with SEC and hence useful data may be obtained with a lower quantity of sample. In this case, however, the sample must be resilient to aggregation and a perfectly matched sample buffer must be obtained, which can be achieved with an overnight dialysis of the sample before data collection.

4. For DU-SAXS a density gradient is required. In this case, buffers and salts will be dissolved in high and low-density solutions, which can be layered on the bottom and the top half of an ultracentrifuge tube respectively. To form the density gradient, the ultracentrifuge can be gently tilted around 80° and gently rotated for around 2 min before returning the tube to an upright position.
5. Data reduction is routine and normally automatically performed by software that is custom-written for each synchrotron beamline however, and can be performed manually if specific parameters are required.
6. It is paramount that a stable baseline is obtained before commencing data collection to ensure that buffer correctly matches the sample. We assess the stability of the buffer baseline by measuring the refractive index of the eluate. To save time, SEC columns are pre-equilibrated into the desired buffer on a separate LC machine offline. Nonetheless, 2–3 column volumes are normally required before a stable baseline is reached.
7. In principle, it is possible to load up to 500 μL in the 23 mL Superdex 200 column; however, our sample volume was by the sample loader on the FPLC machine currently installed at the SAXS/WAXS beamline at the Australian Synchrotron. Ideal flow rates and sample volume will be dependent on the purity and concentration of the sample, as well as the specifications of the SEC column utilized for the SEC-SAXS experiment.
8. These experimental parameters should provide sufficient signal-to-noise ratio without causing measureable radiation damage to protein samples, and ensure sufficiently fine sampling over the elution profile to allow for separation of contaminants, in close proximity to the elution peak. However, ideal data collection parameters will depend on the protein sample and the SAXS beamline setup, and may require optimization. In some instances, especially where there is a detectable baseline drift, selecting appropriate buffer frames may not be trivial and require sampling several regions of the elution profile to find an appropriate solvent blank.

9. A C-shell Unix script was used to automatically subtract the solvent blank from each sample frame independently.
10. ATSAS consists of a graphical user interface known as Primus [21], which automatically overlays these subtracted scattering data and an inverse Fourier Transform of the $P(r)$ profile.
11. In our experience, however, an unrestrained $P(r)$ never intersects exactly 0 at $r = 0$ even with carefully selected solvent blanks from SEC-SAXS. Rather, a well-matched buffer results in an absolute difference between $P(0)$ and 0, of less than 2% of the maximum value in entire the $P(r)$ profile (Fig. 3).
12. In our hands solvent blanks are scaled by a factor of less than 0.1% from successful SEC-SAXS and typically much less or do not require any scaling at all.

Acknowledgments

This work was supported by Australian Research Council Discovery Project Grant (DP130102219) and Discovery Early Career Research Award (DE140100262), as well as by the SAXS/WAXS beamline at the Australian Synchrotron, Victoria, Australia.

References

1. Stock D, Namba K, Lee LK (2012) Nanorotors and self-assembling macromolecular machines: the torque ring of the bacterial flagellar motor. Curr Opin Biotechnol 23:545–554. doi:10.1016/j.copbio.2012.01.008
2. Kojima S, Furukawa Y, Matsunami H et al (2008) Characterization of the periplasmic domain of MotB and implications for its role in the stator assembly of the bacterial flagellar motor. J Bacteriol 190:3314–3322. doi:10.1128/JB.01710-07
3. Roujeinikova A (2008) Crystal structure of the cell wall anchor domain of MotB, a stator component of the bacterial flagellar motor: implications for peptidoglycan recognition. Proc Natl Acad Sci USA 105:10348–10353. doi:10.1073/pnas.0803039105
4. Samatey FA, Matsunami H, Imada K et al (2004) Structure of the bacterial flagellar hook and implication for the molecular universal joint mechanism. Nature 431:1062–1068. doi:10.1038/nature02997
5. Samatey FA, Imada K, Nagashima S et al (2001) Structure of the bacterial flagellar protofilament and implications for a switch for supercoiling. Nature 410:331–337. doi:10.1038/35066504
6. Baker MAB, Hynson RMG, Ganuelas LA et al (2016) Domain-swap polymerization drives the self-assembly of the bacterial flagellar motor. Nat Struct Mol Biol. doi:10.1038/nsmb.3172
7. Rambo RP, Tainer JA (2011) Characterizing flexible and intrinsically unstructured biological macromolecules by SAS using the Porod-Debye law. Biopolymers 95:559–571. doi:10.1002/bip.21638
8. Svergun DI, Koch MHJ, Timmins PA, May RP (2013) Small angle X-ray and neutron scattering from solutions of biological macromolecules. Oxford University Press, Oxford
9. David G, Pérez J (2009) Combined sampler robot and high-performance liquid chromatography: a fully automated system for biologi-

cal small-angle X-ray scattering experiments at the Synchrotron SOLEIL SWING beamline. J Appl Cryst 42:892–900. doi:10.1107/S0021889809029288
10. Hynson RMG, Duff AP, Kirby N et al (2015) Differential ultracentrifugation coupled to small-angle X-ray scattering on macromolecular complexes. J Appl Cryst 48:769–775. doi:10.1107/S1600576715005051
11. Glatter O, IUCr (1977) A new method for the evaluation of small-angle scattering data. J Appl Cryst 10:415–421. doi: 10.1107/S0021889877013879
12. Svergun DI, IUCr (1992) Determination of the regularization parameter in indirect-transform methods using perceptual criteria. J Appl Cryst 25:495–503. doi: 10.1107/S0021889892001663
13. Svergun DI (1999) Restoring low resolution structure of biological macromolecules from solution scattering using simulated annealing. Biophys J 76:2879–2886. doi:10.1016/S0006-3495(99)77443-6
14. Petoukhov MV, Konarev PV, Kikhney AG (2007) ATSAS 2.1-towards automated and web-supported small-angle scattering data analysis. J Appl Crystallogr. 40:S223–S228
15. Jacques DA, Guss JM, Svergun DI, Trewhella J (2012) Publication guidelines for structural modelling of small-angle scattering data from biomolecules in solution. Acta Crystallogr D Biol Crystallogr 68:620–626. doi:10.1107/S0907444912012073
16. Volkov VV, Svergun DI, IUCr (2003) Uniqueness of ab initio shape determination in small-angle scattering. J Appl Cryst 36:860–864. doi: 10.1107/S0021889803000268
17. Tria G, Mertens HDT, Kachala M, Svergun DI (2015) Advanced ensemble modelling of flexible macromolecules using X-ray solution scattering. IUCrJ 2:207–217. doi:10.1107/S205225251500202X
18. Pelikan M, Hura GL, Hammel M (2009) Structure and flexibility within proteins as identified through small angle X-ray scattering. Gen Physiol Biophys 28:174–189
19. Zhu S, Takao M, Li N et al (2014) Conformational change in the periplamic region of the flagellar stator coupled with the assembly around the rotor. Proc Natl Acad Sci USA 111:13523–13528. doi:10.1073/pnas.1324201111
20. O'Neill J, Xie M, Hijnen M, Roujeinikova A (2011) Role of the MotB linker in the assembly and activation of the bacterial flagellar motor. Acta Crystallogr D Biol Crystallogr 67:1009–1016. doi:10.1107/S0907444911041102
21. Konarev PV, Volkov VV, Sokolova AV et al (2003) PRIMUS: a Windows PC-based system for small-angle scattering data analysis. J Appl Cryst 36:1277–1282. doi:10.1107/S0021889803012779

Chapter 9

Structural Study of the Bacterial Flagellar Basal Body by Electron Cryomicroscopy and Image Analysis

Akihiro Kawamoto and Keiichi Namba

Abstract

The bacterial flagellum is a large assembly of about 30 different proteins and is divided into three parts: filament, hook, and basal body. The machineries for its crucial functions, such as torque generation, rotational switch regulation, protein export, and assembly initiation, are all located around the basal body. Although high-resolution structures of the filament and hook have already been revealed, the structure of the basal body remains elusive. Recently, the purification protocol for the MS ring, which is the core ring of the basal body, has been improved for the structural study of the MS ring by electron cryomicroscopy (cryoEM) and single particle image analysis. The structure of intact basal body has also been revealed in situ at a resolution of a few nanometers by electron cryotomography (ECT) of minicells. Here, we describe the methods for the MS ring purification, *Salmonella* minicell culture, and cryoEM/ECT data collection and image analysis.

Key words Bacterial flagellum, CryoEM, Single-particle image analysis, Electron cryotomography, Minicell

1 Introduction

Bacteria actively swim in liquid environments by rotating long, helical, filamentous organelle called the flagellum. At the base of each filament, there is a rotary motor powered by proton or sodium ion motive force across the cytoplasmic membrane. The flagellar motor can rotate at a speed of up to 300 Hz (in the case of the H^+-driven *Salmonella* motor) and can switch its rotational direction quickly, which are the properties that identify it as an elaborate biological nanomachine [1].

The flagellum is a large molecular assembly composed of nearly 30 different proteins and can be roughly divided into the basal body and the tubular axial structures, hook and filament. The basal body, which is embedded in the cell envelope, works as a reversible rotary motor composed of a rotor and multiple stator units. The basal body is composed of the LP ring, MS ring, and C ring. The MS ring is composed of a single-membrane protein, FliF, and acts as the

Tohru Minamino and Keiichi Namba (eds.), *The Bacterial Flagellum: Methods and Protocols*, Methods in Molecular Biology, vol. 1593, DOI 10.1007/978-1-4939-6927-2_9, © Springer Science+Business Media LLC 2017

base for flagellar assembly [2]. The C ring is composed of three proteins, FliG, FliM, and FliN, regulates the reversal frequency of the motor, and facilitates flagellar protein export for flagellar assembly [3, 4]. The LP ring is a bushing for the rod to function as a drive shaft. The MS ring and C ring form the rotor, which is surrounded by stator units, and torque is generated by the interaction between the rotor and stator units [5]. Spontaneous mutants of FliF-FliG fusion were found to be functional, indicating that the C terminus of the FliF is located close to the N terminus of the FliG and the stoichiometry of FliF and FliG is 1:1 in the native flagellum [6]. The rotational symmetry of the MS ring is not clearly visible, but previous biochemical and structural analyses showed that about 26 FliF molecules form the MS ring [7, 8]. In contrast, a cryoEM study has revealed the rotational symmetry of the C ring to be 34-fold [9]. Thus, there is a symmetry mismatch between the MS ring and the C ring. To understand their interactions over this symmetry mismatch, we need to reveal a high-resolution structure of the MS ring and the interactions of these rings.

Single-particle cryoEM image analysis circumvents the requirement of well-ordered crystal for structure determination and therefore represents a transformative approach for studying membrane proteins that are refractory to crystallization. Using this method, proteins in their native conformations are embedded in vitreous ice at liquid nitrogen temperature and are imaged directly by an electron cryomicroscope. Images of identical or similar molecules in random orientations are recorded at low electron doses to minimize radiation damage of biological samples. This leads to an extremely low signal-to-noise ratio (SNR) in each individual image, making it necessary to align and average tens of thousands of particle images to increase the SNR and to reconstruct a 3D image at high resolution. Recently, single-particle cryoEM image analysis has achieved atomic resolution for membrane protein structure [10] and large protein complex [11] by the use of a novel technology, direct electron detector camera. Its extremely high frame rate allows single-electron counting to minimize the noise and motion correction for dose fractionation movie images to correct for electron beam-induced movement that otherwise blurs the particle images. Moreover, low molecular weight polymers and new detergents have been developed for purification and stabilization of membrane proteins in solution to facilitate their structural studies by cryoEM image analysis at atomic resolution [12–14]. Here, we describe methods for the purification of the MS ring and its single-particle cryoEM image analysis. Using the methods described here, the resolution of the MS ring has been much improved to reveal the rotational symmetry.

ECT is an imaging technique that directly provides 3D structure of cells and molecular complexes in their cellular environment at nanometer resolution. EM images are 2D projections of 3D objects. To obtain the structures of specimens, the specimen grid is tilted incrementally around an axis perpendicular to the electron beam, e.g., from −70° to +70° with 2° increments, and images are taken at each tilt angle. Each image represents a different view of the specimen. The images of a tilt series are aligned and back-projected to generate a 3D image (tomogram) of the specimen. In order to compare the structures of in situ and isolated basal body, we carried out the ECT analysis of the intact flagellar basal body in *Salmonella* cells. However, the resolution is limited due to the thickness of *Salmonella* cells, which is approximately 1 μm and is too thick even for 300 keV electrons to pass through. We therefore constructed *Salmonella* minicell by overproducing FtsZ, which assembles into a structure known as the Z ring at the future site of cytokinesis, to make the cell size small enough for ECT to visualize the cellular structures in detail [15]. We were able to prepare *Salmonella* minicells with a diameter of less than 0.5 μm, which allowed the 3D structure of intact basal body to be revealed in situ. Here, we also describe the methods for the preparation of *Salmonella* minicell and data collection for ECT.

2 Materials

Prepare all solutions using ultrapure water (resistance =18 MΩ at 25 °C) and analytical grade reagents. Prepare all reagents at room temperature. Store all reagents at 4 °C. Particular attention should be paid to the adequate disposal of material contaminated with *E. coli* and *Salmonella*.

2.1 Bacterial Stain and Plasmid

1. *E. coli*: BL21(DE3) carrying pLysS [2].
2. SJW1103: *Salmonella* wild-type strain [16].
3. pKOT105: full length of *Salmonella* FliF on an expression vector pET3b [2].
4. pYVM031: *Salmonella* FtsZ on an expression vector pBAD24 [15].

2.2 Cell Culture and Solution

1. *E. coli* carrying pKOT105, frozen stock with 10% dimethyl sulfoxide (DMSO), kept at −80 °C.
2. SJW1103 carrying pYVM031, frozen stock with 10% DMSO, kept at −80 °C.
3. Lysogeny Broth (LB) medium: 2 L of LB medium containing 10 g tryptone, 5 g yeast extract, 5 g NaCl and is sterilized by autoclaving at 121 °C for 30 min.
4. Cell culture medium: LB containing 50 μg/mL of ampicillin and 30 μg/mL of chloramphenicol.
5. LB agar plate.

6. Isopropyl β-D-1-thiogalactopyranoside (IPTG).
7. Arabinose.
8. Protease inhibitor cocktail (Complete, EDTA-free).
9. French press buffer: 50 mM Tris–HCl, pH 8.0, 5 mM EDTA-NaOH, pH 8.0, 50 mM NaCl.
10. Alkaline buffer: 50 mM CAPS-NaOH, pH 11.0, 5 mM EDTA-NaOH, pH 11.0, 50 mM NaCl, 1% TritonX-100.
11. Buffer C: 10 mM Tris–HCl, pH 8.0, 5 mM EDTA-NaOH, pH 8.0, 1% TritonX-100.
12. Buffer S: 25 mM Tris–HCl, pH 8.0, 1 mM EDTA-NaOH, pH 8.0, 50 mM NaCl, 0.1% TritonX-100.
13. 15% (w/w) and 40% (w/w) sucrose solution in Buffer C.
14. 12.5% precast polyacrylamide gel.
15. 2× Sample buffer solution: 0.125 M, Tris–HCl pH 6.8, 4% (w/v) SDS, 20% (w/v) Glycerol, 0.002% (w/v) Bromophenol Blue.
16. 2-mercaptoethanol.
17. Running Buffer: 1 L of Running buffer contains Tris 3 g, Glycine 14.4 g, SDS 1 g.
18. 2% (w/v) Uranyl acetate.
19. 0.5% (w/v) phosphotungstic acid.
20. M9 medium: 1 L of M9 medium contains $Na_2HPO_4{\cdot}12H_2O$ 17.1 g, KH_2PO_4 3 g, NaCl 0.5 g, NH_4Cl 1 g, 0.2% glycerol, 1% tryptone, 1 mM $MgSO_4$.
21. 10 nm colloidal gold particles solution (MP Biomedicals, CA, US).

2.3 Equipment and Electron Microscope

1. 30 °C incubator.
2. 37 °C incubator.
3. 37 °C heated shaker.
4. 30 or 37 °C heated orbital shaker.
5. 5 L shaking Erlenmeyer flask with baffles.
6. Spectrophotometer (OD readings at wavelength 600 nm).
7. Sterile conical tubes of 50 mL capacity.
8. French pressure cell press.
9. 1 mL plastic syringes with 20-gauge needle.
10. Protein gradient fractionator.
11. Model 2110 Fraction collector.
12. Micro centrifuge.
13. Centrifuge.
14. Ultracentrifuge: Optima XE-90 (BECKMAN COULTER, US).

15. JLA-10.500 rotor (BECKMAN COULTER, US).
16. 50.2Ti rotor (BECKMAN COULTER, US).
17. 45Ti rotor (BECKMAN COULTER, US).
18. SW 32Ti swinging bucket rotor (BECKMAN COULTER, US).
19. Tweezer.
20. Ion coater.
21. Thin layer of continuous carbon coated copper 200 mech EM grid (Nisshiin-EM, Tokyo, Japan).
22. Desiccator (SANPLATEC, Osaka, Japan).
23. JEM-1011 transmission electron microscope (JEOL, Tokyo, Japan).
24. TVIPS TemCam-F415MP 4 k × 4 k CCD camera (TVIPS, Gauting, Germany).
25. Quantifoil molybdenum 200 mesh R0.6/1.0 grid (Quantifoil Micro Tools, Jena, Germany).
26. Vitrobot Mark III (FEI Company, Eindhoven, The Netherlands).
27. Titan Krios FEG transmission electron microscope (FEI Company, Eindhoven, The Netherlands).
28. FEI Falcon II 4 k × 4 k CMOS direct electron detector camera (FEI Company, Eindhoven, The Netherlands).

2.4 Program for Data Collection and Structural Analysis by Single-Particle Image Analysis and ECT

1. EPU software (FEI Company, Eindhoven, The Netherlands).
2. Motioncorr [17].
3. CTFFIND3 [18].
4. Relion 1.4 [19].
5. ResMap [20].
6. UCSF Chimera [21].
7. Xplore 3D software package (FEI Company, Eindhoven, The Netherlands).
8. IMOD software package [22].

3 Methods

3.1 Purification of FliF

1. *E. coli* carrying pKOT105 is streaked from a frozen stock to a LB plate and then the cells are grown in a 37 °C incubator overnight.
2. A single-bacterial colony is inoculated and the cells are grown overnight in 20 mL of the cell culture medium in a 37 °C heated shaker.

3. The 20 mL of overnight culture is added to 2 L cell culture medium, and then the cells are grown in a 37 °C orbital shaking incubator (100 rpm) until the culture density reaches an OD_{600} of 0.5 ~ 0.7 (*see* **Note 1**).
4. A final concentration of 0.5 mM IPTG is added to 2 L cell culture medium and the growth is continued in a 30 °C orbital shaking incubator (100 rpm) for 4 h (*see* **Note 2**).
5. The cells are collected by centrifugation (4600 × *g*, 10 min, 4 °C).
6. Harvested cells are resuspended in a 40 mL volume of the French press buffer containing protease inhibitor cocktail (*see* **Note 3)** and are disrupted using a French press at a pressure level of 10,000 psi.
7. After cell debris and undisrupted cells are removed by centrifugation (20,000 × *g*, 20 min, 4 °C), a crude membrane fraction is isolated by ultracentrifugation (90,000 × *g*, 60 min, 4 °C, 50.2Ti).
8. The pellet is suspended in a 40 mL solution of Alkaline buffer (mechanical shearing with 1 mL plastic syringes with 20-gauge needle), and the solution is incubated at 4 °C for 1 h.
9. After insoluble materials are removed by centrifugation (20,000 × *g*, 20 min, 4 °C), solubilized proteins are collected by ultracentrifugation (90,000 × *g*, 60 min, 4 °C, 50.2Ti).
10. The pellet is resuspended in a 3 mL solution of buffer S (mechanical shearing with 1 mL plastic syringes with 20-gauge needle), and the solution is loaded onto a 15–40% (w/w) continuous sucrose density gradient in Buffer C and is subjected to centrifugation (49,100 × *g*, 13 h, 4 °C, SW 32Ti) (*see* **Note 4)**.
11. Fractions of the sucrose density gradient, each containing a 700 μL volume, around the opaque band are collected by a gradient fractionator and fraction collector (*see* **Note 5)**.
12. The collected fractions are analyzed by SDS-PAGE to examine the amount of FliF and contaminants in each fraction (*see* **Note 6**; Fig. 1).
13. Peak fractions for the amount of FliF are collected and diluted by 6–7 times volume of buffer S (e.g., main fractions 700 μL × 4 = 2800 μL + 19.2 mL of buffer S = 22 mL). Then the MS rings are enriched by ultracentrifugation (90,000 × *g*, 60 min, 4 °C, 45Ti).
14. The supernatant is removed, and the pellet containing the MS ring is resuspended in a 30 μL solution of buffer S.
15. The MS ring concentration can be measured by the Lowry protein assay.

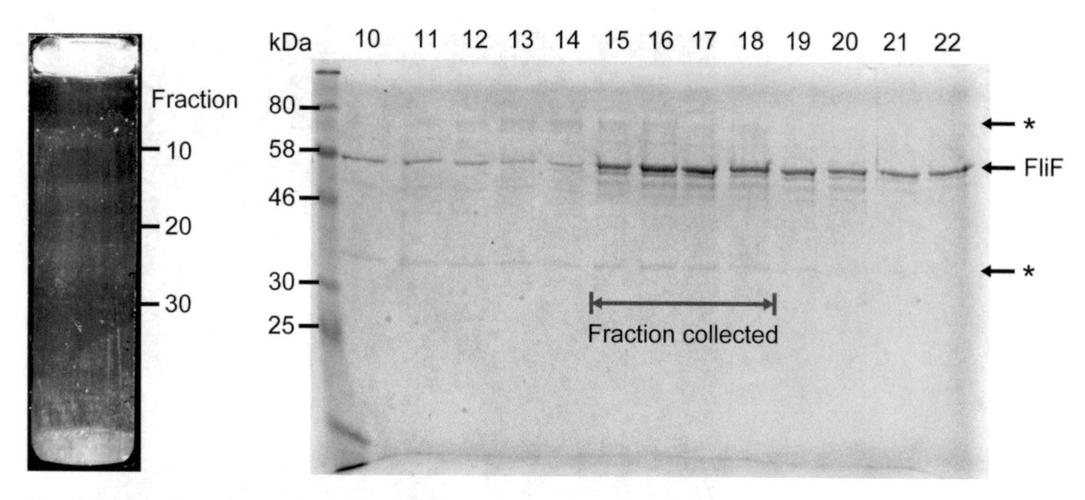

Fig. 1 Purification of the MS ring by 15–40% (w/w) sucrose gradient. *Left panel* shows the bands in the sucrose gradient. Numbers on the right indicate the fractions collected by the gradient fractionator and fraction collector. *Right panel* shows an SDS-PAGE of each fraction collected from the sucrose gradient. The band position of FliF is labeled. Asterisks indicate contaminating bands

3.2 Preparation and Data Collection of Negative Stained EM

1. Continuous carbon-coated EM grids are glow-discharged on a grass slide for 20 s.
2. A 3 μL aliquot of the sample solution is applied onto the EM grid, and then the extra solution is removed by a fan-shaped strip of a filter paper.
3. Three droplets of the 2% (w/v) uranyl acetate solution are placed on a parafilm, and then a sample grid is stained by putting the grid on each droplet and removing the extra solution over the three droplets.
4. The stained EM grid is dried on a filter paper in desiccator at least for 30 min.
5. Negative stained EM images are observed with JEM-1011 transmission electron microscope operating at 100 kV, and EM images are recorded using a TVIPS TemCam-F415MP CCD camera (Fig. 2).

3.3 Preparation and Data Collection of CryoEM

1. Quantifoil molybdenum 200 mesh R0.6/1.0 holey carbon grids are glow discharged on a grass slide for 20 s.
2. A 2.6 μL aliquot of the sample solution is applied onto the grid, is blotted by filter papers for 6 s at 100% humidity and 4 °C, and the grid is quickly frozen by rapidly plunging it into liquid ethane using Vitrobot Mark III.
3. The grids are inserted into a Titan Krios FEG transmission electron microscopy operated at 300 kV with a cryo specimen stage cooled with liquid nitrogen. CryoEM images are recorded with a FEI Falcon II 4 k × 4 k CMOS direct electron

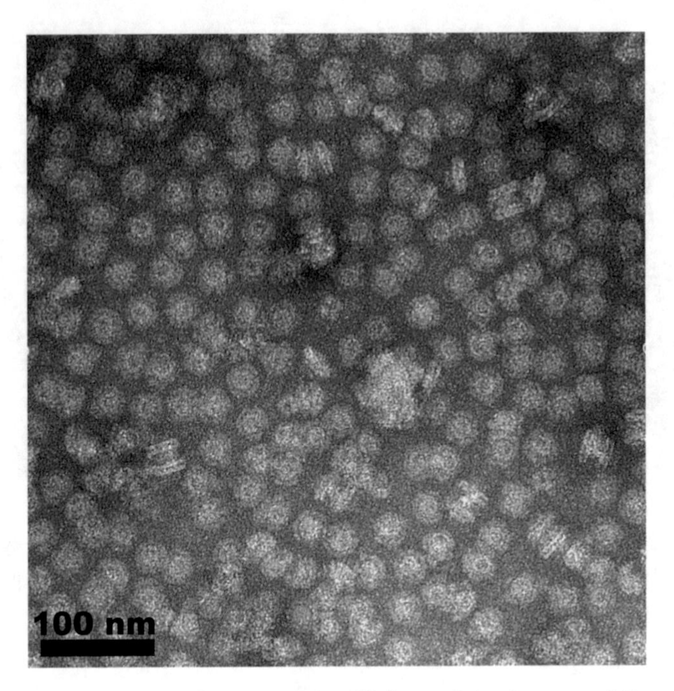

Fig. 2 Negatively stained EM image of the MS ring particles. A representative negatively stained EM image of the MS ring particles from fractions 15–18 in Fig. 1

detector at a nominal magnification of ×75,000, corresponding to an image pixel size of 1.07 Å, and a defocus range of 1.0–2.5 μm. The EPU software is used for automated data acquisition.

4. Images are acquired by collecting seven movie frames with a dose rate of 45 $e^-/Å^2/s$ and an exposure time of 2 s.

3.4 Image Processing

1. The seven movie frames are subsequently aligned to correct for beam-induced movement and drift using the Motioncorr program [17].
2. The parameters for the contrast transfer function (CTF) are estimated using CTFFIND3 [18].
3. About 5000 particle images are manually picked from several hundreds of cryoEM images, and then a reference-free two-dimensional (2D) classification is carried out using Relion [19]. The particle images are classified into 20 classes.
4. "Good" class-averaged images are selected as reference images, and more particle images are automatically picked from all of the cryoEM images using the autopicking program of Relion.
5. To avoid the effect of radiation damage, only the first six movie frames of each image (a total dose of ~30 $e^-/Å^2$) are used for 2D and three-dimensional (3D) classification.
6. To accelerate image analysis, particle images selected from motion-corrected images are binned 2 × 2.

7. 2D classification is carried out using the recommended procedures of Relion, and then only "good" class-averaged images are selected for 3D classification.
8. 3D classification is done using the recommended procedures of Relion. A previous cryoEM density map of the MS ring [8] is used as the initial model after low-pass filtering to 60 Å resolution and without applying any rotational symmetry. Particle images are classified into three classes, and only good class images are selected for the next step.
9. 3D classification is recalculated using only good class-averaged maps to be used as an initial model. The initial model is low-pass filtered to 60 Å resolution, and a rotational symmetry that is observed in the reference-free 2D classification images is applied. The particle images are classified into three classes, and good classes are selected.
10. 3D auto-refinement is done by Relion. Relion's postprocessing procedure with a soft auto-mask estimates the resolution of the 3D density map by the "gold standard" Fourier shell correlation (FSC) at a 0.143 criterion and a B-factor.
11. For visualization of the 3D density map and atomic model building, the map is sharpened and low-pass filtered by Relion postprocessing using the above-mentioned B-factor and resolution. The local resolution is calculated by ResMap [20] using the two cryoEM maps independently refined from each of the two halves of the data set.
12. The 3D density maps are visualized using UCSF Chimera [21].

3.5 Culture Conditions of *Salmonella* Minicell for ECT

1. SJW1103 carrying pYVM031 is streaked from a frozen stock to a LB plate containing 50 μg/mL of ampicillin, and then the cells are grown in a 30 °C incubator overnight.
2. A single-bacterial colony is inoculated into a 5 mL solution of LB medium containing 50 μg/mL of ampicillin and the cells are grown overnight in a 30° heated shaker.
3. A 50 μL aliquot of the overnight culture is added to a 5 mL solution of the M9 medium containing 50 μg/mL of ampicillin, and then the cells are grown in a 30 °C heated shaker until the culture density reaches an OD_{600} of 0.5.
4. Arabinose is added to the 5 mL M9 culture medium to a final concentration of 0.2% for the induction of FtsZ overproduction, and the cell growth is continued at 30 °C until the culture density reaches an OD_{600} of 1.0.
5. After large size cells and cell debris are removed by centrifugation (4400 × *g*, 5 min, 4 °C), the minicells are collected by centrifugation (15,300 × *g*, 5 min, 4 °C).
6. The pellet is resuspended in a 50 μL solution of the M9 medium.

3.6 Preparation and Data Collection of Negative Stained EM

1. Continuous carbon-coated EM grids are glow-discharged on a grass slide for 20 s.
2. A 3 μL aliquot of the sample solution is applied onto the EM grid, and then the extra solution is removed by a fan-shaped strip of a filter paper.
3. Three droplets of the 0.5% (w/v) phosphotungstic acid solution are placed on a parafilm, and then a sample grid is stained by putting the grid on each droplet and removing the extra solution over the three droplets.
4. The stained EM grid on filter paper is dried in desiccator at least 30 min.
5. Negative stained EM images are observed with JEM-1011 transmission electron microscope operating at 100 kV, and EM images are recorded using a TVIPS TemCam-F415MP CCD camera (Fig. 3).

3.7 ECT Data Collection and Tomogram Reconstruction

1. Quantifoil molybdenum 200 mesh R0.6/1.0 holey carbon grids are glow-discharged on a grass slide for 5 s.
2. A 3.0 μL solution containing 10 nm colloidal gold particles is applied onto the grid, and the grid is dried up at room temperature (*see* **Note 7**).

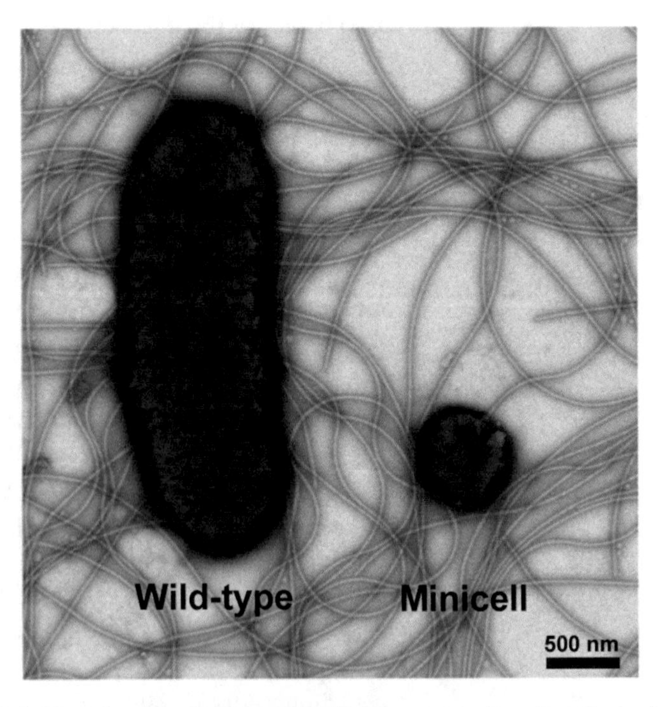

Fig. 3 *Salmonella* minicell produced by FtsZ overproduction. Negatively stained EM image of a Salmonella wild-type cell and a minicell with a size of ~0.5 μm

3. The pretreated grids are glow-discharged again on a grass slide for 20 s.
4. A 2.6 μL aliquot of the sample solution is applied onto the pretreated grid, is blotted by a filter paper for 7 s at 100% humidity and 22 °C, and the grid is quickly frozen by rapidly plunging it into liquid ethane using Vitrobot Mark III.
5. The grids are inserted into a Titan Krios FEG transmission electron microscope operated at 300 kV with a cryo specimen stage cooled with liquid nitrogen. CryoEM images are recorded with a FEI Falcon II 4 k × 4 k direct electron detector at a nominal magnification of ×29,000, corresponding to an image pixel size of 5.7 Å. Single-axis tilt series are collected covering an angular range from −70° to +70° with a nonlinear Saxton tilt scheme at 4–6 μm underfocus using the Xplore 3D software package. A cumulative dose of 100 e^{-}/Å^{2} or less is used for each tilt series. Images are generally binned twofolds, and 3D image reconstructions are calculated using the IMOD software package [22].

4 Notes

1. For efficiency of cell growth, we highly recommend the use of the erlenmeyer flasks with baffles. The baffles make aeration more efficient to facilitate cell growth and allow us to obtain a large amount of *E. coli* cells more quickly and efficiency than flasks without baffles.
2. In case of *E. coli* carrying pKOT105, the expression level of FliF is higher at 30 °C than at 37 °C. Before induction of the FliF expression by IPTG, the temperature condition of the orbital shaker needs to be changed from 37 °C to 30 °C.
3. Before resuspension of harvested cells in the French press buffer, one tablet of the protease inhibitor cocktail (Complete, EDTA-free) should be added to the French press buffer and dissolved using a magnetic stirrer.
4. For making a sucrose gradient, the 40% (w/w) sucrose solution in buffer C is poured into a plastic centrifuge tube to its half volume, and then the 15% (w/w) sucrose solution is poured onto the 40% (w/w) sucrose solution to the full volume of tube. A continuous sucrose density gradient is made by a gradient program-installed fractionator. Alternatively, the tube can be tilted to the horizontal orientation, kept in this position for 4 h, and then tilted back very slowly to stand vertically. The sucrose density gradient should be kept at 4 °C.
5. Set the piston down speed to 0.09 and the fraction collect time to 25 s.

6. The sample fractions and its twice volume of the sample buffer solution containing 5% 2-mercaptoethanol are mixed at the 1:1 ratio. After boiling the sample solutions at 95 °C for 3 min, the boiled samples are centrifuged for a short time to spin down the condensate. The sample solutions and protein standard makers are loaded to each lane, and electrophorese is carried out at a current of 28 mA/gel until the sample fractions reach the bottom of the gel. Following the electrophoresis, proteins in the gel are fixed and stained by Coomassie Brilliant Blue (CBB).
7. The 10 nm colloidal gold particle solution should be concentrated 1.5 times before it is applied onto holey carbon EM grids.

Acknowledgments

The research study described in this chapter was supported by JSPS KAKENHI Grant Number JP25000013 to K.N.

References

1. Berg HC (2003) The rotary motor of bacterial flagella. Annu Rev Biochem 72:19–54
2. Ueno T, Oosawa K, Aizawa SI (1992) M ring, S ting and proximal rod of the flagellar basal body of *Salmonella typhimurium* are composed of subunits of a single protein FliF. J Mol Biol 227:672–677
3. Berg HC (2000) Constraints on models for the flagellar rotary motor. Philos Trans R Soc London B Biol Sci 335:491–502
4. González-Pedrajo B, Minamino T, Kihara M et al (2006) Interactions between C ring proteins and export apparatus components: a possible mechanism for facilitating type III protein export. Mol Microbiol 60:984–998
5. Zhou J, Lloyd SA, Blair DF (1998) Electrostatic interactions between rotor and stator in the bacterial flagellar motor. Proc Natl Acad Sci U S A 95:6436–6441
6. Francis NR, Irikura VM, Yamaguchi S et al (1992) Localization of the *Salmonella typhimurium* flagellar switch protein FliG to the cytoplasmic M-ring face of the basal body. Proc Natl Acad Sci U S A 89:6304–6308
7. Jones CJ, Macnab RM, Okino H et al (1990) Stoichiometric analysis of the flagellar hook-(basal-body) complex of *Salmonella typhimurium*. J Mol Biol 212:377–387
8. Suzuki H, Yonekura K, Namba K (2004) Structure of the rotor of the bacterial flagellar motor revealed by electron cryomicroscopy and single-particle image analysis. J Mol Biol 337:105–113
9. Thomas DR, Francis NR, Xu C et al (2006) The three-dimensional structure of the flagellar rotor from a clockwise-locked mutant of *Salmonella enterica serovar typhimurium*. J Bacteriol 188:7039–7048
10. Gao Y, Cao E, Julius D et al (2016) TRPV1 structures in nanodiscs reveal mechanisms of ligand and lipid action. Nature 534:347–351
11. Taylor NM, Prokhorov NS, Guerrero-Ferreira RC et al (2016) Structure of the T4 baseplate and its function in triggering sheath contraction. Nature 533:346–352
12. Frauenfeld J, Löving R, Armache JP et al (2016) A saposin-lipoprotein nanoparticle system for membrane protein. Nat Methods 13:345–351
13. Liao M, Cao E, Julius D et al (2013) Structure of the TRPV1 ion channel determined by electron cryo-microscopy. Nature 504:107–112
14. Hauer F, Gerle C, Fischer N et al (2015) GraDeR: membrane protein complex preparation for single-particle cryo-EM. Structure 23:1769–1775
15. Kawamoto A, Morimoto VY, Miyata T et al (2013) Common and distinct structural features of *Salmonella* injectisome and flagellar basal body. Sci Rep 3:3369

16. Yamaguchi S, Fujita H, Ishihara A et al (1986) Subdivision of flagellar gens of *Salmonella typhimurium* into regions responsible for assembly, rotation and switching. J Bacteriol 166:187–193
17. Li X, Mooney P, Zheng S et al (2013) Electron counting and beam-induced motion correction enable near-atomic-resolution single-particle cryo-EM. Nat Methods 10:584–590
18. Mindell JA, Grigorieff N (2003) Accurate determination of local defocus and specimen tilt in electron microscopy. J Struct Biol 142:334–347. doi:10.1016/S1047-8477(03)00069-8
19. Scheres SH (2012) RELION: implementation of a Bayesian approach to cryo-EM structure determination. J Struct Biol 180:519–530
20. Kucukelbir A, Sigworth FJ, Tagare HD (2014) Quantifying the local resolution of cryo-EM density maps. Nat Methods 11:63–65
21. Pettersen EF, Goddard TD, Huang CC et al (2004) UCSF Chimera—a visualization system for exploratory research and analysis. J Comput Chem 25:1605–1612
22. Kremer JR, Mastronarde DN, McIntosh JR (1996) Computer visualization of three-dimensional image data using IMOD. J Struct Biol 116:71–76

Chapter 10

Structure of the MotA/B Proton Channel

Akio Kitao and Yasutaka Nishihara

Abstract

Flagellar motors utilize the motive force of protons and other ions as an energy source. To elucidate the mechanisms of ion permeation and torque generation, it is essential to investigate the structure of the motor stator complex; however, the atomic structure of the transmembrane region of the stator has not been determined experimentally. We recently constructed an atomic model structure of the transmembrane region of the *Escherichia coli* MotA/B stator complex based on previously published disulfide cross-linking and tryptophan scanning mutations. Dynamic permeation by hydronium ions, sodium ions, and water molecules was then observed using steered molecular dynamics simulations, and free energy profiles for ion/water permeation were calculated using umbrella sampling. We also examined the possible ratchet motion of the cytoplasmic domain induced by the protonation/deprotonation cycle of the MotB proton binding site, Asp32. In this chapter, we describe the methods used to conduct these analyses, including atomic structure modeling of the transmembrane region of the MotA/B complex; molecular dynamics simulations in equilibrium and in ion permeation processes; and ion permeation-free energy profile calculations.

Key words Structure modeling, Transmembrane prediction, Distance restraint, Molecular dynamics simulation, Steered molecular dynamics, Umbrella sampling, Proton transfer

1 Introduction

1.1 Background

Flagellar motors utilize the motive force of protons and other ions as an energy source [1–7]. To elucidate the mechanisms of ion permeation and torque generation, it is essential to investigate the structure of the motor stator complex; however, the atomic structure of the transmembrane (TM) region of the stator has not been determined experimentally.

Flagellar motors are reversible. In other words, they can rotate in both the clockwise (CW) and counter-clockwise (CCW) directions, if viewed from the outside of bacteria. Reversal of the direction of motor rotation regulates the swimming pattern of bacteria [8]. *Escherichia coli* has several proton-driven flagellar motors. The FliG, FliM, FliN, MotA, and MotB proteins in *E. coli* are involved in torque generation [8, 9]. FliG, FliM, and FliN constitute the

Tohru Minamino and Keiichi Namba (eds.), *The Bacterial Flagellum: Methods and Protocols*, Methods in Molecular Biology, vol. 1593, DOI 10.1007/978-1-4939-6927-2_10, © Springer Science+Business Media LLC 2017

flagellar rotor and are also involved in CW/CCW switching. Each rotor is typically surrounded by ten stators that comprise MotA and MotB.

The stator complex of the flagellar motor is embedded in the bacterial inner membrane. In *E. coli*, the stators (MotA/B complex) act as a proton channel [1, 3–6]. *Vibrio alginolyticus* has a polar flagellum powered by sodium ions [2]. Instead of MotA/B, PomA and PomB function as the stator in *V. alginolyticus*. The MotA/MomB7 motor functions under a Na^+ gradient in *V. alginolyticus*, where MomB7 is a chimera of the TM N-terminal region of MotB from *E. coli* and the periplasmic C-terminal domain of PomB from *V. alginolyticus* [10]. Analyses of this chimera suggested that the TM region of MotA/B utilizes both protons and sodium ions as energy sources. In *Bacillus alcalophilus*, the stator consists of MotP/S and is driven by rubidium (Rb^+), potassium (K^+), and sodium ions (Na^+) [7]. The MotP/S stator can be converted to a Na^+-driven motor by a single mutation. Therefore, bacterial flagella can be considered multi-fuel engines that convert the motive force of distinct ions to molecular motor rotation.

The MotA/B complex of *E. coli* consists of four MotA and two MotB proteins. Each MotA protein consists of 295 residues and contains four TM alpha-helical segments (A1–A4), two short loops (between A1–A2 and A3–A4) in the periplasm, and two long segments (residues 61–160 between A2–A3 and a C-terminal segment composed of residues 228–295) in the cytoplasm [3, 11]. The MotB protein (308 residues) is composed of a short N-terminal cytoplasmic segment, one TM helix (B), and a large C-terminal periplasmic domain [4, 5]. The B segment is expected to form a proton channel together with A3 and A4 [12]. Interestingly, only a few polar residues have been identified in the predicted TM segments of MotA/B [13, 14]. Therefore, the channel surface is expected to be relatively hydrophobic. Asp32 of MotB is conserved across species and considered to be the most plausible proton binding site [11]. This residue is situated near the cytoplasmic end of the B segment.

The generation of torque is hypothesized to originate from conformational changes in the MotA cytoplasmic domain upon proton association/dissociation at the carboxyl group of Asp32 in MotB and by interaction with FliG in the rotor [15–18]. The periplasmic region of MotB contains a peptidoglycan-binding motif that acts to anchor the stator complex to the peptidoglycan layer around the rotor [5, 19]. The atomic structure of the periplasmic domain was determined for MotB of *Helicobactor pylori* [20–22] and *Salmonella typhimurium* [23] and PomB of *V. alginolyticus* [24]. Overall electron microscopic molecular images of isolated PomA/B of *V. alginolyticus* at 21.3 (wild type) and 30.1 Å resolutions (PomB C-terminal deletion mutant) were reported by Yonekura et al. [25].

As mentioned above, the atomic structure of the TM region of the MotA/B complex has not been determined experimentally. However, systematic mutagenesis studies of *E. coli* MotA/B conducted by Blair and coworkers have provided essential information on the structure and function of the flagellar motor. Trp scanning mutagenesis of MotA [13] and MotB [14] showed the effects of mutations on relative swarming rates and identified the lipid-facing regions of the helices and their arrangements. Disulfide cross-liking analyses of MotB [26] revealed the arrangement of the MotB dimer in the MotA/B complex. Subsequent cross-linking studies of the A3 and A4 segments of MotA and the B segment of MotB [12] revealed the helical arrangement of the complex core. Cys mutagenesis of A1 and A2 [27] led to the determination of the overall arrangement of the MotA/B complex.

Based on these experimental data [12–14, 26, 27], we recently constructed a model of the atomic structure of the TM region of the MotA/B complex (Fig. 1) [28] and investigated the mechanism of proton permeation involving MotA/B. The dynamics of hydronium ion, sodium ion, and water molecule permeation through the channel formed in MotA/B were studied using steered molecular dynamics (SMD) simulations [30–32]. During the SMD simulations, we observed that Leu46 of MotB acts as the gate for

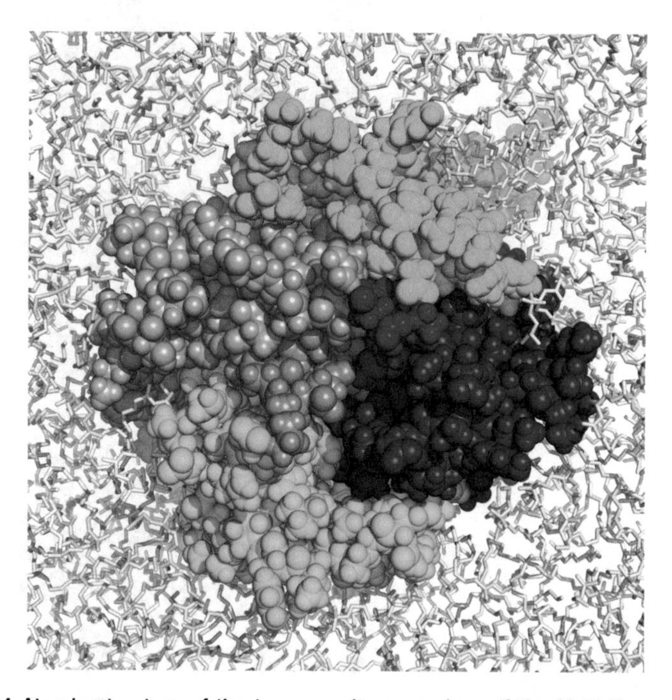

Fig. 1 Atomic structure of the transmembrane regions of the MotA/B complex viewed from the periplasmic side after SMD1 (Subheading 3.5) [28]. Four MotA (*green*, *purple*, *cyan*, and *pink*) and two MotB (*blue* and *orange*) molecules are shown. Two types of lipid molecules, POPG and POPE, are shown in *white* and *pale cyan*, respectively. The image was created with PyMOL [29]

hydronium ion permeation, which induces the formation of a water wire that may mediate the transfer of a proton to Asp32 of MotB (Fig. 2). We also calculated free energy profiles for ion permeation using umbrella sampling. The free energy barrier for H_3O^+ permeation was consistent with the proton transfer rate deduced from the flagellar rotational speed and number of protons transferred per rotation [34–36], suggesting that gating is the rate-limiting step. The structure and dynamics of MotA/B with non-protonated and protonated Asp32, Val43Met, and Val43Leu MotB mutants were also investigated using molecular dynamics (MD) simulations. A narrowing of the channel consistent with size-dependent ion selectivity was observed in the mutants [7]. In MotA/B with non-protonated Asp32, the A3 segment of MotA maintained a kink, whereas protonation of Asp32 induced a straighter conformation. Assuming that the cytoplasmic domain that was not included in the atomic model moves as a rigid body, these data suggest that protonation/deprotonation of Asp32 induces a ratchet-like motion of the cytoplasmic domain, which may be correlated with the motion of the flagellar rotor.

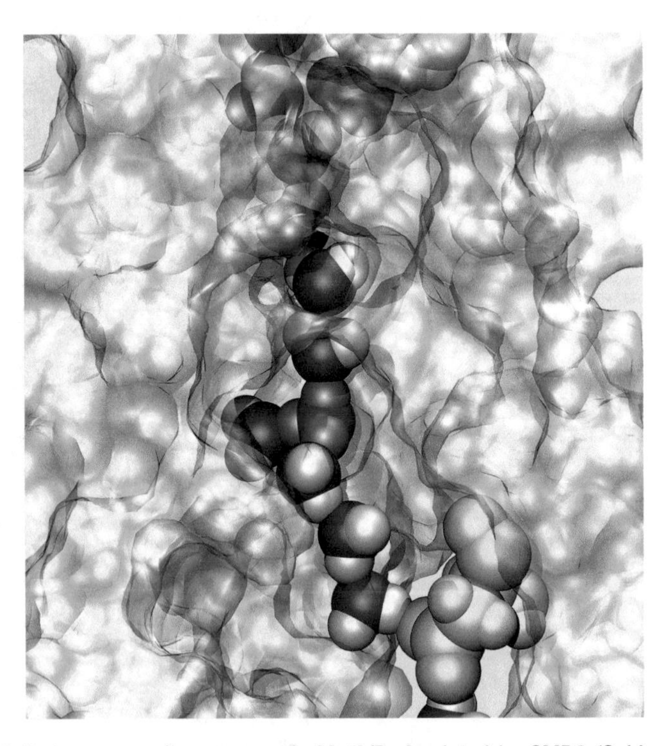

Fig. 2 Proton permeation process in MotA/B simulated by SMD2 (Subheading 3.7) [28]. Protons are permeated by the diffusion of hydronium ion (*blue*) through a gate of the channel (*green*. Leu46 of MotB), which induces the formation of water wire (*red* and *white*) that mediates the proton transfer to a proton binding site (*yellow*, Asp32 of MotB). The image was created with VMD [33]

1.2 Overview of the Methods

This chapter describes the protocol we used to model the atomic structure of the TM regions of MotA/B and elucidate the proton permeation mechanism [28]. The protocol involved computational structure modeling based on disulfide cross-linking and Trp scanning mutations, evaluation and selection of the generated model structures, observation of dynamic behavior in equilibrium using MD simulations, SMD simulations of ion permeation processes, and free energy profile calculations using umbrella sampling.

In the first part of the chapter (Subheadings 3.1–3.4), we describe the computational methods used to construct the atomic model structure of the TM region of MotA/B and investigate the ion permeation mechanisms. In Subheading 3.1, we illustrate the protocol for constructing the initial atomic model based on the prediction of the TM regions of MotA and MotB and information regarding disulfide cross-linking mutations, and then generate the candidate structures using MD simulations. The methods used for the first evaluation and selection of the generated structures are described in Subheading 3.2. The steps described in Subheading 3.2 were conducted based on consistency with the experimental data, the distribution of the generated structure ensemble, and their convergence. The next step, shown in Subheading 3.3, is all-atom refinement of the MotA/B complex embedded in a lipid membrane and in solution. Finally, Subheading 3.4 describes the evaluation of the refined structure based on structural properties and the consistency of the structure with the experimental data. Based on these results, the best atomic model structure was selected.

The second part of the chapter (Subheadings 3.5 and 3.6) describes further examinations of the channels with water permeation using SMD simulations (Subheading 3.5) and subsequent conformational sampling using MD simulations to investigate the dynamic properties of the MotA/B complex in equilibrium (Subheading 3.6). By conducting MD simulations for both protonated and non-protonated Asp32 (the essential proton binding site), the effect of protonation on the structure and dynamics of MotA/B was elucidated. The structure and dynamics of MotA/B mutants can be also investigated in this step.

The third part of the chapter (Subheading 3.7) describes methods for observing H_3O^+, H_2O, and Na^+ permeation through the channels of the MotA/B complex using SMD simulations. SMD simulations are useful for examining the atomic details of the permeation process, including the transient formation/breakage of atomic interactions and channel opening mechanisms.

The last part of the chapter (Subheading 3.8) describes the protocol for calculating the free energy profiles for the permeation of H_3O^+, H_2O, and Na^+ using an umbrella sampling technique. With this information, the energetics of ion transfer can be investigated.

2 Materials

2.1 Transmembrane Regions of MotA/B

1. Determine the TM regions of the MotA/B complex of *E. coli* to be considered in the modeling. We predicted the TM regions using multiple TM helix prediction methods (e.g., TopPred [37], TMpred [38], SOSUI [39], TMHMM 2.0 [40, 41], and MEMSAT 1.5 [42]) and also multiple alignment techniques (e.g., ClustalX [43]). Examine the degree of consensus within the predicted results and determine the TM regions considered in the modeling. We adopted the results of ClustalX: A1, 1–26; A2, 31–56; A3, 165–188; A4, 204–233; B, 24–49. The loops between A1–A2 and A3–A4 were considered in the modeling. We also added five additional residues to both the ends of the helices. The model constructed included residues 1–61 and 160–228 of MotA and 19–54 of MotB.

2.2 Definition of S-S Restraint Pairs

1. Classify the disulfide cross-linking experimental data into several groups. We classified the experimental data into five groups (100–80, 80–60, 60–40, 40–20, and 20–0%) based on the values of the yields of dimers or tetramers [12, 26, 27].
2. Set the distance restraints (S-S restraints. *See* **Note 1**) to the C^{β}-C^{β} atom pairs (C^{α} for Gly) that are identified as being close according to the disulfide cross-linking experiment. We applied the S-S restraints to the first four groups, with $R = 6.0$, 6.5, 7.0, and 7.5 Å, respectively (+1.0 Å per Gly).
3. Consider all possible combinations of the TM segment pairs for the dimer-forming S–S restraints. We generated eight distinct initial models (groups 1–8).

2.3 Definition of Hydrogen Bond–Restrained Pairs

1. Apply the hydrogen bond (H-bond) restraints (*see* **Note 1**) to the predicted TM segments (Subheading 2.1). We used $R = 3.0$ Å.

2.4 Definition of E-E Restrained Pairs

1. Impose the distance restraints between both ends of the A1–A4 and A2–A3 segments (E-E restraints. *See* **Note 1**). We used $R = 20.0$ Å.

3 Methods

3.1 Structure Generation in Vacuo

1. Generate the initial models, with the primarily experimental information consisting of the S-S and H-bond and E-E restraint energies. We initially arranged the predicted helical regions based on the arrangement reported by Kim et al. [27]. Consider only bond, angle, and torsion energies and the repulsive component of the Lennard-Jones energy in vacuo to simplify the potential energy surface and facilitate more efficient conformational sampling. We used GROMACS software [44] with the CHARMM36 force field [45].

2. Set the cut-off distance for non-bonding interactions. We used a 14-Å cutoff.
3. Minimize the energy until the initial interatomic contacts are relaxed.
4. Select the method for controlling the temperature in the MD simulations. In this case, a Nosé-Hoover thermostat was selected.
5. Apply the LINCS algorithm to bonds involving hydrogen atoms with a 1.0-fs time step.
6. Sample the conformational space at high temperature. We conducted MD simulations for 0.5 ns by gradually heating the system to 1000 K and maintained that temperature for 5 ns.
7. Conduct simulated annealing. We repeated 0.5-ns cooling and 0.5-ns equilibration steps four times (800, 600, 400, and 310 K) until the system temperature reached 310 K.
8. Equilibrate the system at 310 K for 5 ns.
9. Generate the candidate structures using additional 25-ns MD simulations.
10. Store snapshots every 1 ps and use them for evaluation (Subheading 3.2).

3.2 First Evaluation and Selection of Generated Models

1. Examine the convergence and distribution of the generated structures using principal component analysis [46]. We confirmed that the probability distribution along the principal components was almost Gaussian.
2. Calculate the root-mean-square fluctuations of MotA/B from the average structures. We confirmed that the generated model structures exhibited good convergence without non-bonded attractive interactions.
3. Confirm that the S-S restraint energy is low.
4. Determine whether the inter-residue distances not used for the S-S restraints are consistent. We referred to previously published disulfide cross-linking studies [12, 26, 27].
5. Calculate the Trp contact number (*see* **Note 2**) and confirm the consistency between the generated model and results of Trp scanning mutations. We referred to the Trp scanning mutations reported by Sharp et al. [13, 14].
6. Select the most plausible models for use in subsequent structure refinement (Subheading 3.3). We selected the three best snapshots from three groups (groups 3, 5, and 6).

3.3 Structure Refinement MD Simulations in a Lipid Membrane and in Solution

1. Embed the selected models in a lipid membrane using the program g_membed [47]. In this case, the MotA/B complex was embedded into a lipid bilayer consisting of 306 1-palmitoyl-2-oleoyl-sn-phosphatidylethanolamine (POPE) and 102 1-pal mitoyl-2-oleoyl-sn-phosphatidylglycerol (POPG) molecules (a 3:1 POPE:POPG ratio). The *z*-axis of the system was set to be perpendicular to the membrane surface.
2. Solvate the MotA/B/lipid model. In this case, the system was solvated with 30,783 TIP3P water molecules [48], and chloride and sodium ions were added to create a 0.15 M physiological concentration and neutralize the system. We used the NAMD program package [49] for the following MD simulations.
3. Gradually equilibrate the system. Keep the MotA/B structure rigid for the initial 1 ns to fit the lipids and solvent with the MotA/B molecules. Next, apply the positional restraints to MotA/B and gradually weaken them by decreasing the force constants. This process required 6 ns.
4. Control the temperature of the simulated system. We kept the system temperature at 310 K using Langevin dynamics.
5. Maintain the pressure of the system at 1 bar in the direction perpendicular to the membrane (NPAT ensemble simulations) using the Nosé-Hoover Langevin piston.
6. Use the particle-mesh Ewald algorithm for long-range electrostatic interactions and a cut-off distance of 14 Å for other short-range nonbonding interactions.
7. Apply the SHAKE algorithm to the bonds involving hydrogen atoms using a time step 2.0 fs.
8. Equilibrate the entire system without any restraints. We performed 100-ns MD simulations.
9. Save the conformations for further analyses. We stored the conformations every 10 ps and used the last 75-ns trajectories for the analyses.

3.4 Second Evaluation and Selection of Generated Models

1. Confirm that the S-S restraint energy is low.
2. Determine whether the inter-residue distances not used for the S-S restraints are consistent. We referred to previously published disulfide cross-linking studies [12, 26, 27].
3. Calculate the Trp contact number (*see* **Note 2**) and evaluate the consistency between the generated model and the Trp scanning mutations. We referred to the Trp scanning mutations reported by Sharp et al. [13, 14].
4. Estimate the deformation from the initial model during the refinement.

5. Examine the interactions among the MotA/B helices.
6. Examine the interactions between MotA/B and the lipid and solvent molecules.
7. Identify whether a channel exists.
8. Evaluate the twofold symmetry of the MotA/B model.
9. Select the best model from the structures from the distinct groups based on the aforementioned evaluation methods. We selected the best model from group 6.

3.5 Steered Molecular Dynamics Simulations of Water Permeation to Equilibrate the System

1. Conduct SMD simulations (SMD1) as follows.
2. Use H_2O as a pulled molecule (referred to as the SMD molecule).
3. Set the SMD molecule to pass through one of the two channels according to the constant velocity pulling scheme.
4. Pull a dummy atom connected to the center-of-mass of the SMD molecule having a harmonic spring with a force constant of 10 $k_B T/\text{Å}^2 = 6.1574$ kcal/mol/Å^2 along the direction of the z-axis, with a velocity of 0.05 Å/ps.
5. Conduct the H_2O permeation simulation starting from the best snapshot among the refinement MD trajectories (Subheading 3.4). Pull the SMD molecule up 50 Å from the starting position along the z-axis. We repeated the forward (from the periplasmic to cytoplasmic side) SMD simulation of H_2O 20 times, alternating between the two channels. A snapshot [28] is shown in Fig. 1.
6. Use the relaxed structure for further analyses.

3.6 Molecular Dynamics Simulations for Conformational Sampling in Equilibrium

1. Use the last snapshot of SMD1 (Subheading 3.5).
2. Conduct the sampling MD simulation using the conditions described in Subheading 3.3. We performed 100-ns MD simulations and used the results for the analyses.
3. Conduct another MD simulation with protonated MotB Asp32. We also simulated Val43Met and Val43Leu MotB mutants to examine the mutational effects on channel size.
4. Analyze the differences in the structure and dynamics between the non-protonated and protonated Asp32 models.

3.7 Steered Molecular Dynamics Simulations of Water and Ion Permeation to Investigate the Permeation Processes

1. Perform SMD simulations for H_3O^+, H_2O, and Na^+ as SMD molecules (SMD2) using methods similar to those described in Subheading 3.5, with the following conditions.
2. Select distinct snapshots as the initial structures for SMD2. We selected the top ten snapshots that best satisfied the S-S restraints from the refinement MD simulation (Subheading 3.3).

3. Generate ten forward permeation paths starting from each snapshot. An example of SMD snapshot for H_3O^+ [28] is shown in Fig. 2.

3.8 Calculation of Free Energy Profiles for H_3O^+, H_2O, and Na^+ Permeation Using Umbrella Sampling

1. Conduct the following umbrella sampling.
2. Determine the range of ion movement along the z-axis (perpendicular to the membrane), and determine the number of windows for the umbrella sampling. We employed 41 windows to cover the range along the z-axis from −20 to 20 Å.
3. Select initial coordinates for each umbrella sampling. We adopted 41 snapshots from the SMD1 simulations (Subheading 3.5).
4. Apply the umbrella potential, $U(z) = (k/2)(z - z_0)^2$, to the center of mass of H_3O^+, H_2O, or Na^+, where z_0 represents the expected peak position of the local probability density of the position. We used the force constant $k = 2.0$–5.0 kcal/mol/Å^2.
5. Perform the MD simulations with $U(z)$ in each window. We conducted 1-ns MD simulations.
6. Combine the histograms calculated from the simulations with $U(z)$ at different locations using the weighted histogram analysis method [50–52] to obtain the free energy profiles. We used the program distributed by Grossfield [53].

4 Notes

1. The energy terms for the S-S, H-bond, and E-E distance restraints (Subheadings 2.2–2.4) between the i- and jth atoms have a common functional form [28],

$$E_{ij} = \begin{cases} 0 & (r_{ij} < R) \\ (1/2)k(r_{ij} - R)^2 & (R \le r_{ij} < R + \Delta R) \\ (1/2)k(\Delta R)^2 + k\Delta R(r_{ij} - R - \Delta R) & (R + \Delta R \le r_{ij}) \end{cases}$$

where rij represents the interatomic distance between the i- and jth atoms, k represents the force constant, 64 kcal/mol/Å^2, and ΔR = 0.5 Å. The R values for the H-bond and E-E restraints are 3.0 and 20 Å, respectively, and those for the S-S restraints depend on the residue pairs, as shown in Subheading 2.1. The parabolic function started from R turns into a continuous linear function beyond the upper bound ($R + \Delta R$). It should be noted that the value of ΔR is typically 1–2 Å in NMR structure determinations, but we employed ΔR = 0.5 Å to allow a larger deviation from R compared to typical NMR studies [28].

2. The total Trp contact number is introduced as follows [28]. For each Trp mutant, generate nine Trp rotamers by changing two dihedral angles, χ_1 (N-CA-CB-CG) and χ_2 (CA-CB-CG-CD1). Calculate the van der Waals (vdW) energies of the Trp rotamers with the surrounding residues as

$$V_{lmk} = \min_{\text{rotamers}} \sum_{i \in l, j \in m} V_{ijk}^{vdW},$$

where i and j indicate atom indices of the lth Trp and the mth surrounding residues in the kth snapshot, respectively. $Vijk^{\text{vdW}}$ represents the vdW energy between the ith and jth atoms in the kth snapshot. $Vlmk$ represents the minimum vdW energy between the lth Trp rotamer and the mth surrounding residue in the kth snapshot. The cutoff length is 5.0 Å. Do not consider hydrogen atoms and main-chain atoms. The contact probability, pl, is introduced as

$$p_l = \frac{1}{T} \sum_k n_{lk}$$

$$n_{lk} = \begin{cases} 1 & (\text{if } V_{lmk} > 0 \text{ in } \mathit{any}\ m) \\ 0 & (\text{others}) \end{cases},$$

where l, m, and k indicate the index of the Trp mutation index, the residue index on the stator, and the trajectory snapshot index, respectively, and T represents the number of snapshots. The total Trp contact number is defined as

$$N_{\text{contact}} = \sum_{l,k} n_{lk}.$$

Acknowledgements

This research was supported by a Grant-in-Aid for Science Research in Innovative Areas (No. JP25104002) and by Grants-in-Aid for Science Research B (No. JP23370066 and No. JP15H04357) from the Japan Society for The Promotion of Science (JSPS) and Ministry of Education, Culture, Sports, Science and Technology of Japan (MEXT) to A.K.. This work was also financially supported by Innovative Drug Discovery Infrastructure through Functional Control of Biomolecular Systems, Priority Issue 1 in Post-K Supercomputer Development (Project ID: hp150270) to A.K. The computations were partially performed using the supercomputers at the RCCS, National Institute of Natural Science and ISSP, The University of Tokyo.

References

1. Larsen SH, Adler J, Gargus JJ, Hogg RW (1974) Chemomechanical coupling without ATP: the source of energy for motility and chemotaxis in bacteria. Proc Natl Acad Sci U S A 71:1239–1243
2. Hirota N, Kitada M, Imae Y (1981) Flagellar motors of alkalophilic bacillus are powered by an electrochemical potential gradient of Na^+. FEBS Lett 132:278–280
3. Dean GE, Macnab RM, Stader J, Matsumura P, Burks C (1984) Gene sequence and predicted amino acid sequence of the motA protein, a membrane-associated protein required for flagellar rotation in *Escherichia coli*. J Bacteriol 159:991–999
4. Stader J, Matsumura P, Vacante D, Dean GE, Macnab RM (1986) Nucleotide sequence of the *Escherichia coli* motB gene and site-limited incorporation of its product into the cytoplasmic membrane. J Bacteriol 166:244–252
5. Chun SY, Parkinson JS (1988) Bacterial motility: membrane topology of the *Escherichia coli* MotB protein. Science 239:276–278
6. Khan S, Dapice M, Reese TS (1988) Effects of mot gene expression on the structure of the flagellar motor. J Mol Biol 202:575–584
7. Terahara N, Sano M, Ito M (2012) A bacillus flagellar motor that can use both Na+ and K+ as a coupling ion is converted by a single mutation to use only Na+. PLoS One 7:e46248
8. Yamaguchi S, Fujita H, Ishihara A, Aizawa SI, Macnab RM (1986) Subdivision of flagellar genes of *Salmonella-typhimurium* into regions responsible for assembly, rotation, and switching. J Bacteriol 166:187–193
9. Enomoto M (1966) Genetic studies of paralyzed mutants in *Salmonella*. 2. Mapping of 3 mot loci by linkage analysis. Genetics 54:1069–1076
10. Asai Y, Yakushi T, Kawagishi I, Homma M (2003) Ion-coupling determinants of Na+-driven and H+-driven flagellar motors. J Mol Biol 327:453–463
11. Zhou JD, Fazzio RT, Blair DF (1995) Membrane topology of the mota protein of *Escherichia coli*. J Mol Biol 251:237–242
12. Braun TF, Al-Mawsawi LQ, Kojima S, Blair DF (2004) Arrangement of core membrane segments in the MotA/MotB proton-channel complex of *Escherichia coli*. Biochemistry 43:35–45
13. Sharp LL, Zhou JD, Blair DF (1995) Features of Mota proton channel structure revealed by tryptophan-scanning mutagenesis. Proc Natl Acad Sci U S A 92:7946–7950
14. Sharp LL, Zhou JD, Blair DF (1995) Tryptophan-scanning mutagenesis of Motb, an integral membrane-protein essential for flagellar rotation in *Escherichia coli*. Biochemistry 34:9166–9171
15. Berry RM (1993) Torque and switching in the bacterial flagellar Motor—an electrostatic model. Biophys J 64:961–973
16. Elston TC, Oster G (1997) Protein turbines. 1. The bacterial flagellar motor. Biophys J 73:703–721
17. Walz D, Caplan SR (2000) An electrostatic mechanism closely reproducing observed behavior in the bacterial flagellar motor. Biophys J 78:626–651
18. Kojima S, Blair DF (2001) Conformational change in the stator of the bacterial flagellar motor. Biochemistry 40:13041–13050
19. Demot R, Vanderleyden J (1994) The C-terminal sequence conservation between OmpA-related outer membrane proteins and MotB suggests a common function in both gram-positive and gram-negative bacteria, possibly in the interaction of these domains peptidoglycan. Mol Microbiol 12:333–336
20. Roujeinikova A (2008) Crystal structure of the cell wall anchor domain of MotB, a stator component of the bacterial flagellar motor: implications for peptidoglycan recognition. Proc Natl Acad Sci U S A 105:10348–10353
21. O'Neill J, Xie M, Hijnen M, Roujeinikova A (2011) Role of the MotB linker in the assembly and activation of the bacterial flagellar motor. Acta Crystallogr D Biol Crystallogr 67:1009–1016
22. Reboul CF, Andrews DA, Nahar MF, Buckle AM, Roujeinikova A (2011) Crystallographic and molecular dynamics analysis of loop motions unmasking the peptidoglycan-binding site in stator protein MotB of flagellar motor. PLoS One 6:e18981
23. Kojima S, Imada K, Sakuma M, Sudo Y, Kojima C, Minamino T, Homma M, Namba K (2009) Stator assembly and activation mechanism of the flagellar motor by the periplasmic region of MotB. Mol Microbiol 73:710–718
24. Zhu SW, Takao M, Li N, Sakuma M, Nishino Y, Homma M, Kojima S, Imada K (2014) Conformational change in the periplamic region of the flagellar stator coupled with the assembly around the rotor. Proc Natl Acad Sci U S A 111:13523–13528
25. Yonekura K, Maki-Yonekura S, Homma M (2011) Structure of the flagellar motor protein complex PomAB: implications for the torque-generating conformation. J Bacteriol 193:3863–3870
26. Braun TF, Blair DF (2001) Targeted disulfide cross-linking of the MotB protein of *Escherichia coli*: evidence for two H+ channels in the stator complex. Biochemistry 40:13051–13059

27. Kim EA, Price-Carter M, Carlquist WC, Blair DF (2008) Membrane segment organization in the stator complex of the flagellar motor: implications for proton flow and proton-induced conformational change. Biochemistry 47:11332–11339
28. Nishihara Y, Kitao A (2015) Gate-controlled proton diffusion and protonation-induced ratchet motion in the stator of the bacterial flagellar motor. Proc Natl Acad Sci U S A 112:7737–7742
29. The PyMOL Molecular Graphics System, Version 1.7 Schrödinger, LLC
30. Isralewitz B, Baudry J, Gullingsrud J, Kosztin D, Schulten K (2001) Steered molecular dynamics investigations of protein function. J Mol Graph Model 19:13–25
31. Isralewitz B, Gao M, Schulten K (2001) Steered molecular dynamics and mechanical functions of proteins. Curr Opin Struct Biol 11:224–230
32. Jensen MO, Park S, Tajkhorshid E, Schulten K (2002) Energetics of glycerol conduction through aquaglyceroporin GlpF. Proc Natl Acad Sci U S A 99:6731–6736
33. Humphrey W, Dalke A, Schulten K (1996) VMD: visual molecular dynamics. J Mol Graph Model 14:33–38
34. Block SM, Berg HC (1984) Successive incorporation of force-generating units in the bacterial rotary motor. Nature 309:470–472
35. Meister M, Lowe G, Berg HC (1987) The proton flux through the bacterial flagellar motor. Cell 49:643–650
36. Samuel AD, Berg HC (1996) Torque-generating units of the bacterial flagellar motor step independently. Biophys J 71:918–923
37. Vonheijne G (1992) Membrane-protein structure prediction—hydrophobicity analysis and the positive-inside rule. J Mol Biol 225:487–494
38. Hofmann K, Stoffel W (1993) TMbase—a database of membrane spanning proteins segments. Biol Chem Hoppe Seyler 374:166
39. Hirokawa T, Boon-Chieng S, Mitaku S (1998) SOSUI: classification and secondary structure prediction system for membrane proteins. Bioinformatics 14:378–379
40. Sonnhammer ELL, Gv H, Krogh A (1998) A hidden Markov model for predicting transmembrane helices in protein sequences. Proc Int Conf Intell Syst Mol Biol 6:175–182
41. Krogh A, Larsson B, von Heijne G, Sonnhammer ELL (2001) Predicting transmembrane protein topology with a hidden Markov model: application to complete genomes. J Mol Biol 305:567–580
42. Jones DT (2007) Improving the accuracy of transmembrane protein topology prediction using evolutionary information. Bioinformatics 23:538–544
43. Larkin MA, Blackshields G, Brown NP, Chenna R, McGettigan PA, McWilliam H, Valentin F, Wallace IM, Wilm A, Lopez R, Thompson JD, Gibson TJ, Higgins DG (2007) Clustal W and clustal X version 2.0. Bioinformatics 23:2947–2948
44. Hess B, Kutzner C, van der Spoel D, Lindahl E (2008) GROMACS 4: algorithms for highly efficient, load-balanced, and scalable molecular simulation. J Chem Theory Comput 4:435–447
45. Klauda JB, Venable RM, Freites JA, O'Connor JW, Tobias DJ, Mondragon-Ramirez C, Vorobyov I, MacKerell AD, Pastor RW (2010) Update of the CHARMM all-atom additive force field for lipids: validation on six lipid types. J Phys Chem B 114:7830–7843
46. Kitao A, Hirata F, Go N (1991) The effects of solvent on the conformation and the collective motions of protein—normal mode analysis and molecular-dynamics simulations of melittin in water and in vacuum. Chem Phys 158:447–472
47. Wolf MG, Hoefling M, Aponte-Santamaria C, Grubmuller H, Groenhof G (2010) g_membed: efficient insertion of a membrane protein into an equilibrated lipid bilayer with minimal perturbation. J Comput Chem 31:2169–2174
48. Jorgensen WL, Chandrasekhar J, Madura JD, Impey RW, Klein ML (1983) Comparison of simple potential functions for simulating liquid water. J Chem Phys 79:926–935
49. Phillips JC, Braun R, Wang W, Gumbart J, Tajkhorshid E, Villa E, Chipot C, Skeel RD, Kale L, Schulten K (2005) Scalable molecular dynamics with NAMD. J Comput Chem 26:1781–1802
50. Ferrenberg AM, Swendsen RH (1989) Optimized Monte-Carlo data-analysis. Phys Rev Lett 63:1195–1198
51. Kumar S, Bouzida D, Swendsen RH, Kollman PA, Rosenberg JM (1992) The weighted histogram analysis method for free-energy calculations on biomolecules. 1. The method. J Comput Chem 13:1011–1021
52. Souaille M, Roux B (2001) Extension to the weighted histogram analysis method: combining umbrella sampling with free energy calculations. Comput Phys Commun 135:40–57
53. Grossfield A WHAM: the weighted histogram analysis method 2.0.6. http://membrane.urmc.rochester.edu/content/wham

Chapter 11

Mechanism of Stator Assembly and Incorporation into the Flagellar Motor

Seiji Kojima

Abstract

In many cases, conformational changes in proteins are related to their functions, and thereby inhibiting those changes causes functional defects. One way to perturb such conformational changes is to covalently link the regions where the changes are induced. Here, I introduce an example in which an intramolecular disulfide crosslink in the stator protein of PomB, introduced based on its crystal structure, reversibly inhibits the rotation of the flagellar motor, and I detail how we analyzed that phenotype. In this Chapter, first I describe how we monitor the motility inhibition and restoration by controlling disulfide bridge formation, and secondly how we detect intramolecular disulfide crosslinks, which are sometimes difficult to monitor by mobility shifts on SDS-PAGE gels.

Key words Bacterial flagellum, Motility, Flagellar motor, Stator, Torque generation, PomA, PomB, Disulfide crosslink, Biotin maleimide

1 Introduction

1.1 Background

Many motile bacteria swim by rotating their motility organ, the flagellum [1, 2]. Rotation of the flagellum is driven by a motor embedded in the cell surface at the base of each flagellum. The flagellar motor is harnessed by a specific ion motive force (H^+ or Na^+), and energy conversion occurs at the multiple stator units that surround the rotary part of the motor (rotor): ion translocation through the stator units is coupled with the rotor-stator interaction that generates motor torque [3–5]. Each stator unit is composed of two membrane protein subunits: MotA(PomA) and MotB(PomB). The H^+-driven *Escherichia coli* and *Salmonella* motors use the MotA/MotB complex [6], and the Na^+-driven *Vibrio* and *Shewanella* motors use the PomA/PomB complex [3, 7]. MotA and PomA, MotB and PomB are orthologs, respectively, and form an A4B2 hetero-hexamer complex that functions as an ion-conducting channel [8–11]. MotA(PomA) has four transmembrane segments and cytoplasmic regions that interact with the

Tohru Minamino and Keiichi Namba (eds.), *The Bacterial Flagellum: Methods and Protocols*, Methods in Molecular Biology, vol. 1593, DOI 10.1007/978-1-4939-6927-2_11, © Springer Science+Business Media LLC 2017

rotor, whereas MotB(PomB) has a single transmembrane segment with a large periplasmic region that contains an OmpA-like domain with the peptidoglycan (PG)-binding motif, at which the stator unit is anchored at the PG layer (Fig. 1a) [12]. In the functioning motor, smooth rotation is achieved by the proper placement and anchoring of the stator units around the rotor, which is mediated by the B subunits of the stator complex.

Therefore, incorporation of the stator units into the motor is a critical step to complete the functional motor assembly. Because overproduction of the stator does not arrest cell growth, the ion-conductivity of the stator is activated only when it is incorporated into the motor (Fig. 1a) [13]. To investigate how the stator assembly around the rotor is coupled with the activation of ion conductivity, structure-based functional studies were performed.

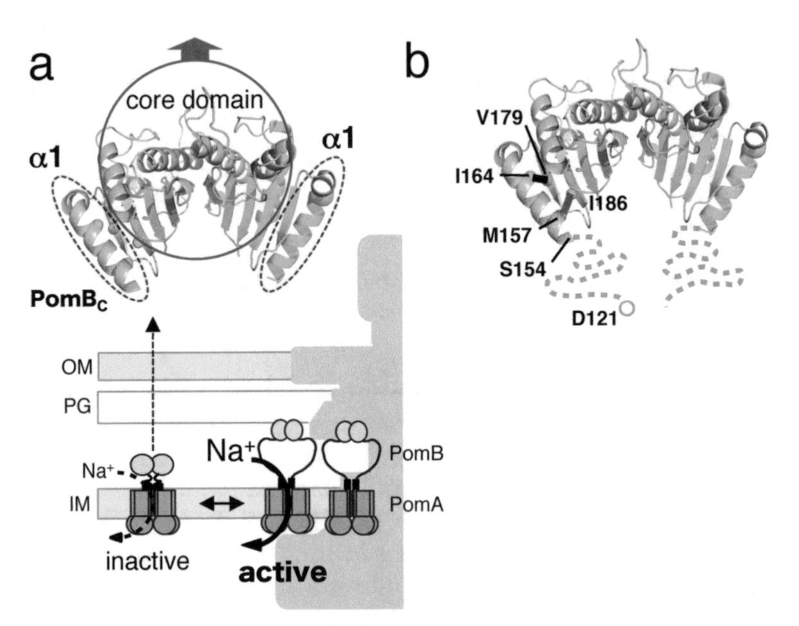

Fig. 1 A model for the conformational change in the periplasmic region of PomB and the design of crosslinking pairs to test the model. (**a**) In the *Vibrio alginolyticus* polar flagellar motor, the stator units are composed of PomA and PomB. The PomA/B complex becomes active for Na^+ conduction coupled with torque generation only when it is incorporated into the motor. Based on the structural information and functional assay, a model is proposed for the periplasmic C-terminal region of PomB ($PomB_C$), where its crystal structure has been solved (shown above), would change its conformation during the incorporation of stator units around the motor. The hydrophobic interactions between helix α1 and the core domain would be weakened by the conformational change, and the released core domain can reach the peptidoglycan (PG) layer. IM inner membrane, OM outer membrane. (**b**) The designed crosslinking pairs to test the model. The M157C-I186C and I164-V179C pairs are shown as representative pairs. In the crystal structure of $PomB_C$ dimer (PDB code: 3WPW), the region between residues D121 to S154 is disordered (shown as a dotted line)

A series of deletion studies of *Salmonella* MotB identified a functionally minimum MotB protein (MotBΔL), whose *p*eriplasmic region contains the entire region *e*ssential for *m*otility [14]. This region, called PEM, was crystallized and revealed an unexpectedly compact structure [15]. Since MotBΔL is functional, it was suggested that a large conformational change must be induced, presumably around the N-terminal region of PEM (helix α1, α2 and strand β1), which is a characteristic of MotB family proteins, to reach the PG layer when incorporated into the motor [15]. In support of this model, a point mutation (L119P) in the middle of helix α1 was found to improve incorporation of the MotA/MotBΔL complex into the motor. Importantly, overproduction of the MotA/MotBΔL(L119P) complex arrested cell growth, suggesting that this mutant stator mimics an active conformation for proton conduction which is also competent for PG anchoring. Therefore, a model was proposed in which the strand β1 was extended to cause disruption of hydrophobic interactions between helix α1 and the PG-binding core, leading to the release of the PG-binding core to anchor the stator (Fig. 1a) [15].

Analogous studies of the Na^+-driven motor protein PomB were conducted [16], and the crystal structure of the PEM region of PomB showed a very similar fold to its MotB counterpart (Fig. 1a), suggesting that the proposed conformational changes during the assembly-coupled stator activation might be a common feature regardless of the coupling ion or species [17]. To test this idea, an in vivo disulfide crosslinking approach in *Vibrio* PomB was employed. It was expected that an intramolecular crosslink between helix α1 and the PG-binding core would impair the conformational change proposed above (Fig. 1), and thereby abolish motility. The results showed that the M157C-I186C disulfide crosslink reversibly impairs motility but the I164C-V179C crosslink still allows motility (Figs. 1 and 2), suggesting that a conformational change is induced at the N-terminal two-thirds of helix α1 and connecting an N-terminal disordered region of PEM [17]. Therefore, although the assembly-coupled conformational change in the stator B subunits seems to be a common mechanism of the stator, the actual conformational change may be different from the one proposed for MotB. In the next step, alternative approaches that directly detect the conformational changes must be developed to clarify exactly what happens in vivo during the assembly-coupled stator activation.

1.2 Overview of the Methods

In this chapter, I introduce how our team investigated the conformational change in the Na^+-driven flagellar stator protein PomB, based on the crystal structure of the PEM region of PomB [17]. If a similar conformational change proposed for MotB occurs in the *Vibrio* stator, a crosslink between helix α1 and the PG-binding core (Fig. 1) would impair motility. Therefore, we generated a series of double cysteine replacements (one in helix α1 and the other in the

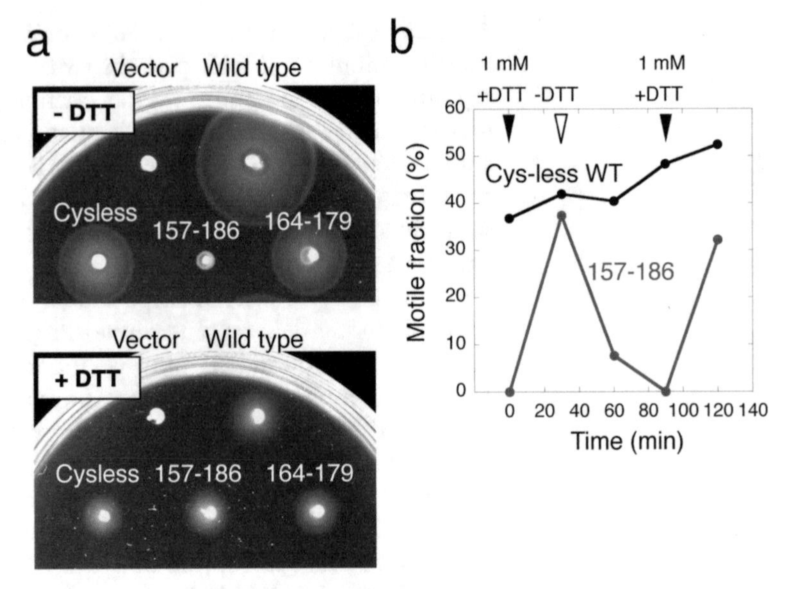

Fig. 2 Reduction of the disulfide bridge restores motility. (**a**) Motility assay in soft agar plates. Overnight cultures of NMB191 cells (Δ*pomAB*) harboring the vector, a plasmid pHFAB (expressing the wild-type PomA/PomB complex), pHFAB Cysless (expressing the cysteine-less PomA/PomB complex), or pHFAB Cysless with double cysteine substitutions (157C-186C or 164C-179C) were inoculated on VPG soft agar plates and incubated at 30 °C. Addition of the reducing agent DTT (5 mM) to the agar restored the motility of the PomB(M157C-I186C) mutant. The *upper panel* shows motility without DTT (incubated for 6 h) and the *lower panel* shows motility with 5 mM DTT (incubated for 7 h). (**b**) Motility assay under dark-field microscopy. Cells were grown until late-log phase, and then were suspended in TMN500 medium. At time zero, DTT (final concentration of 1 mM) was added to the suspension and motility was observed using dark-field microscopy and was recorded. The addition of DDT increased the motile fraction of M157C-I186C mutant cells but its removal (at 30 min) decreased the motile fraction, indicating that the intramolecular crosslink in the periplasmic region of PomB reversibly inhibits motor function

PG-binding core, *see* Fig. 1b), and examined the effects of a disulfide crosslink on motility in soft agar plates (Fig. 2a) and in aqueous medium using dark-field microscopy (Fig. 2b). The disulfide crosslink was confirmed by thiol-specific labeling (Fig. 3). These two procedures, using the marine bacterium *Vibrio alginolyticus*, which has a Na^+-driven flagellar motor powered by the PomA/PomB stator complex, are described here.

The first section is divided into two subsections (Subheadings 3.1 and 3.2). In Subheading 3.1, I describe how we designed cysteine replacements to form disulfide crosslinks (*see* **Note 1**), and then explain the experiments used to test their motility in soft agar plates. This simple assay is convenient and easy if you test multiple mutants at a time. We can detect the restoration of motility when the reducing agent dithiothreitol (DTT) is included in the medium. The

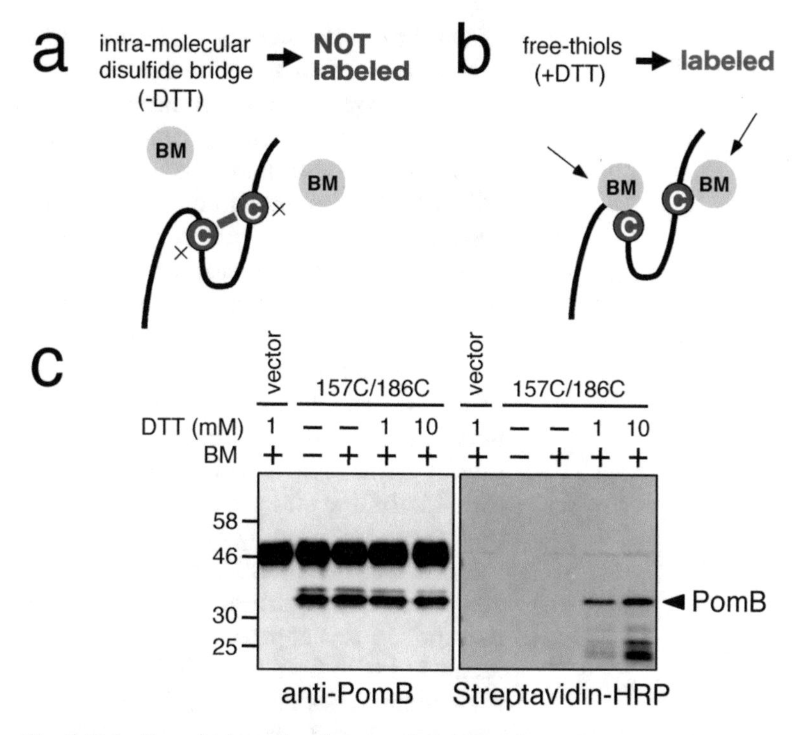

Fig. 3 Detection of intramolecular crosslink formation by free-cysteine labeling with biotin maleimide. Schematic representation of the labeling experiment (**a** and **b**). As shown in (**a**), biotin maleimide cannot react with crosslinked cysteine residues, but free thiol groups of cysteine generated by the cleavage of the disulfide crosslink can react with biotin maleimide (**b**). Therefore, if biotinylated PomB appears only after DTT treatment, it is likely that a disulfide crosslink is formed between the designed double cysteine pairs. (**c**) Detection of biotinylated PomB using streptavidin-conjugated HRP and chemiluminescence using the in vivo labeling method (Subheading 3.3). Depending on the concentration of DTT, the band of biotinylated PomB appeared (*right panel*). BM, biotin maleimide. The same samples were analyzed by an anti-PomB immunoblot to show the loading control (*left panel*). A similar band pattern can be obtained by the in vitro labeling method

effects of a crosslink on motility can be evaluated by comparing the sizes of the motility rings (Fig. 2a). In Subheading 3.2, I describe the assay of detecting swimming cells using microscopy. Although the soft agar assay is simple and convenient, the motility defect in soft agar could be attributed to reasons unrelated to torque generation (e.g., chemotaxis and cell growth). Therefore, direct observation of cell motility using a microscope must be performed. Cells are cultured until the late logarithmic phase, then are harvested, washed, and suspended in the motility medium. Cells expressing mutant PomB with an intramolecular disulfide crosslink do not swim at this step, but the addition of DTT to the suspension gradually restores their motility, which can be evaluated as the motile fraction (percentage of motile cells among the observed total cells, *see* Fig. 2b). The effect of crosslinking on motility is reversible, since the removal of DTT from the suspension again impairs motility because of the re-formation of the disulfide crosslink.

The second section also has two subsections (Subheadings 3.3 and 3.4). In this section, I describe the method used to detect intramolecular disulfide crosslinks. Since an intramolecular crosslink in PomB did not significantly alter its mobility on SDS-PAGE gels, we could not confirm the crosslink by a mobility shift of PomB on the gel. Instead, we use in vivo labeling of free cysteine residues by a thiol-reactive reagent, biotin maleimide, to evaluate the formation of the disulfide crosslink. Biotin maleimide irreversibly modifies the free thiol group exposed to the aqueous surface of the protein. If an intramolecular disulfide crosslink is formed between the designed double cysteines, biotin maleimide cannot modify them (Fig. 3a). However, the addition of DTT cleaves a disulfide bridge and produces free cysteines that are now competent to modification by biotin maleimide (Fig. 3b). Therefore, we can evaluate the crosslink formation by the detection of biotinylated PomB. In Subheading 3.3, I describe the in vivo labeling method. In this approach, cells treated with or without DTT are incubated with biotin maleimide. The cells are then disrupted and their membranes are isolated and solubilized. PomB is immunoprecipitated and separated by SDS-PAGE, transferred onto a PVDF membrane, and biotinylated PomB is detected with streptavidin-conjugated HRP and chemiluminescence (Fig. 3c). The in vivo method is straightforward and simple, but if the cysteine residues are buried in the protein core, biotin maleimide would not be able to access them and labeling would not be efficient. In Subheading 3.4, I describe an alternative in vitro method that can overcome this accessibility problem. In this approach, the proteins are denatured to allow biotin maleimide to access free cysteines that are otherwise buried inside. After treating the cells with or without DTT, cells are mixed with SDS and boiled, cooled down to room temperature, and then reacted with biotin maleimide to label free cysteine. PomB is immunoprecipitated and biotinylated PomB is detected as described above. These two methods can be used to detect any proteins that form an intramolecular disulfide crosslink, especially cases that are difficult to evaluate by the mobility shift of protein on SDS-PAGE gels.

2 Materials

2.1 Motility Assay in Soft Agar Plates

All media should be autoclaved. Chloramphenicol, DTT, and L-arabinose are added to the autoclaved medium after it has cooled down to ~60 °C.

1. *Vibrio alginolyticus* strain NMB191 (*laf* Δ*pomAB*) harboring the plasmid pBAD33 or pHFAB-Cysless (encoding *pomA* and

cysteine-free *pomB* in an operon placed under the arabinose-inducible promoter) with or without desired mutations in *pomB* (*see* **Note 1**).

2. VC medium: 0.5% (w/v) polypeptone, 0.5% (w/v) yeast extract, 3% (w/v) NaCl, 0.4% (w/v) K_2HPO_4, 0.2% (w/v) glucose.
3. VPG soft agar plate: 1% (w/v) polypeptone, 0.5% (w/v) glycerol, 3% (w/v) NaCl, 0.4% (w/v) K_2HPO_4, 0.3% (w/v) bacto-agar (*see* **Note 2**).
4. Chloramphenicol (2.5 mg/mL stock in ethanol, store at −20 °C).
5. Dithiothreitol (DTT) (100 mM stock in MilliQ water, store at −20 °C).
6. L-Arabinose (20% (w/v) stock in MilliQ water, store at −20 °C).
7. 30 °C incubator for bacterial plates.
8. Sterile petri dishes for soft agar plates.

2.2 Swimming Motility Assay Using Dark-Field Microscopy

Use the same materials (**items 1, 2, 4–6**) as listed in Subheading 2.1, and additional materials as listed below. All media should be autoclaved.

1. VPG medium: 1% (w/v) polypeptone, 0.5% (w/v) glycerol, 3% (w/v) NaCl, 0.4% (w/v) K_2HPO_4.
2. TMN500 medium: 50 mM Tris–HCl, pH 7.5, 5 mM glucose, 5 mM $MgCl_2$, 500 mM NaCl.
3. L-Serine (1 M stock in MilliQ water, store at 4 °C), filter sterilized.
4. 30 °C shaker for bacterial culture.
5. 30 °C water bath.
6. Ice.
7. Eppendorf tubes.
8. Microcentrifuge.
9. Test tubes.
10. Slide glasses.
11. Dark-field microscope equipped with a CCD camera and a video image recording system.
12. Image analysis software (*see* **Note 3**).

2.3 Detection of Intramolecular Disulfide Crosslink Formation (In Vivo Method)

Use the same materials (**items 1, 2, 4–6**) as listed in Subheading 2.1, and (**items 1–9**) as listed in Subheading 2.2, and additional materials as listed below. All media should be autoclaved.

1. TNET: 50 mM Tris–HCl, pH 8.0, 150 mM NaCl, 5 mM EDTA, 1%(w/v) Triton X-100.

2. Biotin maleimide (Sigma, 40 mM stock in DMSO, store at –20 °C).
3. Protein-A Sepharose CL-4B beads (GE Healthcare).
4. Antibody against PomB periplasmic region [18].
5. Streptavidin-conjugated HRP (GE Healthcare).
6. PVDF membranes for western blotting.
7. TBST buffer: 20 mM Tris–HCl, pH 8.0, 500 mM NaCl, 0.05% (w/v) Tween-20.
8. Skim milk.
9. Sonicator.
10. Tubes for ultracentrifugation.
11. Ultracentrifuge.
12. Rotary wheel in a cold room.
13. Heat block for boiling SDS-PAGE samples.
14. SDS-PAGE system.
15. Semidry blotting transfer system.
16. Chemiluminescence signal detection system.

2.4 Detection of Intramolecular Disulfide Crosslink Formation (In Vitro Method)

Use the same materials as listed in Subheading 2.3, and additional materials as listed below.

1. 2% (w/v) SDS.
2. Cysteine (400 mM stock in MilliQ water, store at –20 °C).

3 Methods

3.1 Motility Assay in Soft Agar Plates

1. Start overnight cultures of the *Vibrio* cells of interest in 3 mL VC medium containing 2.5 μg/mL chloramphenicol at 30 °C, with vigorous lateral shaking at 180 rpm in the shaker.
2. On the same day as the motility analysis, prepare VPG soft agar plates containing 2.5 μg/mL chloramphenicol, 0.02% arabinose, and 5 mM DTT. Also prepare a plate without DTT. Chloramphenicol, arabinose, and DTT are added right before pouring the agar into the petri dish; otherwise, these chemicals are damaged by heat (*see* **Note 4**).
3. Inoculate 2 μL of each overnight culture on the surface of a freshly prepared soft agar plate (with or without DTT). Strains to be compared for motility must be prepared on the same day and inoculated in the same soft agar plate.
4. Incubate the plates at 30 °C in the incubator for an appropriate time, until a suitable size of motility ring is formed (typi-

cally 6 ~ 7 h, *see* Fig. 2a). DTT seems to affect cell growth, so plates with DTT take a little longer to reach an appropriate ring size.

5. Take a photograph of each plate and measure the sizes of the motility rings using a scale.

3.2 Swimming Motility Assay Using Dark-Field Microscopy

1. Start overnight cultures of the *Vibrio* cells of interest in 3 mL VC medium containing 2.5 μg/mL chloramphenicol at 30 °C, with vigorous lateral shaking at 180 rpm in the shaker.
2. In the morning of the next day, inoculate 100 μL of each overnight culture in 3 mL VPG medium (1:30 dilution) containing 2.5 μg/mL chloramphenicol and 0.02% arabinose.
3. Grow cells until late-log phase (typically 4 h) at 30 °C with vigorous lateral shaking at 180 rpm in the shaker.
4. Harvest cells (1 mL culture) at room temperature by microfuge centrifugation (3500 × *g*, 5 min).
5. Remove each supernatant and wash cells with 1 mL TMN500 medium.
6. Resuspend cells in 500 μL TMN500 medium, place them in glass tubes, and then incubate them at 30 °C for 10 min in a water bath. In this step, cells are equilibrated to the buffer condition.
7. DTT is added to the suspension at a final concentration of 1 mM. This sets the time at zero. Cell suspensions should be kept at 30 °C in the water bath.
8. At each time point, 5 μL of the suspension is taken and diluted 100 times with fresh TMN500 medium in a glass tube containing 10 mM serine (final concentration), then immediately observe the motility using a dark-field microscope (*see* **Note 5**). We usually record images of motility for 1 min. We set observation time points at 0, 5, 10, 20, and 40 min.
9. If necessary, DTT can be removed from the cell suspensions by washing cells twice with 1 mL TMN500 medium. For long-range observations of motility to examine the effects of removing DTT, we remove DTT at 30 min after the DTT addition, then observe motility every 30 min after that (i.e., 60, 90 min; *see* Fig. 2b).
10. The motile fraction is defined as the percentage of motile cells among the observed total cells. The video images recorded at each time point are integrated for 1 s using image analysis software (*see* **Note 3**), and the number of swimming tracks (S) of motile cells is counted in the observed field. Simultaneously, the number of all cells (T) in the observed field is counted, and the motile fraction is obtained as S/T × 100 (%). This measurement is performed in at least three different fields, and averaged motile fractions are reported in the figure.

3.3 Detection of Intramolecular Disulfide Crosslink Formation (In Vivo Method)

1. Grow *Vibrio* cells in 3 mL VPG medium containing 2.5 μg/mL chloramphenicol and 0.02% arabinose until late-log phase at 30 °C, the same way as described in Subheading 2.2. If the expression level is low, scale up the culture volume (~20 mL).
2. Harvest all cells at room temperature by microfuge centrifugation (3500 × *g*, 5 min) and wash cells with 1 mL TMN500 medium.
3. Resuspend cells in 1 mL TMN500 medium and divide the suspension into two halves (500 μL each) in Eppendorf tubes.
4. DTT is added to one tube at a final concentration of 1 mM, and the same volume of distilled water is added to the other tube.
5. Incubate both tubes at 30 °C in the water bath for 40 min.
6. Check motility by dark-field microscopy for both the samples. If a disulfide crosslink inhibits motility, the addition of DTT should cleave the crosslink and restore motility 40 min after the addition of DTT.
7. Cells are collected at room temperature by centrifugation (13,500 × *g*, 1 min), and then resuspended in 500 μL TMN500 medium.
8. Add biotin maleimide to the suspension at a final concentration of 0.4 mM.
9. Incubate the suspensions at 30 °C for 10 min in the water bath (or heat block).
10. Wash cells twice with 1 mL TMN500 medium, and resuspend them in 500 μL TMN500 medium.
11. Sonicate to disrupt cells until the suspension becomes transparent.
12. Unbroken cells are removed by centrifugation (6000 × *g*, 5 min).
13. Each cell lysate is subjected to ultracentrifugation (112,000 × *g*, 30 min, 4 °C), using an ultracentrifuge tube.
14. The pellet (membrane fraction) is dissolved in 1 mL TNET.
15. Incubate each membrane suspension on ice for 30 min.
16. Insoluble materials are removed by centrifugation (16,100 × *g*, 10 min, 4 °C).
17. During this centrifugation, equilibrate 5 mg (per tube) Protein-A Sepharose CL-4B beads in 1 mL TNET for at least 5 min. Prepare two sets of beads per sample.
18. Soluble materials obtained from **step 16** are applied to equilibrated Protein-A Sepharose CL-4B beads.
19. Slowly mix the beads and solution on a rotary wheel in a cold room for 30 min. In this step, materials that are nonspecifically bound to the beads are removed.
20. Precipitate beads by flash centrifugation, and resuspend the supernatant in fresh TNET-equilibrated beads (second set).

21. Add anti-PomB antibody (1:1000 dilution) to the bead-solution mixture, and slowly mix them overnight on a rotary wheel in a cold room.
22. Wash beads with 1 mL TNET 3×, and finally resuspend the beads in 50 μL 1× SDS-loading buffer.
23. Boil each sample for 5 min, spin briefly to precipitate the beads, and analyze each supernatant that contains immunoprecipitated PomB by SDS-PAGE.
24. Protein samples are transferred onto a PVDF membrane.
25. Blots are blocked by 3% (w/v) skim milk in TBST, and are then incubated in 1% (w/v) skim milk in TBST containing streptavidin-conjugated HRP (1:5000 dilution) for 1 h.
26. Biotinylated PomB on the blot is detected by the standard chemiluminescence method (*see* **Notes 6** and 7).

3.4 Detection of Intramolecular Disulfide Crosslink Formation (In Vitro Method)

1. *Vibrio* cells are grown as described in Subheading 3.3, and are suspended in 500 μL MN500 medium at a cell concentration equivalent to an optical density (OD) 660 nm of 5 (*see* **Note 8**).
2. Cells are divided into two halves; one half is incubated with 1 mM DTT, the other half is incubated without DTT (just add distilled water) at 30 °C for 40 min in a water bath.
3. Motility is checked by dark-field microscopy.
4. Take 100 μL of each cell suspension from each condition, and mix with 100 μL of 2% (w/v) SDS.
5. Boil each mixture at 95 °C for 10 min, followed by cooling down to room temperature.
6. Biotin maleimide is added to the mixture at a final concentration of 2 mM.
7. Incubate each mixture at 37 °C for 30 min.
8. The reaction is stopped by the addition of cysteine to a final concentration of 4 mM.
9. Add 1100 μL TNET to the mixture, and then centrifuge (16,100 × *g*, 2 min, 4 °C) (*see* **Note 9**).
10. Recover the supernatant, immunoprecipitate with the PomB antibody, and detect biotinylated PomB by following **steps 18–26** of Subheading 3.3.

4 Notes

1. For the crosslinking study, we first selected four residues (M157, L160, I164, and L168) in helix α1. When replaced to cysteine, these residues are expected to face the core domain of the PomB C-terminal region ($PomB_C$, *see* Fig. 1) within the

distance to form a disulfide bridge (6 ~ 7 Å). We measured the distance between the C_β atoms of the selected residues and their neighboring residues, and picked the closest and the second closest residues as crosslink candidates. Finally, we chose pairs of one residue in helix α1 and the other in the core domain [17]. It should be emphasized that we introduced double cysteine substitutions into the cysteine-less PomB mutants. Intact PomB has three native cysteine residues, and we replaced all of them with alanine. This cysteine-less PomB was found to be still functional.

2. To make the soft agar plates, we recommend the use of bactoagar (Difco). Better motility and clear motility rings can be obtained using bactoagar.
3. For the measurement of the motile fraction, swimming tracks for 1 s are analyzed from the recorded video images. For this purpose, we use image analysis software "Move-tr/2D" (Library Co., Tokyo, Japan), but alternatively Image J can be used.
4. We use freshly prepared soft agar plates right before the motility analysis since DTT is easily oxidized. Prepare the DTT-containing plates on the same day (agar becomes solid about 1 h after pouring it into the dish). For short time storage of plates without DTT, put them in a cold room or in an incubator with a beaker containing some water to avoid drying the agar surface.
5. Serine works as an attractant for *Vibrio alginolyticus* polar flagellar motility. The addition of serine to the TMN500 medium suppresses the change of direction for swimming, so that we can analyze the swimming speed and the motile fraction without the complexity of swimming tracks with directional changes [19].
6. To show the loading controls, the amount of the immunoprecipitated PomB in each sample should be analyzed simultaneously by immunoblotting with the anti-PomB antibody (*see* Fig. 3c, left panel).
7. If crosslinking occurs in a designed pair (e.g., M157C-I186C), biotinylated PomB can be detected only when the cells are treated with DTT. Without DTT, a disulfide crosslink does not produce free cysteine, and thereby biotinylated PomB would not appear. On the other hand, if a designed double cysteine pair does not form a disulfide bridge, then biotinylated PomB would be detected regardless of the DTT treatment.
8. In our experience, the in vitro method tends to require more cells than the in vivo method to detect the biotinylated PomB.
9. In our experience, SDS does not precipitate during an overnight mixing in a cold room when diluted 12-fold in TNET.

Acknowledgments

The methods introduced here were developed during the project reported in the manuscript of Zhu et al. [17], and that project was partially supported by the Japan Society for the Promotion of Science KAKENHI Grant JP24657087 (to S. K.).

References

1. Berg HC (2003) The rotary motor of bacterial flagella. Annu Rev Biochem 72:19–54
2. Macnab RM (2003) How bacteria assemble flagella. Annu Rev Microbiol 57:77–100
3. Li N, Kojima S, Homma M (2011) Sodium-driven motor of the polar flagellum in marine bacteria *Vibrio*. Genes Cells 16(10):985–999
4. Terashima H, Kojima S, Homma M (2008) Flagellar motility in bacteria structure and function of flagellar motor. Int Rev Cell Mol Biol 270:39–85
5. Zhu S, Kojima S, Homma M (2013) Structure, gene regulation and environmental response of flagella in *Vibrio*. Front Microbiol 4:410
6. Blair DF (2003) Flagellar movement driven by proton translocation. FEBS Lett 545(1):86–95
7. Thormann KM, Paulick A (2010) Tuning the flagellar motor. Microbiology 156:1275–1283
8. Sato K, Homma M (2000) Functional reconstitution of the Na$^+$-driven polar flagellar motor component of *Vibrio alginolyticus*. J Biol Chem 275(8):5718–5722
9. Sato K, Homma M (2000) Multimeric structure of PomA, a component of the Na$^+$-driven polar flagellar motor of *Vibrio alginolyticus*. J Biol Chem 275(26):20223–20228
10. Braun TF, Al-Mawsawi LQ, Kojima S, Blair DF (2004) Arrangement of core membrane segments in the MotA/MotB proton-channel complex of *Escherichia coli*. Biochemistry 43(1):35–45
11. Kojima S, Blair DF (2004) Solubilization and purification of the MotA/MotB complex of *Escherichia coli*. Biochemistry 43(1):26–34
12. Kojima S (2015) Dynamism and regulation of the stator, the energy conversion complex of the bacterial flagellar motor. Curr Opin Microbiol 28:66–71
13. Stolz B, Berg HC (1991) Evidence for interactions between MotA and MotB, torque-generating elements of the flagellar motor of *Escherichia coli*. J Bacteriol 173(21):7033–7037
14. Muramoto K, Macnab RM (1998) Deletion analysis of MotA and MotB, components of the force-generating unit in the flagellar motor of *Salmonella*. Mol Microbiol 29(5):1191–1202
15. Kojima S, Imada K, Sakuma M, Sudo Y, Kojima C, Minamino T, Homma M, Namba K (2009) Stator assembly and activation mechanism of the flagellar motor by the periplasmic region of MotB. Mol Microbiol 73(4):710–718
16. Li N, Kojima S, Homma M (2011) Characterization of the periplasmic region of PomB, a Na$^+$-driven flagellar stator protein in *Vibrio alginolyticus*. J Bacteriol 193(15):3773–3784
17. Zhu S, Takao M, Li N, Sakuma M, Nishino Y, Homma M, Kojima S, Imada K (2014) Conformational change in the periplamic region of the flagellar stator coupled with the assembly around the rotor. Proc Natl Acad Sci U S A 111(37):13523–13528
18. Terauchi T, Terashima H, Kojima S, Homma M (2011) The critical role of a conserved residue, PomB-F22, in the transmembrane segment of the flagellar stator complex for conducting ions and generating torque. Microbiology 157:2422–2432
19. Nishikino T, Zhu S, Takekawa N, Kojima S, Onoue Y, Homma M (2016) Serine suppresses the motor function of a periplasmic PomB mutation in the *Vibrio* flagella stator. Genes Cells 21(5):505–516

Part III

Dynamics of the Bacterial Flagellar Motor

Chapter 12

Rotation Measurements of Tethered Cells

Yuichi Inoue

Abstract

Direct observation of the rotation of tethered cells using optical microscopy is a simple method to examine dynamics of the bacterial flagellar motor. The rotational speed indicates not only the existence of the rotary motor, but also approximate number of stators that are torque-generating units in a motor. Since "run" and "tumble" of the peritrichous cells as *Escherichia coli* are regulated by the counterclockwise rotation and the clockwise rotation, respectively, rotational direction of the tethered cell is an important clue to understand the chemotactic system of the cells.

Key words Bacterial flagellar motor, Tethered cell, *fliC*-sticy mutant, Speed analysis, Switching of the rotational direction

1 Introduction

Rotation of the bacterial flagellar motor was clearly demonstrated by "tethering" a single flagellum to a glass surface, showing the motor-driven rotation of the entire cell body [1]. The tethered cell assay has been used to examine dynamics of the motors in various bacteria including*E. coli*, *S. enterica typhimurium*, and *Streptococcus species.* One of the most striking observations is that motors switch spontaneously between clockwise (CW) and counterclockwise (CCW) rotations, in which the probabilities of the directional switching are controlled in such a way as to allow the bacteria to perform chemotaxis [2, 3]. Tethered cells typically rotate at a slow speed of ~10 Hz in either direction with a high load of the cell body. Recently, various range of rotational speed with different load conditions is measured at high temporal resolution using 0.30–1.0 μm diameter beads, attached to the flagellar filament stub whose cell body had been fixed on the glass surface (bead assay) [4]. However, tethered cell assay still has a significant role because of its

Electronic supplementary material: The online version of this chapter (doi:10.1007/978-1-4939-6927-2_12) contains supplementary material, which is available to authorized users.

Tohru Minamino and Keiichi Namba (eds.), *The Bacterial Flagellum: Methods and Protocols*, Methods in Molecular Biology, vol. 1593, DOI 10.1007/978-1-4939-6927-2_12, © Springer Science+Business Media LLC 2017

simplicity. Especially, FliCst, which is a mutant flagellin, in which centrally located 57 residues are replaced by six other residues [5], has made this method simpler. FliCst forms a sticky filament that naturally adsorbs on the nonwashed glass surface, obviating the need for the preparations in the previous protocol including pre-treatment of glass surface and anti-flagellin antibody before tethering [6]. Simple observation of the tethered cells would be helpful to optimize the culture conditions as the expression level of stators for the beginner in this field. Also for the experts, the simple protocol of both experiment and image analysis without the custom optics to analyze the rotational speed [7] would be a powerful tool to examine the motor dynamics as well as to combine with other techniques including pressure (Chapter 13), fluorescence imaging (Chapter 17), and temperature control.

2 Materials

2.1 Solutions

1. Prepare all solutions using sterilized Milli-Q water and reagent-grade chemicals.
2. T-broth: 1% Bacto tryptone, 0.5% NaCl.
3. Motility buffer: 10 mM potassium phosphate, 85 mM NaCl, 0.1 mM EDTA, pH 7.0.

2.2 Cell Culture and Sample Preparation for Microscopy

1. Transform *E. coli* strain HCB1271 (*fliC*::Tn*10*, Δ*pil*A′-KnR, *motA448*) with plasmids pDFB36 (*motA*, IPTG inducible, ampicillin resistance) and pDF313Cm (*fliCsticky*, chrolamphenicol resistance) [4], and *E. coli* strain JHC36 (Δ*cheY*, *fliC-sticky*, Δ*pilA*, Δ*motAmotB*) with a plasmids pTH200 (PomA, PotB, inducible by IPTG) [8].
2. Grow *E. coli* cells in 5 mL of T-broth containing 0.05 mM IPTG, 25 μg/mL chrolamphenicol (and 50 μg/mL ampicillin) in a 100 mL conical flask with shaking (180 rps) at 31 °C for 5 h.
3. Pass the cells through 26 gauge needles 60 times to shorten the sticky filament (*see* **Note 1**).
4. Construct a sample chamber composed of two cover glasses (24 × 50 mm and 22 × 26 mm), using double-stick tape (e.g., 665-1-18, 3 M Scotch) as a spacer (*see* **Note 2**).

2.3 Microscopy

1. Inverted microscope.
2. 40× phase-contrast objective lens (NA = 0.75).
3. High-speed camera (IPX-VGA210LMCN or CLB-B0620M-TC, Imperx, FL, USA).

4. 64 bit-windows-based computer installed with frame grabber (PIXCI® EB1, EPIX, Inc., IL, USA) and control software (XCAP-Ltd., EPIX, Inc., IL, USA) (*see* **Note 3**).

2.4 Software for the Image Analysis

1. ImageJ (National Institute of Mental Health, Bethesda, Maryland, USA) and excel (Microsoft Corporation, WA, USA) for basic calculation of rotational speed.
2. Supplemental programs that are executable on windows-based PC (both 32bit and 64bit) at no charge with LabVIEW Run-Time Engine 2016 (32bit version) downloadable from http://www.ni.com (National Instruments, Austin, TX, USA) (*see* **Note 4**).

2.5 Equipment for the Simultaneous Recording of Temperature with a Phase-Contrast Image

1. A thermometer (HA-200E, Anritsu, Japan) with a thin thermocouple probe (ST-21E-010-TS1-ASP, ANRITSU, Japan).
2. A CCD camera (WAT100N, Watec, Yamagata, Japan) to record phase-contrast images of *E. coli* cells, and a digital camera (F-30, Fujifilm, Japan) to record a display of the thermometer.
3. A screen separation device (MTQC-14, Mother tool, Korea) to combine images of *E. coli* cells and those of the display of the thermometer in a picture-in-picture mode.

3 Methods

3.1 Observation of Tethered Cells

Overview of the tethered cell assay is shown in Fig. 1.

1. Apply the cultured cells (~30 μL) into the glass chamber.
2. Wait for 5 min to adsorb *E. coli* cells on the glass surface through their sticky filaments.
3. Perfuse the glass chamber with 200 μL Motility buffer to remove unbound cells.
4. Mount the glass chamber on optical microscopy to observe the tethered cells.
5. Count number of rotating cells if you need to optimize the motor conditions by modulating IPTG concentration and/or culture time.
6. Record phase-contrast images of the rotating cells at frame rate of 200 fps (or faster) into uncompressed AVI files for later analysis (*see* **Notes 5** and **6**).

3.2 Speed Calculation from the Video Images

Basic calculation using ImageJ and Excel is mainly described in **steps 1–13** (Fig. 2a), whereas automatic calculation using an executable program is shown in the **steps 14–17** (Fig. 2b).

1. Open a recorded AVI file with ImageJ (Do not use Virtual Stack, Convert to Grayscale, Do not Flip Vertical).

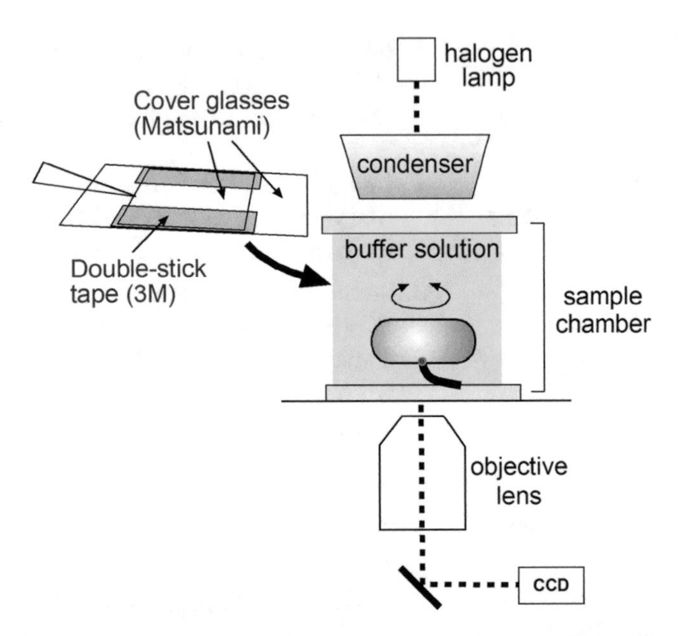

Fig. 1 Illustration of the experimental system for the tethered cell assay. Rotation of the motor is clearly observed by "tethering" a single flagellum to a glass surface [1]. Simple observation and image analysis of the tethered cell is described in this chapter

2. Select an area of single rotating cell by "Rectangular tool" and duplicate the area as a new stack (Image > Duplicate, check to "Duplicate Stack").
3. Use a supplemental movie of "Test-CCW-CW.avi" in the following **steps 4–7** as an example.
4. Set a threshold of the image to define the area of the cell (Image > Adjust > Threshold, 120–255 for Test-CCW-CW.avi).
5. Set parameters for the image analysis (Analyze > Set Measurements, Fit ellipse).
6. Calculate the angle of the tethered cell (Analyze > Analyze Particles) by setting a cellular size as 100-Infinity (pixel^2).
7. Save the analyzed result as "Result.txt" and open with Microsoft excel.
8. Use a supplemental movie of "Speed calculation.xls" in the following **steps 9–13** as an example.
9. Calculate the angle change/frame by subtracting the 1-frame shifted trace (D-sita1 in column D, such as "D3 = B3–C3).
10. Calculate two other candidates for angle change/frame (D-sita2 in column E, such as "E3 = D3 + 180; D-sita3 in column F, such as "F3 = D3–180).

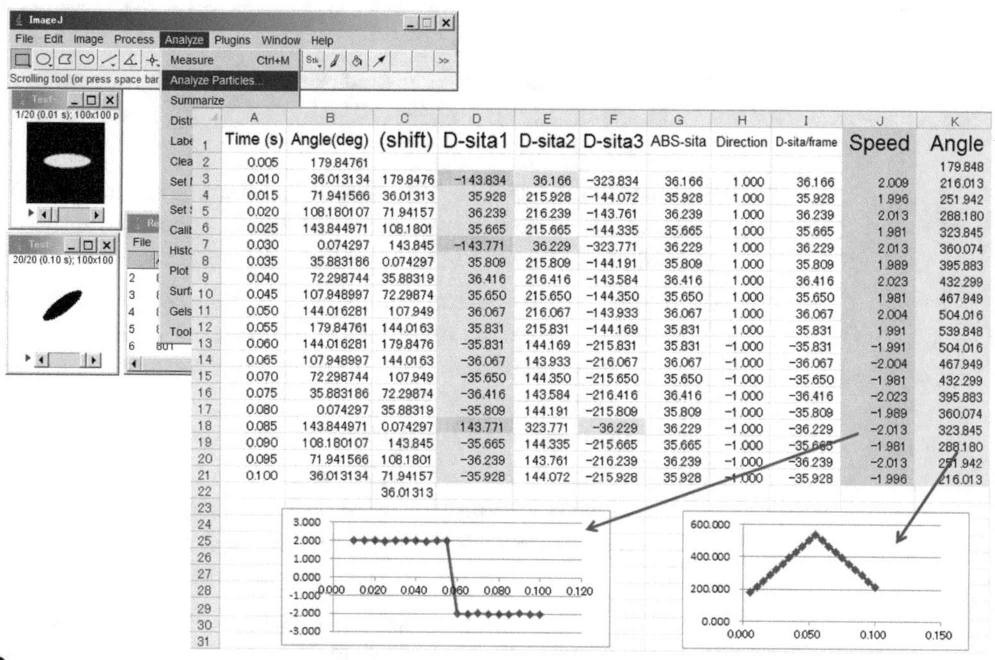

	Time (s)	Angle(deg)	(shift)	D-sita1	D-sita2	D-sita3	ABS-sita	Direction	D-sita/frame	Speed	Angle
2	0.005	179.84761									179.848
3	0.010	36.013134	179.8476	−143.834	36.166	−323.834	36.166	1.000	36.166	2.009	216.013
4	0.015	71.941566	36.01313	35.928	215.928	−144.072	35.928	1.000	35.928	1.996	251.942
5	0.020	108.180107	71.94157	36.239	216.239	−143.761	36.239	1.000	36.239	2.013	288.180
6	0.025	143.844971	108.1801	35.665	215.665	−144.335	35.665	1.000	35.665	1.981	323.845
7	0.030	0.074297	143.845	−143.771	36.229	−323.771	36.229	1.000	36.229	2.013	360.074
8	0.035	35.883186	0.074297	35.809	215.809	−144.191	35.809	1.000	35.809	1.989	395.883
9	0.040	72.298744	35.88319	36.416	216.416	−143.584	36.416	1.000	36.416	2.023	432.299
10	0.045	107.948997	72.29874	35.650	215.650	−144.350	35.650	1.000	35.650	1.981	467.949
11	0.050	144.016281	107.949	36.067	216.067	−143.933	36.067	1.000	36.067	2.004	504.016
12	0.055	179.84761	144.0163	35.831	215.831	−144.169	35.831	1.000	35.831	1.991	539.848
13	0.060	144.016281	179.8476	−35.831	144.169	−215.831	35.831	−1.000	−35.831	−1.991	504.016
14	0.065	107.948997	144.0163	−36.067	143.933	−216.067	36.067	−1.000	−36.067	−2.004	467.949
15	0.070	72.298744	107.949	−35.650	144.350	−215.650	35.650	−1.000	−35.650	−1.981	432.299
16	0.075	35.883186	72.29874	−36.416	143.584	−216.416	36.416	−1.000	−36.416	−2.023	395.883
17	0.080	0.074297	35.88319	−35.809	144.191	−215.809	35.809	−1.000	−35.809	−1.989	360.074
18	0.085	143.844971	0.074297	143.771	323.771	−36.229	36.229	−1.000	−36.229	−2.013	323.845
19	0.090	108.180107	143.845	−35.665	144.335	−215.665	35.665	−1.000	−35.665	−1.981	288.180
20	0.095	71.941566	108.1801	−36.239	143.761	−216.239	36.239	−1.000	−36.239	−2.013	251.942
21	0.100	36.013134	71.94157	−35.928	144.072	−215.928	35.928	−1.000	−35.928	−1.996	216.013
22			36.01313								

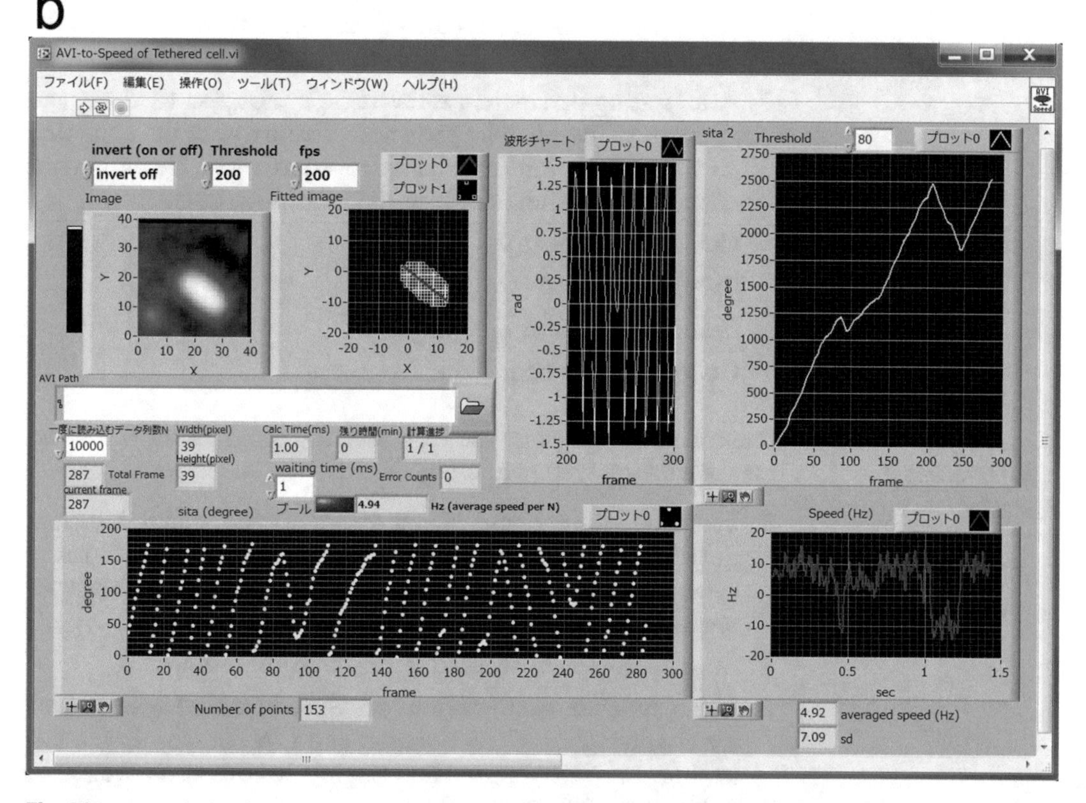

Fig. 2 Image analysis of tethered cells to calculate rotational speed. (**a**) Basic analysis using ImageJ and Excel. See Subheading 3.2 (**steps 1–13**). (**b**) Automatic analysis using a supplemental program developed with LabVIEW. An execution screen of "AVI-to-Speed of Tethered cell.exe" is shown. *See* Subheading 3.2 (**steps 14–17**)

11. Select minimum value for absolute value of the angle change/frame (column G) from sita1-3, such as G3 = MIN(ABS(D3), ABS(E3), ABS(F3)).
12. Make a decision for rotational direction in each frame (column H), such as H3 = IF(OR(D3/G3 = 1,E3/G3 = 1,F3/G3 = 1),1,-1).
13. From columns G to H, determine the angle change/frame (column I) and convert into rotational speed (column J) as well as accumulated angle (column K).
14. If you want to perform automatic analysis listed above, use a supplemental program of "AVI-to-Speed of Tethered cell.exe" with windows-based PC. To execute this program, LabVIEW Run-Time Engine 2016 (32bit version), which is available for free from National Instruments, is required to be installed.
15. After install of LabVIEW Run-Time Engine 2016 (or LabVIEW development system), unzip the supplemental file to make an executable file and data folder. Keep the "data" folder in the same directory of the executable file.
16. Double click "AVI-to-Speed of Tethered cell.exe." The program starts with a sample data "test-movie.avi." To analyze a new AVI file for single tethered cell, drag the AVI file into the file path on the execution screen.
17. After setting parameters as invert on or not, recorded fps, and file path to the AVI file, press the white arrow at the upper left to start calculation. Results are saved as a text file within the same folder of the AVI file in the following order.

    Column 1: Time (s).

    Column 2: Speed (Hz).

    Column 3: Total Angle (degree).

    Column 4: 180-turn Angle (degree).

3.3 Analysis of the Directional Switching

The execution screen of the supplemental program for the directional switching is shown in Fig. 3.

1. To analyze the chemotactic responses of a tethered cell, such as duration of CCW rotation and CW bias, record the AVI files of rotating cells as Subheading 3.1 with or without an environmental stimulus as a chemical attractant.
2. Analyze the speed of rotating cells by Subheading 3.2.
3. Use a supplemental program of "Speed-to-2or3 groups.exe" for classifying the CCW rotation and CW rotation in the following. LabVIEW Run-Time Engine 2016 is also required as Subheading 3.2 (**step 14**).
4. Double click "Speed-to-2or3 groups.exe." The program starts with a sample data "test-speed.txt," waiting for pressing the

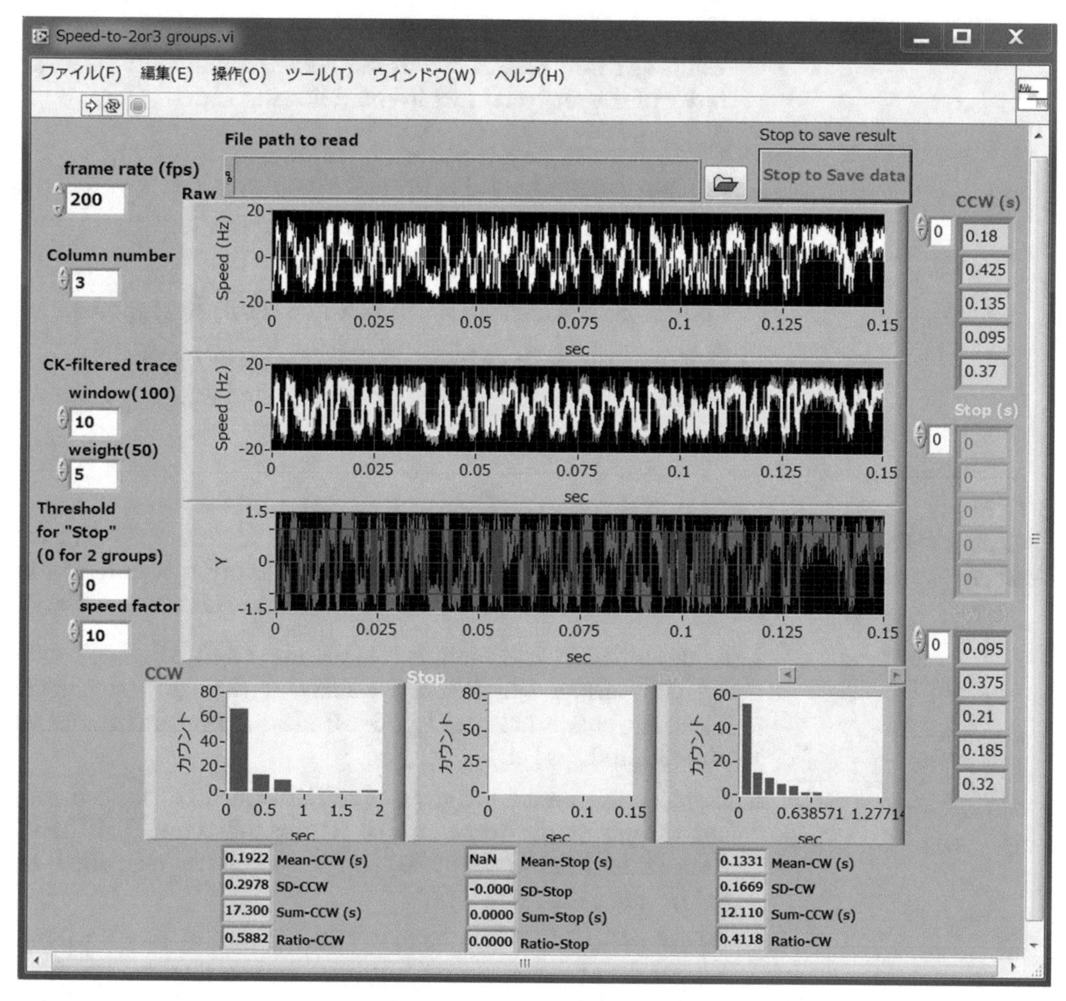

Fig. 3 Analysis of the rotational speed into the two groups of CCW/CW rotations. An execution screen of the supplemental program "Speed-to-2or3 groups.exe" is shown. *See* Subheading 3.3

"stop to save data" button or the red circle at the upper left to stop calculation.

5. To analyze a new text file, drag the text file into the file path on the execution screen. After setting number of the recorded frame rate (fps) and column number for the speed in the text file (3 for "test-speed.txt"), press the white arrow at the upper left to start calculation.

6. Set window size and weight ratio for Chung-Kennedy filter [9], which detects stepwise change of the trace. (Initial values for test-speed.txt are 10 and 5, respectively.)

7. Set threshold = 0 to classify the speed into the two states of CCW/CW rotations. If threshold is not zero, the threshold determines the third state of "stopped rotation."

8. Press "Stop to save data" button to stop calculation and save results as a new text. A new result file is generated in the same folder of the analyzed text in the following order.

   Column 1: Time (s) for original speed.

   Column 2: Speed (Hz) as original trace.

   Column 3: Time (s) for the CK-filtered trace.

   Column 4: the CK-filtered Speed (Hz).

   Column 5: Normalized speed by the 2 (or 3) state model.

   Column 6: Durations of CCW rotation (s) (*see* **Note 7**).

   Column 7: Durations of "stopped rotation" (s).

   Column 8: Durations of CW rotation (s).

3.4 Torque Estimation in the Slow Rotation

Fluctuation theorem (FT) provides a new tool for force measurement of a wide range of biological systems including F_1-ATPase and bacterial flagellar motors [10]. The execution screen of the supplemental program for the torque calculation is shown in Fig. 4.

1. As described in Subheading 3.1, record the AVI files of rotating cells using a high-speed camera at the fast frame rate. Typically, rotational speed of 3–40 Hz requires frame rate of 500–30,000 fps [10].
2. Use a supplemental program of "FT-Torque.exe" for estimating torque of the tethered cell in the following. LabVIEW Run-Time Engine 2016 is also required as described in Subheading 3.2 (**step 14**).
3. Double click "FT-Torque.exe." The program starts with a sample data "test-angle.txt," waiting for pressing the "push to save the selected trace" button or the red circle at the upper left to stop calculation.
4. To analyze a new text file, drag the text file into the file path on the execution screen. Then, press the white arrow at the upper left to start calculation.
5. Set parameters as column number for the angle (degree) in the text file, number of the frame-shift (Δframe), recorded frame rate (column = 3, Δframe = 1–5, fps = 1000 for "test-angle.txt").
6. Drag the two red cursor bars in the white angle trace to select the range for the torque estimation. The range at approximately constant speed (without switching direction) should be selected. If the red cursor bars are out of view, press "Center Cursor."

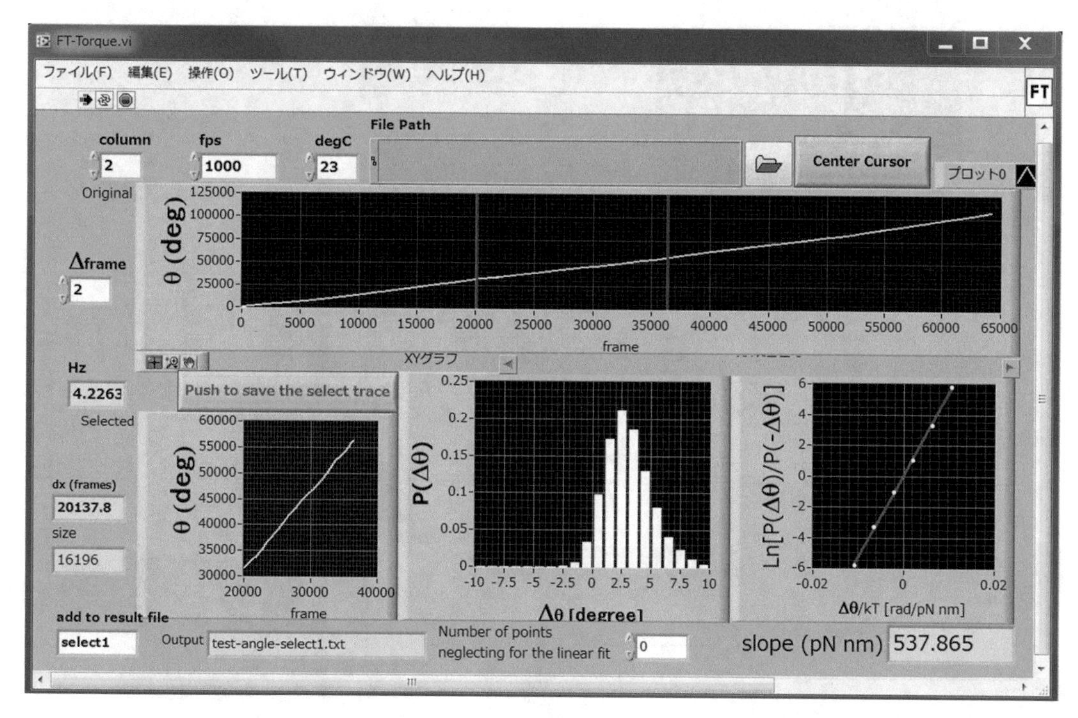

Fig. 4 Analysis of generated torque by the tethered cell. An execution screen of the supplemental program "FT-Torque.exe" is shown. *See* Subheading 3.4

7. Press "Push to select data" button to determine the range. The programs shows Probability distribution, $P(\Delta\theta)$, as well as the plot of $\ln[P(\Delta\theta)/P(-\Delta\theta)]$ with $\Delta\theta/k_BT$. From the linear fit to the plot, estimated slope provides generated torque (450–550 pN nm for "test-angle.txt") (*see* **Note 8**).

3.5 Tethered Cells Under Temperature Control

Because of the simplicity of the tethered cell, this method can be easily combined with other techniques such as temperature control. Fig. 5 shows an overview of the experimental system for the cold response.

1. Fabricate the fluid flow chip for temperature control using two cover glasses (18 × 18 mm) with four glass pipettes as 1 mm-spacer [11]. At the four corners of the chamber, four polyethylene tubes (ϕ_{out} = 1.3 mm) are connected and then the entire chamber is sealed with adhesive agent (e.g., Araldite, Huntsman Advanced Materials, Switzerland). The polyethylene tubes are linked to the silicon tubes (ϕ_{out} = 3.5 mm and ϕ_{in} = 2.0 mm) and connected to Y-type mini-fitting (e.g., VFY306, Isis Co., LTD, Tokyo, Japan).
2. After dry out of the adhesive agent, fill up the fluid flow chip and the connecting tubes with water using a plastic syringe.

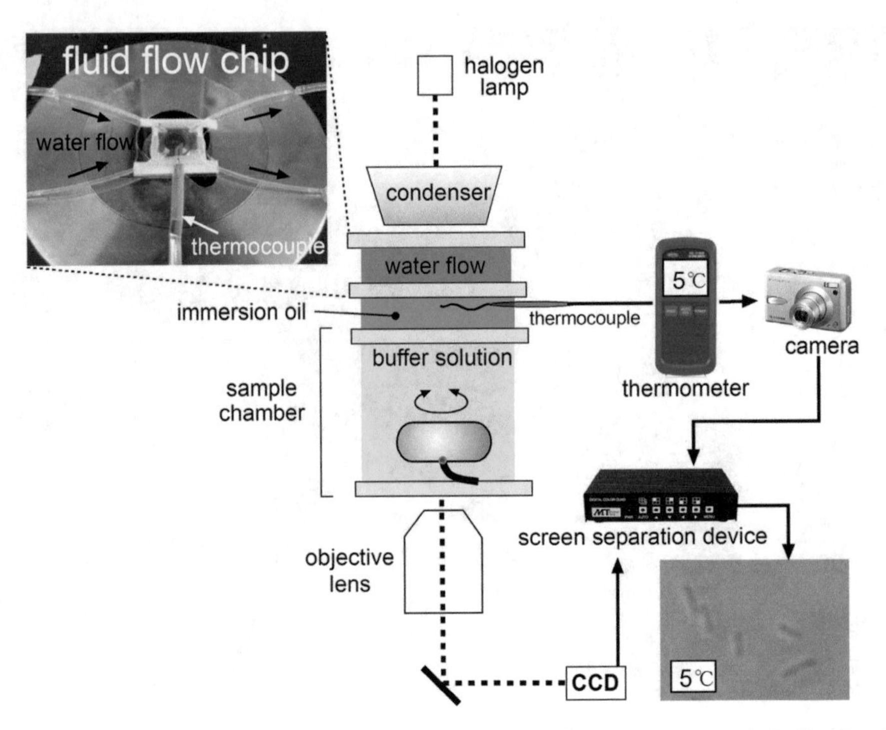

Fig. 5 Overview of the experimental setup for a tethered cell under temperature control. *See* Subheading 3.5

3. Set the fluid flow chip on the sample chamber of the tethered cell which had been mounted on the inverted microscope. Between the fluid flow chip and the sample chamber, insert a thin thermocouple probe and fill with immersion oil (*see* **Note 9**).
4. During the observation of the tethered cells, apply a flow of ice water with a flow rate of 0.003–0.3 mL/s to examine cold response of the motors. For the simultaneous recording of temperature with a phase-contrast image of *E. coli* cells by a CCD camera, a display of the thermometer was imaged using a digital camera to combine with a screen separation device in a picture-in-picture mode.
5. Apply a flow of hot water similarly to examine thermal response of the motors (*see* **Note 10**).

4 Notes

1. A *fliC*st mutation sometimes induces the strain with the very short filament, which cannot support swimming in solution. In this case, this shearing step can be skipped.
2. Some other double-stick tapes inhibit normal chemotactic behavior of bacteria, possibly by leaking chemical(s) that disturb chemotaxis.

3. For broader memory space that determines maximum number of frames to be recorded, 32bit-windows-based PC is not recommended for recording.
4. Supplemental programs are available for both 32bit-PC and 64bit-PC with 32bit version of LabVIEW Run-Time Engine 2016 (http://www.ni.com/download/labview-run-time-engine-2016/6066/en/), but not with 64bit version of LabVIEW Run-Time Engine 2016.
5. Phase-contrast image is not essential for the tethered cell. Although bright field image can be analyzed as well, higher contrast is advantageous for the image analysis.
6. Real-time speed analysis of the motor rotation (before saving images) is possible up to 1250 fps [12], if you use a frame grabber from National Instruments and custom programs developed with LabVIEW 2016 development system.
7. Sequential "0" following the detected durations for the columns 6–8 in the result text should be neglected for further analysis.
8. Unidirectional slow rotation of Test-angle.txt is measured for the Na^+-driven chimeric motor expressed in *E. coli* without CheY [8], so that slow rotation of ~4 Hz driven by the limited stators is observed at low Na^+ concentration. To estimate the maximum torque with 11–13 stators, FT-method requires fast frame rate over 30,000 fps for recording the forward and backward fluctuation in $\theta(t)$, which is essential for the reliable estimation of $P(\Delta\theta)/P(-\Delta\theta)$. If the frame rate is fast enough, the FT method would be also available to estimate the motor torque by bead assays.
9. We confirmed that temperature gradient between the thermocouple and sample solution is negligible for a dry objective (UPlanFI ×40, NA 0.75), whereas temperature gradient for an oil immersion objective is significant [11].
10. To record rotation faster than 10 Hz, the rotation should be recorded by the fast detector as high-speed camera [12, 13] or quadrant photodiode sensor [8] using custom program developed with LabVIEW. In this case, temperature detected by the thermocouple can be simultaneously recorded by the same program [14] using thermocouple measurement device (USB-TC01, National Instruments, Austin, TX, USA).

Acknowledgments

I thank Prof. A. Ishijima, Dr. H. Fukuoka, Dr. Y-S. Che, Dr. T. Sagawa, and H. Takahashi for their key contributions to the protocol.

References

1. Silverman M, Simon M (1974) Flagellar rotation and the mechanism of bacterial motility. Nature 249:73–74
2. Larsen SH, Reader RW, Kort EN, Tso WW, Adler J (1974) Change in direction of flagellar rotation is the basis of the chemotactic response in *Escherichia coli*. Nature 249:74–77
3. Berg HC, Tedesco PM (1975) Transient response to chemotactic stimuli in *Escherichia coli*. Proc Natl Acad Sci U S A 72:3235–3239
4. Ryu WS, Berry RM, Berg HC (2000) Torque-generating units of the flagellar motor of *Escherichia coli* have a high duty ratio. Nature 403:444–447
5. Kuwajima G (1988) Construction of a minimum-size functional flagellin of *Escherichia coli*. J Bacteriol 170:3305–3309
6. Scharf BE, Fahrner KA, Turner L, Berg HC (1998) Control of direction of flagellar rotation in bacterial chemotaxis. Proc Natl Acad Sci U S A 95:201–206
7. Kobayasi S, Maeda K, Imae Y (1977) Apparatus for detecting rate and direction of rotation of tethered bacterial cells. Rev Sci Instrum 48:407–410
8. Inoue Y, Lo CJ, Fukuoka H, Takahashi H, Sowa Y, Pilizota T, Wadhams GH, Homma M, Berry RM, Ishijima A (2008) Torque-speed relationships of Na^+-driven chimeric flagellar motors in *Escherichia coli*. J Mol Biol 376:1251–1259
9. Chung SH, Kennedy RA (1991) Forward-backward non-linear filtering technique for extracting small biological signals from noise. J Neurosci Methods 40:71–86
10. Hayashi K, Ueno H, Iino R, Noji H (2010) Fluctuation theorem applied to F_1-ATPase. Phys Rev Lett 104:218103
11. Baker MA, Inoue Y, Takeda K, Ishijima A, Berry RM (2011) Two methods of temperature control for single-molecule measurements. Eur Biophys J 40:651–660
12. Sagawa T, Kikuchi Y, Inoue Y, Takahashi H, Muraoka T, Kinbara K, Ishijima A, Fukuoka H (2014) Single-cell *E. coli* response to an instantaneously applied chemotactic signal. Biophys J 107:730–739
13. Fukuoka H, Sagawa T, Inoue Y, Takahashi H, Ishijima A (2014) Direct imaging of intracellular signaling components that regulate bacterial chemotaxis. Sci Signal 7:ra32
14. Inoue Y, Baker MA, Fukuoka H, Takahashi H, Berry RM, Ishijima A (2013) Temperature dependences of torque generation and membrane voltage in the bacterial flagellar motor. Biophys J 105:2801–2810

Chapter 13

Tracking the Movement of a Single Prokaryotic Cell in Extreme Environmental Conditions

Masayoshi Nishiyama and Yoshiyuki Arai

Abstract

Many bacterial species move toward favorable habitats. The flagellum is one of the most important machines required for the motility in solution and is conserved across a wide range of bacteria. The motility machinery is thought to function efficiently with a similar mechanism in a variety of environmental conditions, as many cells with similar machineries have been isolated from harsh environments. To understand the common mechanism and its diversity, microscopic examination of bacterial movements is a crucial step. Here, we describe a method to characterize the swimming motility of cells in extreme environmental conditions. This microscopy system enables acquisition of high-resolution images under high-pressure conditions. The temperature and oxygen concentration can also be manipulated. In addition, we also describe a method to track the movement of swimming cells using an ImageJ plugin. This enables characterization of the swimming motility of the selected cells.

Key words Prokaryote, Bacteria, Archaea, Motility machinery, Flagella, High-pressure microscopy, Extreme environment, Tracking system

1 Introduction

Many bacterial species sense the surrounding environment, and move toward favorable habitats. The flagellum is one of the most important organelles for bacteria to swim in solution. The mechanisms involved in the assembly and function of flagella are well studied in bacteria such as *Escherichia coli* and *Salmonella enterica* serovar Typhimurium [1, 2]. These cells have 5–10 flagellar filaments. The rotation of these filaments allows them to join in a bundle and propels the cell forward. The swimming motion of the cells can be monitored under conventional microscopes. The detailed mechanism, including the motor rotations, has been well studied using highly developed microscopy techniques [3].

In contrast, many bacteria preferably reside in harsh environments. Physical and chemical conditions, such as temperature, hydrostatic pressure, and oxygen concentration, differ drastically

Tohru Minamino and Keiichi Namba (eds.), *The Bacterial Flagellum: Methods and Protocols*, Methods in Molecular Biology, vol. 1593, DOI 10.1007/978-1-4939-6927-2_13, © Springer Science+Business Media LLC 2017

from our living environments. Interestingly, the flagellum is an organelle that is conserved across a wide range of bacteria. The bacterial motility machinery seems to adapt to various environmental conditions, and then works well with similar mechanisms. In previous studies, flagellar motility under extreme conditions has not been well studied due to the difficulties in constructing the extreme conditions under a microscope.

Here, we describe a method to monitor the movements of swimming cells under conditions of high hydrostatic pressure, which we have published recently [4, 5]. The developed microscopy system allows acquisition of high-resolution microscopic images including bright-field, dark-field, phase-contrast, and fluorescence images [4, 6]. Hydrostatic pressure could be applied up to 150 MPa, which is about 1.5 times the water pressure at the deepest part of the Mariana Trench, the Challenger Deep (10,900 m). The chamber can be sealed tightly and its inner temperature could be controlled [7, 8]. Therefore, this system could be used for monitoring cells of the hyperthermophilic bacterium, *Aquifex aeolicus* [8].

In addition, we report a method to track the movement of particles using an ImageJ plugin. This enables us to characterize the swimming motility of the selected cells. Cell motility could be characterized by analyzing the fraction and speed of the swimming cells under each condition. This analysis could be extended to other motile systems of bacteria (or archaea) such as swarming, gliding, twitching, or floating [9].

2 Materials

2.1 Pressure and Temperature Apparatuses for the Microscope System

1. Custom-made high-pressure chamber (*see* **Note 1**) (Fig. 1a).
2. Hand pump (HP-150, Syn Corporation, Kyoto, Japan), pressure gauge, and indicator gauge (PG-2TH and WGI-400A-02, Kyowa, Kyoto, Japan) (Fig. 1b).
3. Custom-made separator (*see* **Note 2**) (Fig. 1c).
4. High-pressure tubing and connections (30,000 psi, High Pressure Equipment Company, PA, USA).
5. Thermostat bath (RTE-110, Neslab Instruments Inc., NH, USA).
6. Silicon tubing (*see* **Note 3**).

2.2 Microscope System

1. Inverted microscope (Fig. 1c).
2. Custom-made microscope stage (*see* **Note 4**).
3. Long-working distance objective lens.
4. CCD camera.
5. USB2.0 converter (DFG/USB2pro, Imaging Source, Taiwan).

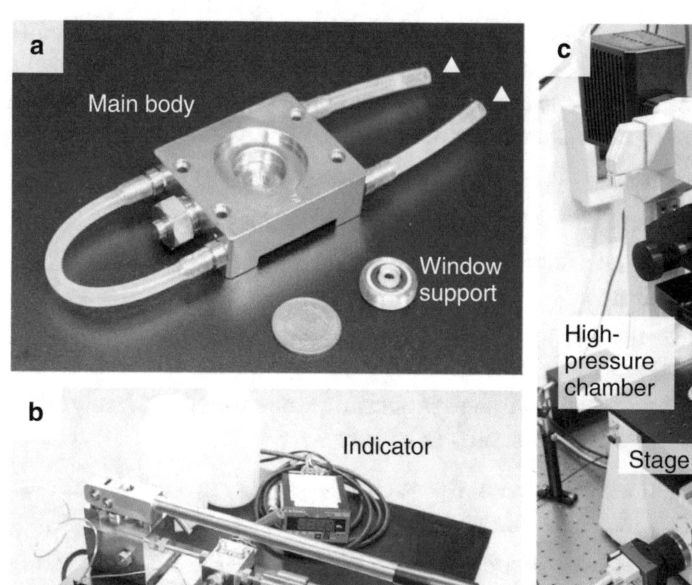

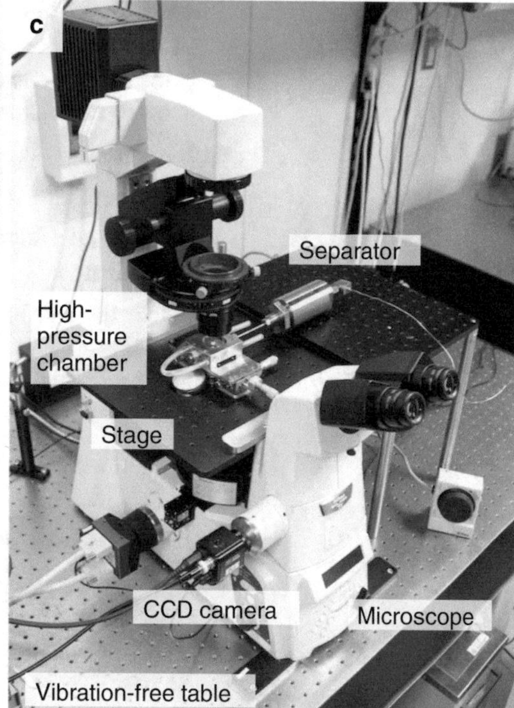

Fig. 1 High-pressure microscope. (**a**) Disassembled components of the high-pressure chamber with a coin (diameter 23.5 mm) shown for size comparison. (**b**) High-pressure pump. (**c**) Microscope system. See text for details

6. Computer installed with control software (IC Capture, Imaging Source).
7. Vibration-free table.

2.3 Culture Media

1. Tryptone Broth (TB) medium: 1% Tryptone, 0.5% NaCl.
2. Motility medium: 10 mM Tris, pH 7, 0.1 mM EDTA (*see* **Note 5**).

3 Methods

3.1 Cell Culture

1. Smooth-swimming *E. coli* strain RP4979 (Δ*cheY*) [10].
2. Store the stocks at −80°C.
3. Dilute the frozen stock in fresh TB medium (1:50).
4. Shake at 30°C.
5. Cultivate to the late logarithmic phase (OD_{600} = 0.6 ~ 0.8).

3.2 Motility Assay

1. Dilute the culture in the logarithmic phase in the motility medium (*see* **Note 6**).
2. Enclose the sample solution in the chamber.

3. Operate the hand pump lever to increase the hydrostatic pressure in the chamber (or open the regulation valve for decreasing the pressure). Focus the microscope on the cells in the chamber and acquire the image over about 20 s.
4. Repeat **step 3** for each pressure condition.
5. Complete all assays within 2 h after culture dilution.
6. Repeat the assays with cells derived from more than five different cultures.
7. Select the results from more than three assays for analysis.

3.3 Required Files for Image Analysis

ImageJ is widely used image processing software and is freely available (in the public domain) (*see* **Note 7**).

1. Download Fiji from the website, http://fiji.sc/. Windows, MacOSX, and Unix versions are available (*see* **Note 8**).
2. Download the plugin (file name: Simple_PTA.jar) for tracking the movement of single particles (*see* **Note 9**).
3. Run Fiji and install the plugin by Plugins menu→ Install Plugin.
4. Restart Fiji. "SimplePTA" can now be found in the Plugins menu.

3.4 Using This Plugin for Tracking Cells

1. Open the analyzed movie file.
2. Set the scale and frame interval by Image menu→ Properties.
3. Check (or convert to Grayscale) by Image menu→ Type → 8-bit (NOT 8-bit color).
4. Start the "Simple PTA" plugin by Plugins menu→ SimplePTA > Simple PTA (Fig. 2a).
5. Set the Threshold value for highlighting pixels in cell images.
6. Play continuously and select an image of the cell for which the translational speed is to be analyzed.
7. Click on the selected cell in the window.

Fig. 2 (continued) maximum size for searching at the next frame. If no cells are found in next frame in this area, the searching process will be stopped. Lower size (pixels) set the lower boundary of cell size. You can ignore noise by using this parameter. 4. Parameters for how the tracked cells are shown. If you choose the "Always" radio button, complete trajectories are shown in any of the frames. On the other hand, the "Growing" radio button makes the trajectory appear from the beginning to the current frame. The "All" check box shows all the tracked cells. "ROI" shows a square ROI at each cell. "Number" indicates the number of tracked cells adjacent to the cells. (**b**) An example of the analysis result. A trajectory of the analyzed cell is superimposed in the window as a blue line. (**c**) Table of the analysis results. Columns indicate the summary of the tracking data. (**d**) Multi-plots of the analyzed data. This window will appear when a row is chosen from the table shown in (**c**). "x–y trajectory" indicates the center positions of x and y. "Intensity," "Square Disp.," "Velocity," "Area," and "Cos-Theta" indicate these instantaneous values of cells. "Cos-Theta" is derived from an inner product using vectors determined by the point at *t*-2, *t*-1, *t* where *t* is the current time point

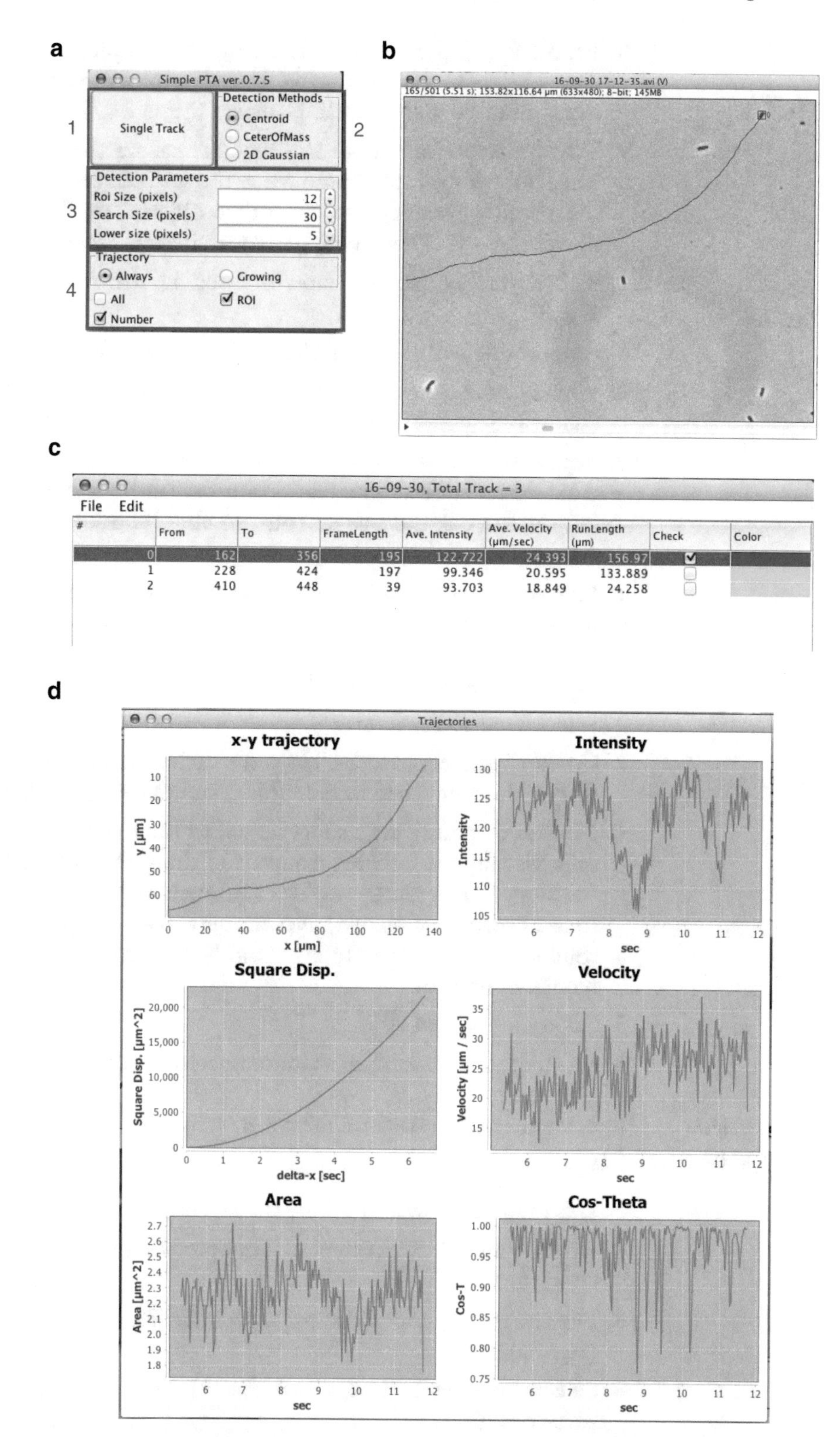

Fig. 2 An ImageJ plugin for tracking single particles. (**a**) SimplePTA main window. 1. Perform tracking after clicking the object. 2. Methods for detecting the center position. 3. Parameters for detecting the object. "ROI Size (pixels)" is used to limit the size of ROI for 2-dimensional Gaussian fittings. "Search Size (pixels)" to use the

8. Press the [Single Track] button on the SimplePTA window. Three methods can be chosen for tracking: centroid, center of mass, and two-dimensional Gaussian fitting.
9. Track automatically until the analysis reaches the final frame. Tracking will be interrupted when the selected cell goes out of the window, overlaps to the other objects, or the software misses the cell in the range of its search area.
10. The trajectory will be superimposed in the movie file (*see* Fig. 2b)
11. A new table will appear in the display (*see* Fig. 2c).
12. When you choose the row in the table, graphs of the tracking data will be generated (*see* Fig. 2d).
13. Repeat **steps 7–12**. The new data will be added to the existing table automatically. The table columns can be sorted by clicking on the column name. You can also "check" the rows to change their color.
14. Save the analysis result as a .csv file from the menu of the table by table → File → Save all (*see* Fig. 2d).

3.5 Classification of the Motility State in Cells into Three Groups

1. Open the analyzed movie file.
2. Select a particular frame.
3. Count all the cells in focus. However, exclude cells with unnatural shapes, those attached to the surface, and those in aggregates.
4. Play continuously for a short period of the movie. Classify the cells visually into three groups: (1) Smoothly swimming in solution, (2) Jiggling and/or rotational motions without any translational motion, (3) No motility (Brownian motion in solution). The same criteria should be applied for each experimental condition.
5. Repeat **steps 2–4**.
6. Calculate the fraction of swimming cells in each condition.

3.6 Motility Analysis of the Swimming Cells

Here, we describe a method for constructing Fig. 3.

1. Open the file "analysis results."
2. Use graphic software to draw the x–y trajectory of the cells.
3. Check and select the typical examples of five cells for each condition.
4. Plot the x and y values at every tenth frame for 150 frames (= 5 s) (Fig. 3a).
5. Select the series of time, t, and $x^2 + y^2$ of 50 cells for each pressure condition.
6. Calculate the average value of $x^2 + y^2$ at every frame.

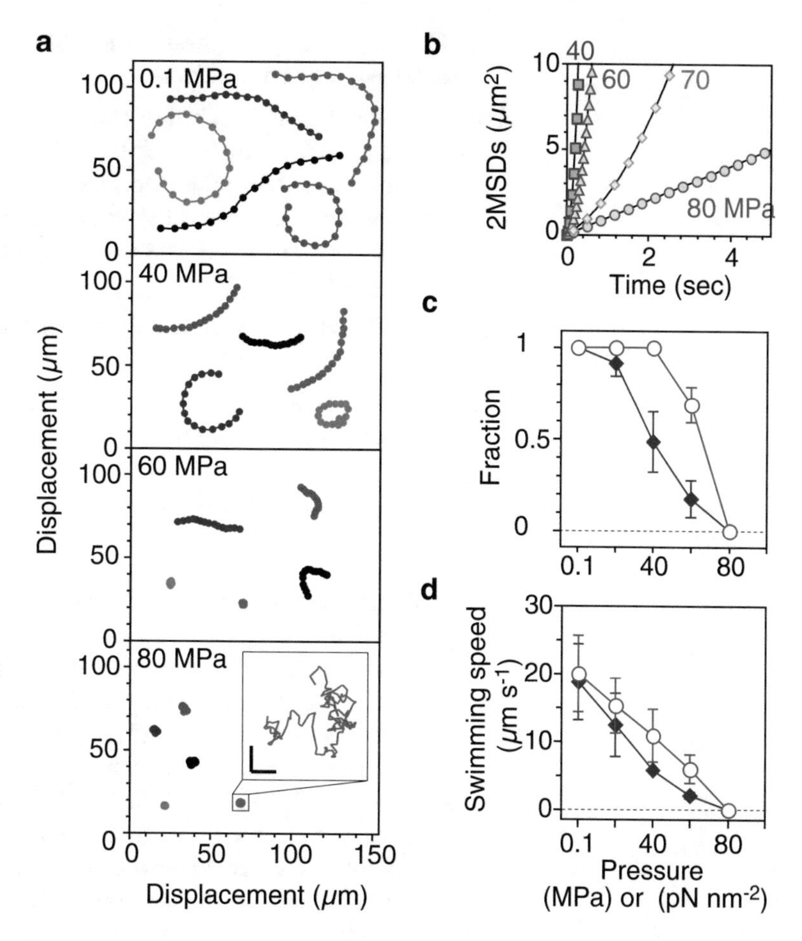

Fig. 3 Motility of swimming cells. (**a**) Trajectories of RP4979 cells. The positions of the cell were plotted at every tenth frame for 5 s. (Inset) Trajectory of a cell at 80 MPa for 5 s on an expanded scale. Scale bar, 2 μm. (**b**) 2MSDs plotted at every frame for 5 s (40 and 60 MPa) or at every 10 frames (70 and 80 MPa) (n = 46–59, total = 202). Data were fitted by $(vt)^2$ and/or $4D_{x\text{-}y}t$, where v is the translational speed, $D_{x\text{-}y}$ is the diffusion constant, and t is time (40 MPa, v = 11 μm s^{-1}; 60 MPa, v = 5.2 μm s^{-1}; 70 MPa, v = 1.0 μm s^{-1} and $D_{x\text{-}y}$ = 0.35 μm^2 s^{-1}; 80 MPa, $D_{x\text{-}y}$ = 0.25 μm^2 s^{-1}). (**c** and **d**) Swimming fraction and speed during the pressurization (solid circles) and depressurization processes (open diamonds). Swimming fractions were based on the number of cells that swam with a speed of >2 μm s^{-1} under each pressure condition. The speed was the average value of the swimming cells in **c**. Error bars represent the SD. Adapted from ref. [4]

7. Plot the time courses of 2-dimensional mean square displacements (2MSDs).
8. Fit $(vt)^2$ and/or $4D_{x\text{-}y}t$ to the data, where v is the translational speed, $D_{x\text{-}y}$ is the diffusion constant, and t is time. The equation should be selected in consideration of the swimming motility of cells (Fig. 3b).

9. (Case #1: Most cells swam smoothly in solution at 40 and 60 MPa) Fit $(vt)^2$ to the data.
10. (Case #2: Some cells swam smoothly, but the others did not show any translational motion at 70 MPa) Fit $(vt)^2 + 4D_{x\text{–}y}t$ to the data. The data could be fitted by this equation when the MSDs would be located between the values in case#1 and #3. It indicates that the cells have very weak motility (*see* Movie S2 and Fig. S3 in Ref. [8]).
11. (Case #3: all cells stopped swimming in solution, at 80 MPa) Fit $4D_{x\text{–}y}t$ to the data.
12. Calculate the average and the standard deviation of the fraction and speed of the swimming cells for each condition (Fig. 3c).

4 Notes

1. A similar high-pressure chamber is commercially available (PMC-100-2-0.6; Syn Corporation, Kyoto, Japan). Briefly, Fig. 1a displays a photograph of the high-pressure chamber [4]. The chamber was composed of a main body and a window support (nickel alloy (Hastelloy C276); Sasahara Giken, Kyoto, Japan), each of which was equipped with an optical window (BK7; Sasahara kogaku, Kyoto, Japan). The window support was screwed onto the main body, and the interior of the chamber was sealed with an O-ring. In addition, the chamber was equipped with two flow paths at the outer edge of the main body.
2. The use of a separator conferred the advantage of reducing the total dead volume of the buffer solution in the pressure line [11]. In addition, we have reported another chamber, equipped with a "built-in" separator [12].
3. The silicon tubes could be connected to the external ports in the chamber, as indicated by arrowheads in Fig. 1a. The inner temperature inside the chamber could be controlled by running temperature-regulated water from the thermostat bath.
4. Complete details of the microscope stage are available to interested parties upon request.
5. Use good buffer solutions, but do not use phosphate buffer for high-pressure experiments. The pH value of the latter is largely changed by applied pressures [11].
6. Adjust the number of cells in a frame for performing the image analysis efficiently.
7. ImageJ is written in Java and runs on the Java Virtual Machine (JVM). Most operating systems including Windows, Mac, and

Unix are compatible with JVM. Therefore, ImageJ users can easily communicate with other users. In addition, the functions of ImageJ can be extended using plugins that are also written in Java. Many plugins are available and several ImageJ derivatives already have many plugins. Fiji is one such ImageJ-based software that is especially designed for image analysis of biological data. Here, we describe a method to perform image analysis using a custom-made plugin for ImageJ.

8. If your operating system is Windows 7 or higher, you should install (or extract the .zip file) to the folder which is free from User Account Control (UAC) for example, to the Document folder. Because Fiji is frequently updated automatically, extraction of Fiji to Program Folder will result in UAC suffering from such processes. Do not choose folders whose names contain 2-bytes characters such as Kanji.
9. Download the latest version of the plugin from the web site, https://github.com/arayoshipta/SimplePTA. This plugin is free, but its copyright is not abandoned. We take NO responsibility for any damage from the use of this plugin.

Acknowledgments

We thank Yoshie Harada and Takeharu Nagai for technical support. We thank Taishi Kasai and Tomofumi Sakai for reviewing the Image J plugin and for valuable comments. We acknowledge support from the Grant-in-Aid for Scientific Research (Nos. JP15H01319 and JP16K04908), Takeda Science Foundation, Research Foundation for Opto-Science and Technology, and the Nakatani Foundation for advancement of measuring technologies in biomedical engineering (to M.N.). M.N. developed the microscope system; Y.A. developed the Image J plugin; M.N. and Y.A. wrote the manuscript.

References

1. Macnab RM (1996) Flagella and motility. In: Neidhardt FC (ed) *Escherichia coli* and *Salmonella*: cellular and molecular biology, 2nd edn. Washington, D.C. ASM Press, pp 123–145
2. Berg HC (2004) *E. coli* in motion, Biological and medical physics biomodeical engineering. Springer, New York
3. Sowa Y, Berry RM (2008) Bacterial flagellar motor. Q Rev Biophys 41(2):103–132. doi:10.1017/S0033583508004691
4. Nishiyama M, Sowa Y (2012) Microscopic analysis of bacterial motility at high pressure. Biophys J 102(8):1872–1880. doi:10.1016/j.bpj.2012.03.033
5. Nishiyama M (2015) High-pressure microscopy for studying molecular motors. Subcell Biochem 72:593–611. doi:10.1007/978-94-017-9918-8_27
6. Watanabe TM, Imada K, Yoshizawa K, Nishiyama M, Kato C, Abe F, Morikawa TJ,

Kinoshita M, Fujita H, Yanagida T (2013) Glycine insertion makes yellow fluorescent protein sensitive to hydrostatic pressure. PLoS One 8(8):e73212. doi:10.1371/journal.pone.0073212

7. Nishiyama M, Sowa Y, Kimura Y, Homma M, Ishijima A, Terazima M (2013) High hydrostatic pressure induces counterclockwise to clockwise reversals of the *Escherichia coli* flagellar motor. J Bacteriol 195(8):1809–1814. doi:10.1128/jb.02139-12
8. Takekawa N, Nishiyama M, Kaneseki T, Kanai T, Atomi H, Kojima S, Homma M (2015) Sodium-driven energy conversion for flagellar rotation of the earliest divergent hyperthermophilic bacterium. Sci Rep 5:12711. doi:10.1038/srep12711
9. Jarrell KF, McBride MJ (2008) The surprisingly diverse ways that prokaryotes move. Nat Rev Microbiol 6(6):466–476. doi:10.1038/nrmicro1900
10. Scharf BE, Fahrner KA, Turner L, Berg HC (1998) Control of direction of flagellar rotation in bacterial chemotaxis. Proc Natl Acad Sci U S A 95(1):201–206
11. Nishiyama M, Kimura Y, Nishiyama Y, Terazima M (2009) Pressure-induced changes in the structure and function of the kinesin-microtubule complex. Biophys J 96(3):1142–1150. doi:10.1016/j.bpj.2008.10.023
12. Nishiyama M, Kojima S (2012) Bacterial motility measured by a miniature chamber for high-pressure microscopy. Int J Mol Sci 13(7):9225–9239. doi:10.3390/ijms13079225

Chapter 14

Measurements of the Rotation of the Flagellar Motor by Bead Assay

Taishi Kasai and Yoshiyuki Sowa

Abstract

The bacterial flagellar motor is a reversible rotary nano-machine powered by the ion flux across the cytoplasmic membrane. Each motor rotates a long helical filament that extends from the cell body at several hundreds revolutions per second. The output of the motor is characterized by its generated torque and rotational speed. The torque can be calculated as the rotational frictional drag coefficient multiplied by the angular velocity. Varieties of methods, including a bead assay, have been developed to measure the flagellar rotation rate under various load conditions on the motor. In this chapter, we describe a method to monitor the motor rotation through a position of a 1 μm bead attached to a truncated flagellar filament.

Key words Bacterial flagellar motor, Rotation speed, Torque, Microsphere, High-speed camera

1 Introduction

Many species of swimming bacteria rotate their flagella that extend from the cell body for generating their propulsion force. Each flagellum consists of a long, thin helical filament turned like a screw by a rotary motor. The motor embedded in the bacterial cell envelope is driven by the flux of ions across the cytoplasmic membrane [1, 2].

The rotation of bacterial flagella was demonstrated by tethered cell assay in the 1970s [3, 4]. Cells were tethered to a microscope slide via one of their filaments and rotation of the cell body, driven by the motor, was observed under a conventional video microscope. The rotation speed of tethered cells, however, was limited below 20 Hz (revolutions per second), because of high load to the motor. To observe the motor rotation under smaller load, i.e., natural conditions, varieties of techniques, such as high-intensity dark-field microscopy, laser dark-field microscopy, and fluorescent microscopy, have been developed and applied [5–7]. Laser dark-field microscopy revealed that the maximum speed of flagellar rotation reached 270 and 1700 Hz for *Escherichia coli* and for *Vibrio alginolyticus*, respectively [6, 8].

Tohru Minamino and Keiichi Namba (eds.), *The Bacterial Flagellum: Methods and Protocols*, Methods in Molecular Biology, vol. 1593, DOI 10.1007/978-1-4939-6927-2_14, © Springer Science+Business Media LLC 2017

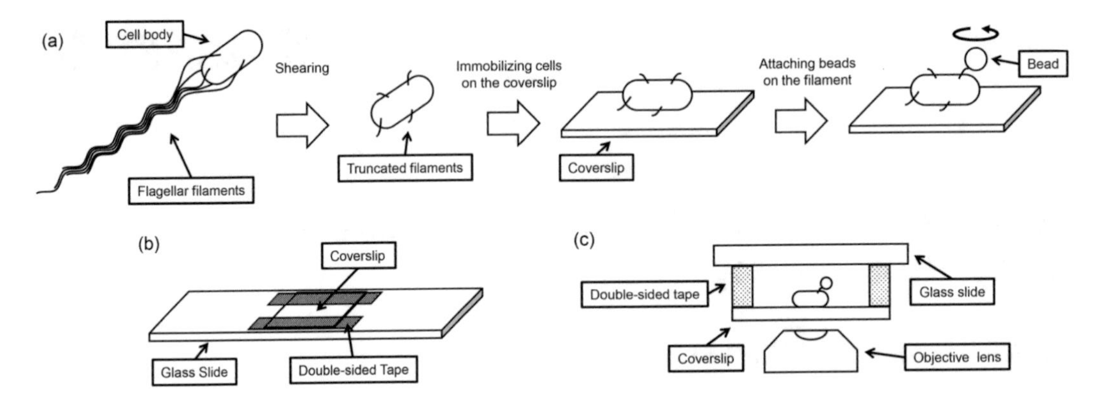

Fig. 1 Schematics of a bead assay. (**a**) Flagellar filaments extending from the cell body can be truncated by "shearing." A bead attached to a truncated filament of a cell immobilized to coverslip is followed by a light microscope equipped with a high-speed camera. (**b**) A flow cell assembly. (**c**) The cross section of the flow cell mounted on an inverted microscope

Recent preferable approach to measure the rotation rate of the flagellar motor is a bead assay [9]. A sub-micrometer microsphere (bead) is attached to truncated flagellar filaments or hooks of immobilized cells as a marker. The rotation of the bead can be followed by a quadrant photodiode (QPD) or high-speed camera with high temporal and spatial resolution. Further analysis of bead centers gives motor speed or fine angular movements in one revolution [9–12]. Because the Reynolds number for the rotating flagellar motor is much smaller than 1, the inertia is negligible; the generated torque of the motor simply can be calculated as its rotational frictional drag coefficient multiplied by the angular velocity (= $2\pi \times$ rotational speed) [13]. The load to the motor can be varied by using different sizes of microspheres, by changing viscosity of media or both; therefore, the torque-speed relationship can be measured in a wide range by a bead assay [9, 14, 15]. Fitting experimentally obtained torque-speed relationship of the flagellar motor under a wide range of conditions is an important test for theoretical models of the mechanochemical cycle of the motor [16–21].

In this chapter, we describe how to measure the speed and torque of the *E. coli* flagellar motor using a 1 μm bead as a marker by combining a conventional light microscope and a high-speed camera (Fig. 1).

2 Materials

Prepare all solutions using ultrapure water.

2.1 Cell Culture

1. *E. coli* strains carrying *fliCst* allele (*see* **Note 1**).
2. T-broth media: 10.0 g/L Bacto™ Tryptone and 5.0 g/L sodium chloride.

3. LB media: 10.0 g/L Bacto™ Tryptone, 5.0 g/L Bacto™ Yeast Extract, and 5.0 g/L sodium chloride.
4. Dimethyl sulfoxide (DMSO).
5. Glass test tube and flask.
6. Shaking incubator.

2.2 Sample Preparation

1. Motility Buffer: 10 mM potassium phosphate, pH 7.0, 0.1 mM EDTA, and 85 mM sodium chloride.
2. 0.01% Poly-L-lysine solution (P4707, Sigma-Aldrich).
3. Beads solution: 1/20 dilution of 2.5% Solids-Latex 1.0 μm-diameter Polystyrene Beads (Polysciences, Inc.) in Motility Buffer (*see* **Note 2**).
4. Glass slide: 1.0 mm thick, 76 × 26 mm.
5. Cover slip: 0.17 mm thick, 18 × 18 mm.
6. Double-stick Scotch tape.
7. Clean tissue.
8. Humidity chamber (box with a soaked tissue).
9. Syringe with 26 G needle.

2.3 Microscope Setup

1. Inverted microscope equipped with 100× phase-contrast objective lens.
2. High-speed camera and computer installed with camera control software (*see* **Note 3**).
3. Vibration-free table.
4. Analysis software; ImageJ [22] or LabVIEW (National Instruments).

3 Methods

All procedures are done at room temperature, unless indicated otherwise.

3.1 Frozen Stock Preparation

1. Inoculate an *E. coli* colony from a fresh agar plate into 10 mL of LB media. Incubate in a shaking incubator overnight at 30 °C.
2. Add 1 mL DMSO to the overnight culture and mix gently.
3. Make 100 μL aliquots and keep them at −80 °C freezer.

3.2 Cell Cultivation

1. Add 60 μL of thawed aliquot to 3 mL of T-broth medium. Incubate in a shaking incubator overnight at 30 °C.
2. Add 60 μL of overnight culture to fresh 3 mL of T-broth medium. Incubate in a shaking incubator for 4–5 h at 30 °C. Optical density of the cell culture at 600 nm is approximately 0.7.

3. Transfer the cell culture to an Eppendorf tube. Pass the cell culture through 26 G needle using syringe 50 times. This procedure truncates flagellar filaments by viscous shear, called shearing process.
4. Centrifuge at 3500 × *g* for 2 min.
5. Discard the supernatant, add 1 mL of fresh Motility Buffer, and resuspend the pellet gently.
6. Repeat **steps 4** and **5** twice.
7. Centrifuge at 3500 × *g* for 2 min.
8. Discard the supernatant, add 0.5 mL of fresh Motility Buffer, and resuspend the pellet gently.

3.3 Flow Cell Assembly

1. Place a glass slide on a clean tissue.
2. Apply two pieces of double-stick Scotch tape onto the slide such that there is about a 5 mm gap between the tapes.
3. Place a coverslip on the top of the tapes and press down over the double-stick tape gently.

3.4 Sample Preparation

1. Flow Poly-L-lysine solution into the flow cell.
2. Immediately rinse the flow cell with 100 μL of Motility Buffer.
3. Flow 10–15 μL of cell solution.
4. Place the flow cell in a humidity chamber such that its coverslip side faces to the downside, and incubate for 5–10 min.
5. Flush 100 μL of Motility Buffer to remove unattached cells.
6. Flow 10–15 μL of Beads Solution.
7. Repeat **step 4**.
8. Flush 100 μL of Motility Buffer to remove unattached beads.

3.5 Microscope Observation and Analysis

1. Mount the sample slide on the microscope stage.
2. Adjust the focus and find spinner beads under a phase-contrast microscopy.
3. Set appropriate camera parameters and record image sequences.
4. Analyze the bead's centroid for each frame by ImageJ or LabVIEW.
5. Convert the data from frame numbers to time and from pixels to micrometer. Measure the bead eccentric rotation radius (Fig. 2).
6. Calculate speed from power spectra of tracked *X*- *Y*-signals, using data windows of length 1 s at 0.1 s intervals (Fig. 3, *see* **Note 4**).
7. Calculate the generated torque from obtained parameters (*see* **Note 5**).

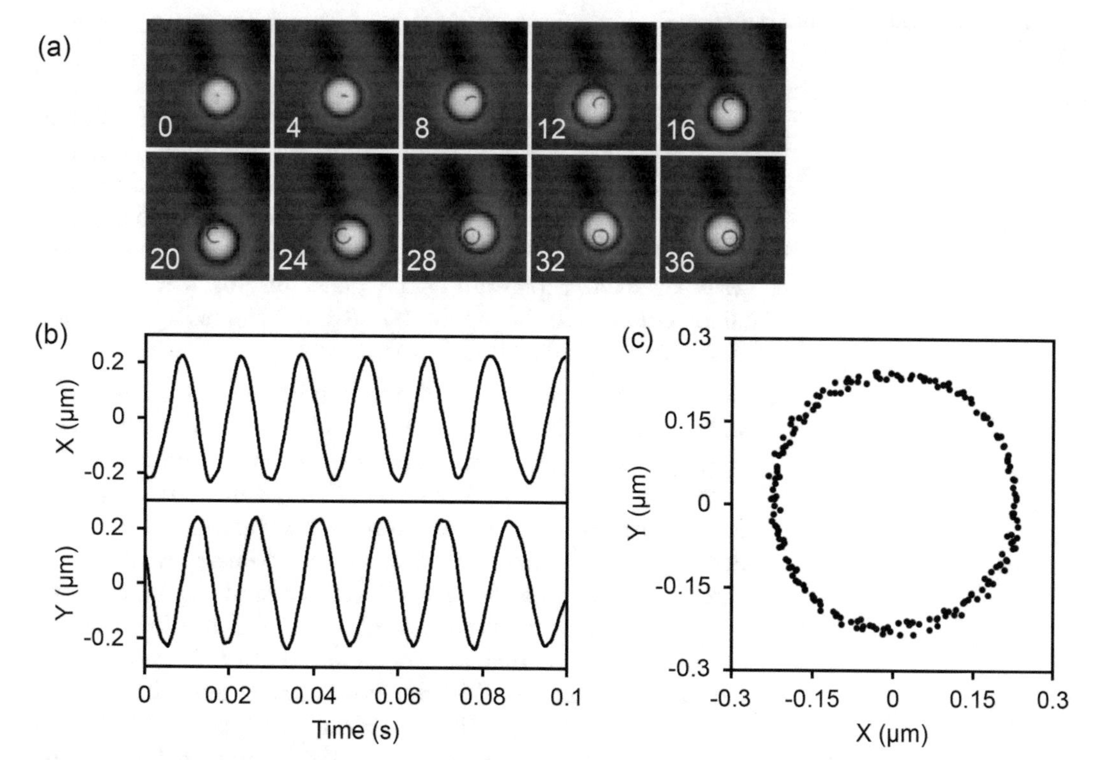

Fig. 2 Typical data of a 1 μm bead assay. (**a**) An example of sequential images of a rotating bead attached to a truncated flagellar filament with calculated bead centers superimposed. The original video was recorded at 2000 fps. The time interval between images is 2 ms and the frame number of each image is shown at *bottom-left*. *Red lines* track the calculated bead centers. Each image size is 4 μm square. (**b**) *X–Y* positions of the rotating bead are shown as a function of time. (**c**) A two-dimensional *X–Y* plot of the data shown in (**b**)

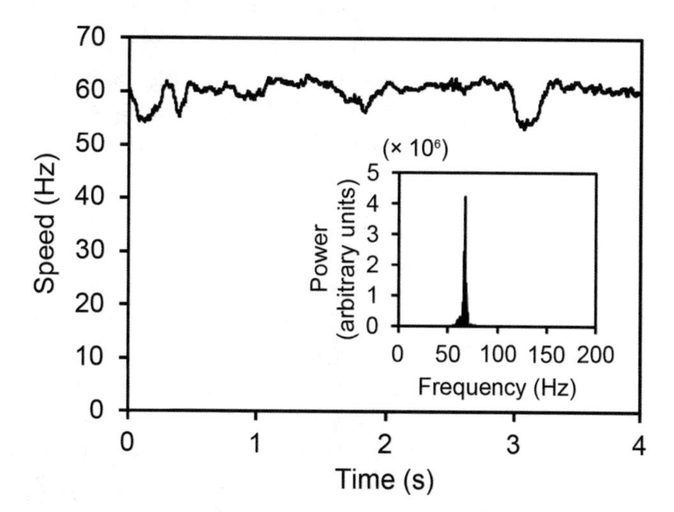

Fig. 3 Speed vs. time trace. The rotation speed of the motor driving a 1 μm bead is shown as a function of time. The speed vs. time trace was smoothed with a median filter using a window of 10 data points. (*Inset*) The power spectrum of the signal from the data of 1.0 s

4 Notes

1. *E. coli* flagellar filament is a tubular polymer of FliC protein (497 residues). FliCst, which is made by deleting amino acids 245–301 from FliC, forms a sticky filament to which a bead with a hydrophobic surface is attached spontaneously and tightly [23, 24]. FliCst can be expressed from chromosomal gene or from a plasmid-borne gene. Strains without *fliCst* allele could also be applicable for a bead assay by using antibody-coated beads [25, 26].
2. Beads of other materials and sizes can be used for a bead assay with some modifications. In general, beads of smaller sizes require a stronger light source or efficient illumination for detection. A method presented in this chapter can be applied to a relatively large latex microsphere (above ~0.5 μm in diameter) [15]. Back-focal plane interferometry with a QPD gives a better signal from small markers and can be used for latex or glass microspheres larger than 0.3 μm in diameter [27]. For even smaller markers, combination of gold-colloidal nanoparticle (60–250 nm in diameter) and either high-intensity dark-field illumination or laser dark-field illumination would give sufficiently high signal from the marker for nanophotometry [25, 28].
3. Any high-speed cameras, which run at a frame rate of more than twice as fast as the expected motor rotation speed, can be used for this protocol.
4. Speeds are obtained from power spectra of combined *X*-*Y* data, using data windows of a time period of 1 s at 0.1 s intervals. This analysis gives a speed resolution of 1 Hz, which is suitable for observing speed changes due to incorporation and dissociation of torque-generating stator units [10, 21, 29]. Note that the window sizes and intervals should be adjusted depending on the purpose of the study.
5. The generated torque by the motor (M) can be calculated as $M = (f_b + f_f)\ \omega$, where f_b and f_f are the rotational frictional drag coefficients of the bead and the filament stub, respectively, and ω is the angular velocity. f_b can be estimated as $8\pi\eta r_b^3 + 6\pi\eta r_b r_r^2$, where r_b and r_r are the radius and the rotational eccentricity of the bead, respectively, and η is the viscosity of the medium. Although f_f is uncertain, f_f is much smaller than f_b and is negligible for analysis when 1 μm beads are used. In the case of Fig. 2, r_b of 0.2 μm was measured, giving M = 1270 pN nm when the motor rotates at 60 Hz. Note that the filament contribution to the drag coefficient is significant, and its correction is required when smaller sizes of beads are used [30].

Acknowledgment

We thank Dr. I. Kawagishi and Dr. M. Nishikawa (Hosei Univ.) for critically reading the manuscript and Dr. Y.-S. Che for *E. coli* strains carrying *fliCst* allele. This work was supported by MEXT KAKENHI Grant Number JP15H01332, JSPS KAKENHI Grant Number JP15K07034, and the MEXT-Supported Program for the Strategic Research Foundation at Private Universities, 2013–2017.

References

1. Berg HC (2003) The rotary motor of bacterial flagella. Annu Rev Biochem 72:19–54
2. Sowa Y, Berry RM (2008) Bacterial flagellar motor. Q Rev Biophys 41:103–132
3. Berg HC, Anderson RA (1973) Bacteria swim by rotating their flagellar filaments. Nature 245:380–382
4. Silverman M, Simon M (1974) Flagellar rotation and the mechanism of bacterial motility. Nature 249:73–74
5. Macnab RM (1976) Examination of bacterial flagellation by dark-field microscopy. J Clin Microbiol 4:258–265
6. Kudo S, Magariyama Y, Aizawa S (1990) Abrupt changes in flagellar rotation observed by laser dark-field microscopy. Nature 346:677–680
7. Turner L, Ryu WS, Berg HC (2000) Real-time imaging of fluorescent flagellar filaments. J Bacteriol 182:2793–2801
8. Magariyama Y, Sugiyama S, Muramoto K, Maekawa Y, Kawagishi I, Imae Y, Kudo S (1994) Very fast flagellar rotation. Nature 371:752
9. Ryu WS, Berry RM, Berg HC (2000) Torque-generating units of the flagellar motor of *Escherichia coli* have a high duty ratio. Nature 403:444–447
10. Reid SW, Leake MC, Chandler JH, Lo CJ, Armitage JP, Berry RM (2006) The maximum number of torque-generating units in the flagellar motor of *Escherichia coli* is at least 11. Proc Natl Acad Sci U S A 103:8066–8071
11. Sowa Y, Rowe AD, Leake MC, Yakushi T, Homma M, Ishijima A, Berry RM (2005) Direct observation of steps in rotation of the bacterial flagellar motor. Nature 437:916–919
12. Nakamura S, Kami-ike N, Yokota JP, Minamino T, Namba K (2010) Evidence for symmetry in the elementary process of bidirectional torque generation by the bacterial flagellar motor. Proc Natl Acad Sci U S A 107:17616–17620
13. Berg HC (1993) Random walks in biology. Princeton University Press, Princeton, NJ
14. Chen X, Berg HC (2000) Torque-speed relationship of the flagellar rotary motor of *Escherichia coli*. Biophys J 78:1036–1041
15. Sowa Y, Hotta H, Homma M, Ishijima A (2003) Torque-speed relationship of the Na^+-driven flagellar motor of *Vibrio alginolyticus*. J Mol Biol 327:1043–1051
16. Oosawa F, Hayashi S (1986) The loose coupling mechanism in molecular machines of living cells. Adv Biophys 22:151–183
17. Läuger P (1988) Torque and rotation rate of the bacterial flagellar motor. Biophys J 53:53–65
18. Berry RM (1993) Torque and switching in the bacterial flagellar motor. An electrostatic model. Biophys J 64:961–973
19. Xing J, Bai F, Berry R, Oster G (2006) Torque-speed relationship of the bacterial flagellar motor. Proc Natl Acad Sci U S A 103:1260–1265
20. Bai F, Lo CJ, Berry RM, Xing J (2009) Model studies of the dynamics of bacterial flagellar motors. Biophys J 96:3154–3167
21. Lo CJ, Sowa Y, Pilizota T, Berry RM (2013) Mechanism and kinetics of a sodium-driven bacterial flagellar motor. Proc Natl Acad Sci U S A 110:E2544–E2551
22. Rasband WS (1997–2016) ImageJ, U.S National Institutes of Health. Bethesda, MD, http://imagej.nih.gov/ij/
23. Kuwajima G (1988) Construction of a minimum-size functional flagellin of *Escherichia coli*. J Bacteriol 170:3305–3309
24. Berg HC, Turner L (1993) Torque generated by the flagellar motor of *Escherichia coli*. Biophys J 65:2201–2216
25. Yuan J, Berg HC (2008) Resurrection of the flagellar rotary motor near zero load. Proc Natl Acad Sci U S A 105:1182–1185
26. Pilizota T, Brown MT, Leake MC, Branch RW, Berry RM, Armitage JP (2009) A molecular

brake, not a clutch, stops the *Rhodobacter sphaeroides* flagellar motor. Proc Natl Acad Sci U S A 106:11582–11587

27. Rowe AD, Leake MC, Morgan H, Berry RM (2003) Rapid rotation of micron and submicron dielectric particles measured using optical tweezers. J Mod Opt 50:1539–1554
28. Sowa Y, Steel BC, Berry RM (2010) A simple back-scattering microscope for fast tracking of biological molecules. Rev Sci Instrum 81: 113704
29. Sowa Y, Homma M, Ishijima A, Berry RM (2014) Hybrid-fuel bacterial flagellar motors in *Escherichia coli*. Proc Natl Acad Sci U S A 111:3436–3441
30. Inoue Y, Lo CJ, Fukuoka H, Takahashi H, Sowa Y, Pilizota T, Wadhams GH, Homma M, Berry RM, Ishijima A (2008) Torque-speed relationships of Na^+-driven chimeric flagellar motors in *Escherichia coli*. J Mol Biol 376:1251–1259

Chapter 15

Measurements of Ion-Motive Force Across the Cell Membrane

Tsai-Shun Lin, Yi-Ren Sun, and Chien-Jung Lo

Abstract

Cells need energy to survive. Ion-motive force (IMF) is one of the most important biological energy formats in bacterial cells. Essentially, the ion-motive force is the sum of electrical and chemical potential differences across the cell membrane. For bacteria, the ion-motive force is involved not only in ATP production but also in flagellar motility. The bacterial flagellar motor is driven either by proton or sodium ion. The ion-motive force measurement therefore requires the measurement of membrane potential, proton concentration, or sodium ion concentration. The bacterial flagellar motor is the most powerful molecular machine we have known so far. To understand the energetic condition of bacterial flagellar motors, together with single-motor torque measurement, methods for single-cell ion-motive force measurement have been developed. Here, we describe fluorescent approaches to measure the components of ion-motive force.

Key words Ion-motive force, Fluorescent indicator, Membrane potential, Intracellular sodium concentration

1 Introduction

1.1 Background

The bacterial flagellar motor (BFM) is a natural rotary molecular machine that rotates a helical flagellar filament to propel a bacterial cell to swim around [1–3]. The BFM is a large protein complex with a diameter of 45 nm and is major part of the flagellar basal body [4, 5]. Two proteins, MotA and MotB, form a proton-driven stator-unit that interacts with the rotor to generate torque. Some bacteria use PomA/B to form a sodium-driven stator-unit to drive the motor [6]. The energy source of the BFM is the ion flux through the stator-unit driven by ion-motive force. It is important to measure the accurate value of ion-motive force in bacteria for understanding the mechanism of chemomechanical coupling between the BFM function and the driving energy [7–9].

The ion-motive force is the sum of electrical and chemical potential across the cell membrane [10, 11]. Bacteria use

Tohru Minamino and Keiichi Namba (eds.), *The Bacterial Flagellum: Methods and Protocols*, Methods in Molecular Biology, vol. 1593, DOI 10.1007/978-1-4939-6927-2_15,

ion-motive force not only for motility but also for several important functions, such as ATP synthesis [12] and active membrane transport [13].

The ion-motive force is expressed as

$$IMF = V_{\mathrm{m}} - Z \times \log \frac{[ion]_{ex}}{[ion]_{\mathrm{in}}} \quad (1)$$

where V_{m} is the membrane potential, $[\mathrm{ion}]_{\mathrm{ex}}$ and $[\mathrm{ion}]_{\mathrm{in}}$ the intracellular and extracelluar ion concentrations, $Z = 2.303 \times RT/F = 2.303 \times k_{\mathrm{B}}T/e$, R the gas constant, T the absolute temperature (K), F the Faradays constant, k_{B} the Boltzmann's constant, and e the electrical charge.

The ion-motive force is an electrochemical potential difference across the bacterial cell membrane (from outside to inside) in the unit of volt. It comprises the electrical potential difference across membrane, V_{m}, and the chemical potential difference of specific ion of interests. To measure the ion-motive force, we need to measure the membrane potential, V_{m}, and the intracellular ion concentration of interest. The recent developments of fluorescent indicators and microscopy tools have made the single-cell ion-motive force measurement possible [7–9, 14]. Here, we describe the measurement of V_{m} using Nernst potential indicators and the measurement of intracellular sodium ion concentration by fluorescent microscopy.

This chapter is divided into two parts. In the first part, we will introduce modern single-cell ion-motive force measurements using optical membrane voltage indicator. The second part will introduce the intracellular sodium concentration measurement.

1.2 Membrane Potential Measurements

Bacteria transport ions across the cell membrane to generate the membrane potential. Nernst potential indicators are charged small fluorescent molecules that diffuse across membrane following Nernst equation. Local electrical-field indicators are molecules inserted into the membrane to sense the local electric files. However, there is no promising local electrical-field indicator for bacteria at present. We only focus on Nernst potential indicator in this chapter.

Nernst potential indicators are typically cationic and membrane permeable molecules. In the equilibrium condition, the distribution of the indicator molecules follows Boltzmann's Law:

$$C_{\mathrm{in}} / C_{ex} = \exp(-qV_{\mathrm{m}} / k_{\mathrm{B}}T) \quad (2)$$

where q is the charge of the indicator, and C_{in} and C_{ex} are the intracellular and extracellular indicator concentrations, respectively.

For typical positive charged indicator molecules and negative membrane potential, the indicator molecules would accumulate inside the cell. For constant external indicator concentration condition in the steady state, the membrane potential can be measured by the Nernst equation

$$\ln(C_{\text{in}} / C_{ex}) = -q(V_{\text{m}}) / k_{\text{B}}T \tag{3}$$

Since ln N ≈ 2.303 log N, the membrane potential can be rewritten in a general form for simpler calculation

$$V_{\text{m}} = -2.303 \frac{k_{\text{B}}T}{e} \log \frac{C_{\text{in}}}{C_{ex}} \tag{4}$$

By measuring the C_{in} in the constant C_{ex} condition, one can measure the membrane potential. The signal changes of the Nernst potential indicator could be as high as 100 times/100 mV. Therefore, fluorescent microscopy is suitable for this kind of single-cell membrane potential measurement.

There are Carbocyanine (cationic) dyes, Oxonol (anionic) dyes, and Rhodamine (cationic) dyes as commercial available Nernst potential indicator. It has been demonstrated that Rhodamin serious dye, TMRM/TMRE, is useful for the membrane potential measurement in *E. coli* and other bacteria at a single-cell level [7, 15]. Any other new Nernst indicators can be tested and used in the same manner. The following protocol is designed for *E. coli* with a sodium-driven chimeric flagellar motor [9].

1.3 Intra-Cellular Sodium Concentration Measurements

The sodium-driven bacterial flagellar motor gives us a wider possibility to probe the BFM energetics. Intracellular sodium concentration measurement is achieved by using cell permeant sodium indicators, Sodium-Green or CoroNa. We will introduce single-cell fluorescence approach to measure the intra-cellular sodium concentration.

2 Materials

All solutions are prepared using deionized pure water. Prepare and store all reagents at room temperature.

2.1 Buffers

1. LB: 1% tryptone, 0.5% Yeast extract, and 0.5% NaCl, autoclaved.
2. TB: 1% tryptone and 0.5% NaCl, autoclaved.
3. Motility buffer: Make 10 mM Potassium phosphate buffer. Add 0.1 mM EDTA and 85 mM NaCl, adjust to pH 7.0 and then autoclave.
4. Cell treating buffer: Make 10 mM Potassium phosphate buffer. Add 10 mM EDTA and 85 mM NaCl, adjust to pH 7.0 and then autoclave.

2.2 Fluidic Chamber

1. Clean cover glass (*see* **Note 1**).
2. Double-sided tapes: Scotch 3 M double-sided tape 136 or other double-sided tapes.
3. Poly-L-lysin: 0.1% poly-L-lysin solution.

2.3 Nernst Potential Indicator

1. TMRM stock solution: 1 mM Tetramethylrhodamine (TMRM) in DMSO.
2. TMRM loading solution: 1 μM TMRM in motility buffer.

2.4 Sodium Indicator

1. Sodium-Green/CoroNa stock solution: 1 mM Sodium-Green/CoroNa cell permeant in DMSO.
2. Sodium-Green/CoroNa loading solution: 40 μM Sodium-Green/CoroNa in motility buffer.

2.5 Calibration Medium

1. Gramicidin: 20 μM Gramicidin in motility buffer.
2. CCCP: 5 μM carbonyl cyanide 3-chlorophenylhydrazone (CCCP) in motility buffer.

2.6 Fluorescent Microscope

1. Fluorescent microscope (Upright or inverted) with high NA (>1.3) 100× objective and epi-fluorescence mode.
2. Fluorescent cube: Choose Excitation (549 nm)/Emission (573 nm) (TRITC) cube for TMRM and Excitation (507 nm)/Emission (532 nm) (GFP/FITC) cube for Sodium-Green and CoroNa.
3. Camera: High-sensitive cameras such as EMCCD, sCMOS or Interline cameras with cooling.
4. Software: Images acquisition software that can control the exposure time. Those come with microscopes or cameras are suitable. Micromanager is also a good option for free software.

2.7 Image Analysis

1. MATLAB, IDL, ImageJ, or compatible softwares are all suitable to analyze the images.

3 Methods

3.1 Fluidic Chamber

1. Use a microscope slide, and two stripes of a double-sided tape and the clean coverglass to make a fluidic chamber.

3.2 Membrane Potential Measurements

All the following procedures are carried out at room temperature.

1. Grow the bacterial culture overnight in LB.
2. Take 50 μL of the overnight culture into 5 mL TB with 0.5% NaCl and grow the culture up to an optical density (OD) of 0.4.

3. Take 1 mL cells culture in a 1.5 mL tube. Centrifuge down the cells and take away the supernatant. Add 1 mL treating buffer into the tube for 10 min at room temperature. Wash the cells three times with 1 mL motility buffer.
4. Centrifuge down the cells and add TMRM-loading solution to adjust the cells density to ~10^8 cells/mL. Leave it for 40 min (*see* **Note 2**).
5. Put 20 μL of 0.1% poly-L-lysine solution into the fluidic chamber and leave it for 10 s. Then wash the chamber three times with motility buffer by capillary effect (add buffer on one side of the chamber and place a tissue paper on the other side). Then add 20 μL cells suspension into the chamber and invert the chamber for 15 min. Wash the non-tethered cells with 100 μL TMRM-loading solution (*see* **Notes 3** and **4**).
6. Take the chamber and mount on the microscope. Observe fluorescent images at an appropriate excitation light intensity. Typically, 100 ms exposure time is sufficient to observe the TMRM signal in cells (*see* **Notes 5** and **6**).
7. Measurement of the point-spread function for calibration curve: Put 20 nm fluorescent beads stuck on a clean coverglass. Take z-stack images of these beads every 50 nm over total 5 μm. After background subtraction and normalization, the averaged point-spread function (PSF), $P(x_i, y_i, z_i)$, of your microscopy system can be obtained from multiple beads data, as shown in Fig. 1b.
8. Create a 3D model of your bacterium in a chamber with its shape on glass surface, as shown in Fig. 1a. The dye concentrations in the model are

$$C_i(x_i,, y_i,, z_i) = \begin{Bmatrix} 0 & : & z \leq 0, \text{glass} \\ C_{ex} & : & \text{medium} \\ C_{\text{in}} & : & \text{cell} \end{Bmatrix} \quad (5)$$

Here, we use a cylinder (2 μm long, 0.9 μm diameter) capped by two hemispheres (0.9 μm diameter) as our *E. coli* cell model. The rest of the volume in the chamber is filled with TMRM-loading solution. Assuming we focus on the middle plane of the cell, a simulated blurred image is calculated by the convolution of the TMRM distribution and the PSF:

$$I_{\text{m}}(x_j, y_j, z_0) = \Sigma[C_i(x_i, y_i, z_i)P(x_j - x_i, y_j - y_i, z_0 - z_i)] \quad (6)$$

where $I_{\text{m}}(x_j, y_j, z_0)$ was the modeled intensity of image pixel j, Fig. 1c lower.

The correction factor $S(F_{\text{in}}/F_{\text{ex}}) = (C_{\text{in}}/C_{\text{ex}})/(F_{\text{in}}/F_{\text{ex}})$ can be found by comparing the images before and after convolution,

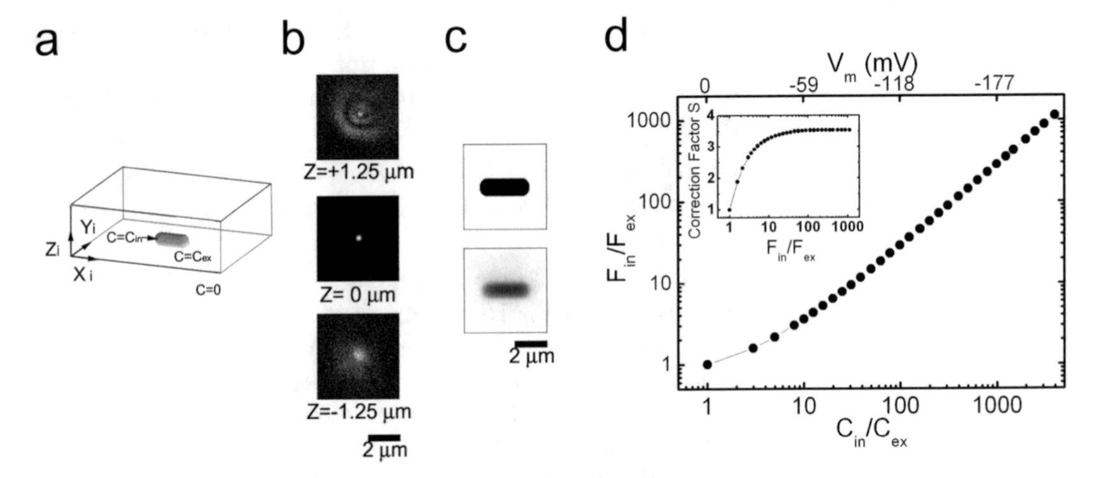

Fig. 1 Correction factor due to microscope convolution. (**a**) A model of a bacterium on a coverglass and dye in the chamber. (**b**) Three selected plans of measured point-spread function of microscope. (**c**) Dye distribution at the midcell middle-plan (Upper) and the corresponding image after convolution (Lower). (**d**) The calculated fluorescence intensity ratio, F_{in}/F_{ex}, versus the concentration ratio C_{in}/C_{ex}. The correction factor, $S(F_{in}/F_{ex}) = (C_{in}/C_{ex})/(F_{in}/F_{ex})$ is shown in the inset ref. 8

as shown in Fig. 1c, d. For specific bacterial shape, one should generate its correction function to obtain the accurate membrane potential.

9. Calculation of membrane potential: The corrected concentration ratio can be found by

$$C_{\text{in}} / C_{ex} = (F_{\text{in}} / F_{ex})S(F_{\text{in}} / F_{ex}) \quad (7)$$

where F_{in} and F_{ex} are measured experimentally and the correction factor $S(F_{\text{in}}/F_{\text{ex}})$ is calculated using the optical convolution model. The membrane potential can be calculated by Eq. (4).

3.3 Intracellular Sodium Concentration Measurements

1. Grow the bacterial culture overnight in LB with 0.5% NaCl.
2. Put 50 μL of overnight culture into 5 mL TB with 0.5% NaCl and grow the cells up to an OD of 0.4.
3. Take 1 mL cells culture in a 1.5 mL tube. Centrifuge down the cells and take away the supernatant. Add 1 mL treating buffer into the tube for 10 min at room temperature. Wash the cells three times with 1 mL motility buffer.
4. Centrifuge down the cells and add Sodium-Green/CoroNa loading solution to adjust the cells density to ~10^8 cells/mL. Leave for 30 min at 37 °C. Wash three times with motility buffer.
5. Take 20 μL 0.1% poly-L-lysine solution into the fluidic chamber for 10 s. Then wash the chamber three times with motility buffer by capillary effect (add buffer on one side of the cham-

ber and place a tissue paper on the other side). Then add 20 μL cells suspension into the chamber and invert the chamber for 15 min. Wash away the non-tethered cells with 100 μL motility buffer.

6. Take the chamber and mount on the microscope. Observe fluorescent images at an appropriate excitation light intensity. Typically, 0.3–1.0 s exposure time is sufficient to observe the cells.
7. Create a calibration curve as follows:

$$[Na]_{in} = K_d \frac{F - F_{min}}{F_{max} - F} \quad (8)$$

F is the average fluorescent intensity of a single cell obtained from the image. Once the fluorescent sodium experiments are done, the calibration curves can be measured in the following steps. The fluorescence intensity (F) should be measured in media of at least four different sodium concentrations containing the ionophores gramicidin (20 μM) and carbonyl cyanide 3-chlorophenylhydrazone (CCCP, 5μM) for 3 min for equilibrium ($[Na]_{in}=[Na]_{ex}$) (*see* **Note 7**). F_{min} is the fluorescent intensity when $[Na]_{in}=[Na]_{ex}$ = 0 mM. K_d and F_{max} can be found by fitting the calibration curves (Fig. 2c)

8. Calculate the intracellular sodium concentration using Eq. (8).

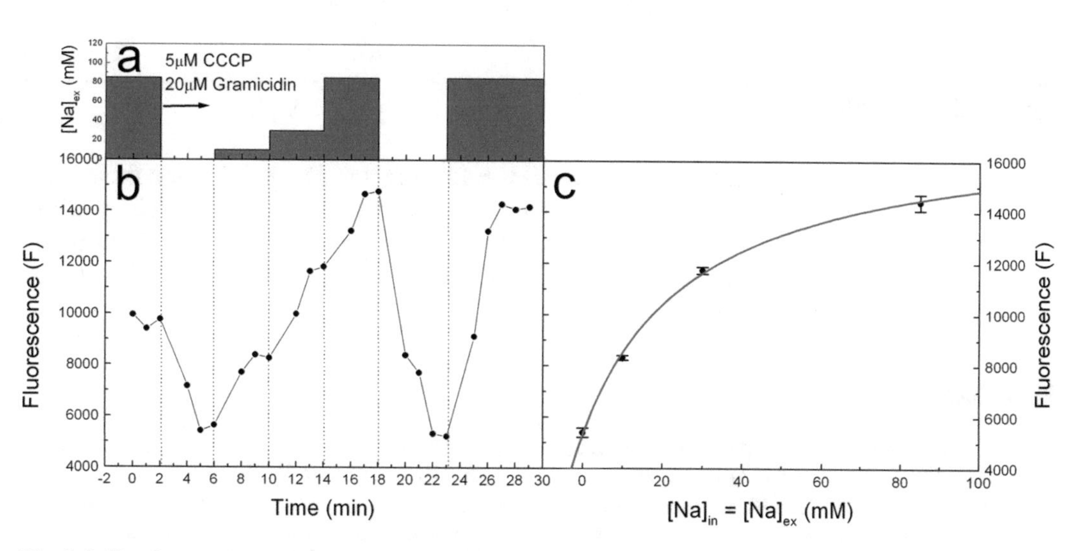

Fig. 2 Calibration method. (**a**) The external sodium concentration, $[Na]_{ex}$, is varied between 0 and 85 mM with gramicidin and CCCP present to equilibrate the sodium concentrations across membrane. (**b**) Fluorescence intensity in response to changes of $[Na]_{ex}$. About 3 min was required for equilibration of $[Na]_{ex}$. (**c**) Steady-state fluorescence intensity versus external sodium concentration, with a fit to Eq. (8). Mean and SD of three successive measurements of F are shown ref. 7

4 Notes

1. Put a coverglass into KOH and ethanol for 10 min. Rinse with clean water and dry it in the clean hood. Then use plasma cleaner to clean it for further 10 min.
2. Typically, Nernst potential indicators are slow (in the order of mins) in response to the membrane potential change. They are suitable for steady-state measurement.
3. Extra care must be taken into the loading process because the membrane structures are different from cell to cell. For bacterial cells, high permeability of Nernst potential indicators is the key to successful membrane potential measurement.
4. The extracellular concentration must be constant. If you are loading in the chamber, multiple dye loading or continuous flow is necessary.
5. The indicator should have low toxicity to cells. Growth curve measurement with indicators is necessary.
6. The indicator should have low binding affinities to the membrane and cellular compartments. After loading, membrane binding or quenching might occur. It is necessary to check the fluorescent intensity of de-energized cell.
7. $[Na]_{ex}$ is varied by mixing sodium motility buffer with potassium motility buffer (10 mM potassium phosphate, 85 mM KCl, 0.1 mM EDTA, pH 7.0), while maintaining a constant ionic strength ([Na] + [K] = 85 mM).

Acknowledgments

I am grateful to Dr. Teuta Pilizota and Dr. Bai Fan for their assistance of single cell protonmotive force measurement. The work in my lab was supported by the Ministry of Science and Technology of the Republic of China under Contract No. MOST-103-2112-M-008-013-MY3.

References

1. Sowa Y, Berry RM (2008) Bacterial flagellar motor. Q Rev Biophys 41(2):103–132
2. Baker M, Berry RM (2009) An introduction to the physics of the bacterial flagellar motor: a nanoscale rotary electric motor. Contemp Phys 50:617–632
3. Xue R, Ma Q, Baker M, Bai F (2015) A delicate nanoscale motor made by nature—the bacterial flagellar motor. Adv Sci:2. doi:10.1002/advs.201500129
4. Minamino T, Imada K (2015) The bacterial flagellar motor and its structural diversity. Trends Microbiol 23(5):267–274
5. Minamino T, Imada K, Namba K (2008) Molecular motors of the bacterial flagella. Curr Opin Struct Biol 18(6):693–701

6. Zhu S, Kojima S, Homma M (2013) Structure, gene regulation and environmental response of flagella in Vibrio. Front Microbiol 4:410
7. Lo CJ, Leake MC, Berry RM (2006) Fluorescence measurement of intracellular sodium concentration in single *Escherichia coli* cells. Biophys J 90:357–365
8. Lo CJ, Leake MC, Pilizota T, Berry RM (2007) Non-equivalence of membrane voltage and ion-gradient as driving forces for the bacterial flagellar motor at low load. Biophys J 93:294–302
9. Lo CJ, Sowa Y, Pilizota T, Berry RM (2013) The mechanism and kinetics of a sodium-driven bacterial flagellar motor. Proc Natl Acad Sci U S A 110:E2544–E2551
10. Mitchell P (1966) Chemiosmotic coupling in oxidative and photosynthetic phosphorylation. Biol Rev Camb Philos Soc 41:445–501. Reprinted (2011) BBA-Bioenergetics 1807:1507–1538
11. Chen MT, Lo CJ (2016) Using biophysics to monitor the essential protonmotive force in bacteria. Adv Exp Med Biol 915:69–79
12. Okuno D, Iino R, Noji H (2011) Rotation and structure of FoF1-ATP synthase. J Biochem 149(6):655–664
13. Shultis DD, Purdy MD, Banchs CN, Wiener MC (2006) Outer membrane active transport: structure of the BtuB:TonB complex. Science 312(5778):1396–1399
14. Martines KA II, Kitko RD, Mershon P et al (2012) Cytoplasmic pH response to acid stress in individual cells of *Escherichia coli* and *Bacillus subtilis* observed by fluorescence ratio imaging microscopy. Appl Environ Microbiol 78:3706–3714
15. Kurre R, Kouzel N, Ramakrishnan K et al (2013) Speed switching of gonococcal surface motility correlates with proton motive force. PLoS One 8(6):e67718

Chapter 16

Stoichiometry and Turnover of the Stator and Rotor

Yusuke V. Morimoto and Tohru Minamino

Abstract

Fluorescence imaging techniques using green fluorescent protein (GFP) and related fluorescent proteins are utilized to monitor and analyze a wide range of biological processes in living cells. Stepwise photobleaching experiments can determine the stoichiometry of protein complexes. Fluorescence recovery after photobleaching (FRAP) experiments can reveal in vivo dynamics of biomolecules. In this chapter, we describe methods to detect the subcellular localization, stoichiometry, and turnovers of stator and rotor components of the *Salmonella* flagellar motor.

Key words Fluorescence imaging, Fluorescent protein, Subcellular localization, FRAP, Stoichiometry, Turnover, Flagellar motor

1 Introduction

Many bacteria have flagella for their motile function. The bacterial flagellum is composed of about 30 different proteins with their copy numbers ranging from one to tens of thousands. The bacteria flagellum has a rotary motor at its base (Fig. 1). The flagellar motor of *Escherichia coli* and *Salmonella enterica* is powered by proton motive force (PMF) across the cytoplasmic membrane. The motor consists of a rotor and a dozen stators. The rotor is composed of four flagellar proteins, FliF, FliG, FliM, and FliN. FliF forms the MS ring in the cytoplasmic membrane. FliG, FliM, and FliN form the C ring at the cytoplasmic side of the MS ring. The C ring proteins are also called the switch proteins since they are responsible for switching the direction of motor rotation. Phosphorylated CheY (CheY-P) binds to FliM and FliN in the C ring, thereby inducing conformational changes of the FliG ring to allow the motor to switch the rotational direction from counter-clockwise (CCW) to clockwise (CW). The stator is a transmembrane heterohexamer complex consisting of four copies of MotA and two copies of MotB and acts as a proton channel to couple the proton flow through the channel with torque generation. Torque is generated

Tohru Minamino and Keiichi Namba (eds.), *The Bacterial Flagellum: Methods and Protocols*, Methods in Molecular Biology, vol. 1593, DOI 10.1007/978-1-4939-6927-2_16, © Springer Science+Business Media LLC 2017

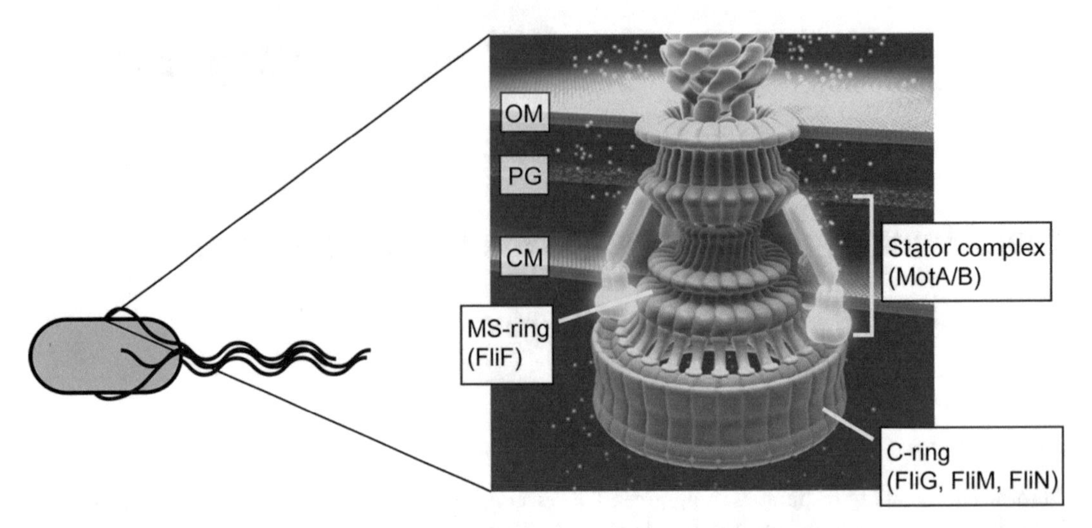

Fig. 1 Schematic illustration of the proton-driven bacterial flagellar motor. A reversible rotary motor is embedded in the cell envelop. The flagellar motor consists of a rotor made of FliF, FliG, FliM, and FliN and a transmembrane stator complex consisting of MotA and MotB. FliF forms the MS ring within the cytoplasmic membrane. FliG, FliM, and FliN form the C ring on the cytoplasmic face of the MS ring. *OM* outer membrane, *PG* peptidoglycan layer, *CM* cytoplasmic membrane

by electrostatic interactions between MotA and FliG [1–6]. N-terminally GFP-tagged MotB (GFP-MotB), which is fully functional, is visualized to be localized to the flagellar base. In mutant strains defective in the electrostatic interactions between MotA and FliG, the efficiency of GFP-MotB localization to the motor is decreased significantly. This suggests that the MotA-FliG interactions are important not only for torque generation but also for efficient stator assembly around the motor [7–10].

The *E. coli* flagellar motor can accommodate about ten stators around a rotor. High-resolution single-molecule fluorescent imaging techniques have revealed that GFP-MotB shows rapid exchanges between the motor and the membrane pool while the motor is rotating [11]. The average rate of GFP-MotB turnover shows that ~0.44 stator units exchange per second [11]. The number of stators in the functional motor is regulated in response to changes in PMF and external loads [12–16]. These suggest that the stators are not permanently fixed in place around the rotor.

FRAP experiments have shown that FliM and FliN alternate rapidly between localized and freely diffusing forms in a CheY-P-dependent manner while the motor is running. This indicates that the C ring is a highly dynamic structure [17–22]. In contrast, FliF and FliG show no exchanges, indicating that both the FliF and FliG rings are static structures [17]. The binding of CheY-P to FliM and FliN induces the dissociation of FliM and FliN molecules from the C ring, while the motor switches the rotational direction from CCW to CW [18–22]. As a result, the copy numbers of FliM and FliN in the CCW motor are ~1.29 times larger than those in the CW motor [20–22]. This suggests that remodeling of the C ring structure is induced by motor switching. It has been shown that this adaptive remodeling of the C ring is important for fine tuning of the chemotactic response to temporal changes in environments.

Thus, fluorescent imaging techniques are powerful tools to investigate the localization dynamics of each component of the bacterial flagellar motor. In this chapter, we provided detail protocols to detect the subcellular localization, stoichiometry and turnover of stator and rotor proteins labeled with a fluorescent protein.

2 Materials

2.1 Tunnel Slide

1. 24 mm × 32 mm coverslip (thickness: 0.12–0.17 mm).
2. 18 mm × 18 mm coverslip (thickness: 0.12–0.17 mm).
3. Double-sided tape.

2.2 Fluorescence Microscope

1. Inverted fluorescence microscope.
2. Electron multiplying CCD (EMCCD) camera (iXon^{EM} + 897-BI).
3. Andor Solis software.
4. 130 W mercury light source system.
5. 150 mW gas laser with a wavelength of 514 nm.
6. High-speed mechanical shutter.
7. Electrical stimulator.

2.3 Culture Media and Buffers

1. L-broth (LB): 10 g Bactotryptone, 5 g Yeast extract, 5 g NaCl per liter.
2. Motility buffer: 10 mM potassium phosphate pH 7.0, 0.1 mM EDTA, 10 mM L-sodium lactate.

2.4 *Salmonella* Strains

1. SJW1103 (wild type for motility and chemotaxis) [23].
2. YVM003 (*gfp-motB*) [7].
3. YVM004 (*gfp-fliG*) [7].
4. MRK5 (*yfp-fliF*) [24].

3 Methods

Carry out procedures at ca. 23 °C unless otherwise specified.

3.1 Preparation of Bacterial Samples

1. Inoculate 50 μL of overnight culture of *Salmonella* cells expressing a fluorescent protein-tagged flagellar motor component

(e.g., YVM003 (*gfp-motB*) or YVM004 (*gfp-fliG*)) into 5 mL of fresh LB and incubate at 30 °C with shaking for 5 h (*see* **Note 1**).

2. Collect the cells from 500 μL of the culture by centrifugation (6000 × *g*, 2 min).
3. Suspend the cell pellet in 1.0 mL of motility buffer.
4. Centrifuge at 6000 × *g* for 2 min.
5. Discard supernatant.
6. Repeat **steps 3–5** twice.
7. Resuspend the cell pellet in 500 μL of the motility buffer.
8. Make a tunnel slide by sandwiching double-sided tape between 24 mm × 32 mm coverslip (bottom side) and 18 mm × 18 mm coverslip (top side) (Fig. 2) (*see* **Note 2**).
9. Add the cell suspension to the tunnel slide and leave for 5–10 min to attach the *Salmonella* cells onto the coverslip surface.
10. Wash out unbound cells by supplying 100 μL of the motility buffer. Absorb an excess amount of the buffer by a filter paper.

3.2 Staining of Flagellar Filaments with a Fluorescent Dye

To observe the localization of GFP-MotB or GFP-FliG to the flagellar base, flagellar filaments are stained with a red fluorescent dye following Subheading 3.1, **step 10**.

1. Mix 1 μL of polyclonal anti-FliC serum with 50 μL of motility buffer in a 1.5 mL tube.
2. Apply 50 μL of the anti-FliC serum mixture to *Salmonella* cells attached to the cover slip.
3. Wash twice with 100 μL of the motility buffer.

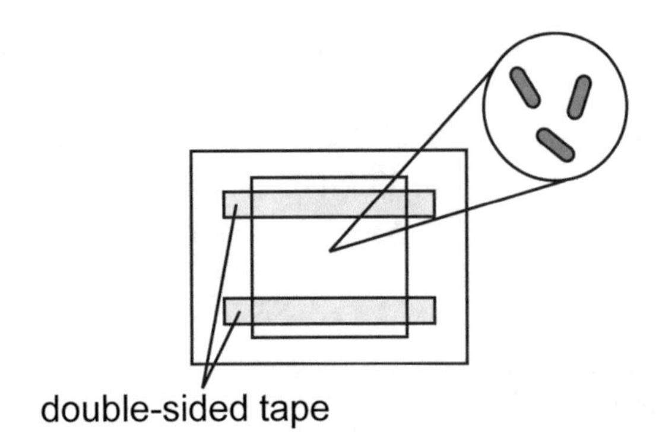

Fig. 2 Tunnel slide for observation of bacteria cells under optical microscopy. Double-sided tapes are sandwiched between a 24 mm × 32 mm coverslip (*bottom side*) and an 18 mm × 18 mm coverslip (*top side*). Cells are added to the space between these two coverslips and then attached onto the coverslip surface. Observation area is magnified in a *circle*

4. Add 1 μL of anti-rabbit IgG conjugated with Alexa Fluor 594 to 50 μL of the motility buffer in a new 1.5 mL tube.
5. Apply 50 μL of the Alexa Fluor 594 mixture to the cells attached to the cover slip.
6. Wash twice with 100 μL of the motility buffer.

3.3 Observation of Subcellular Localization by Fluorescence Microscope

1. Select a 100× objective (*see* **Note 3**).
2. Excite GFP by a 130 W mercury lump and detect fluorescence with a fluorescence mirror unit (Excitation BP 460–480; Emission BP 495–540).
3. Acquire GFP fluorescence images by an EMCCD camera under control of the camera control software (Andor Solis) (e.g., Sensitivity: 300, Exposure time: 50 ms).
4. Excite Alexa Fluor 594 by the 130 W mercury lump and detect the fluorescence with a fluorescence mirror unit (Excitation BP 565–585; Emission BP 600–690).
5. Acquire Alexa Fluor 594 fluorescence images by the EMCCD camera (e.g., Sensitivity: 0, Exposure time: 50 ms) (*see* **Note 4**).
6. Observe a fluorescent spot at the center of the rotational tethered cell (Fig. 3) (*see* **Note 5**).
7. Open the fluorescence images by an image analysis software ImageJ.

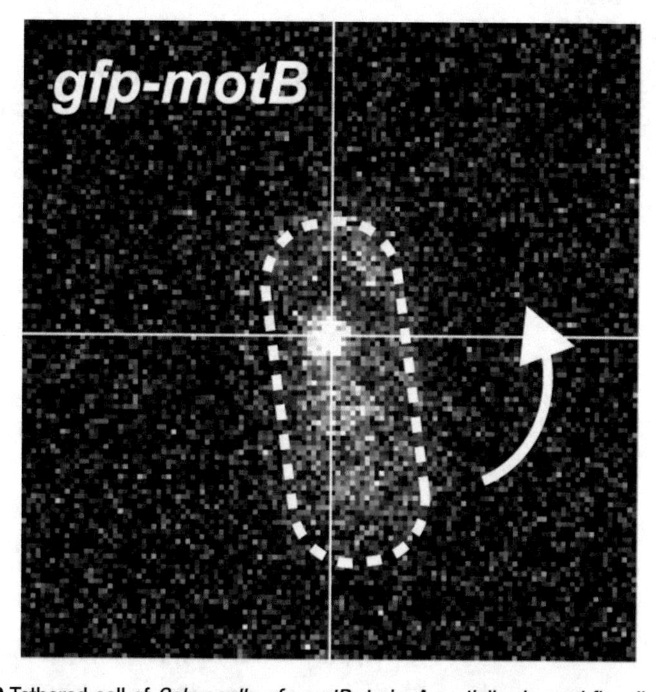

Fig. 3 Tethered cell of *Salmonella gfp-motB* strain. A partially sheared flagellar filament is attached on the coverslip surface and hence the cell body (*dashed line*) rotates. GFP-MotB is localized at the center of the rotational cell as a fluorescent spot

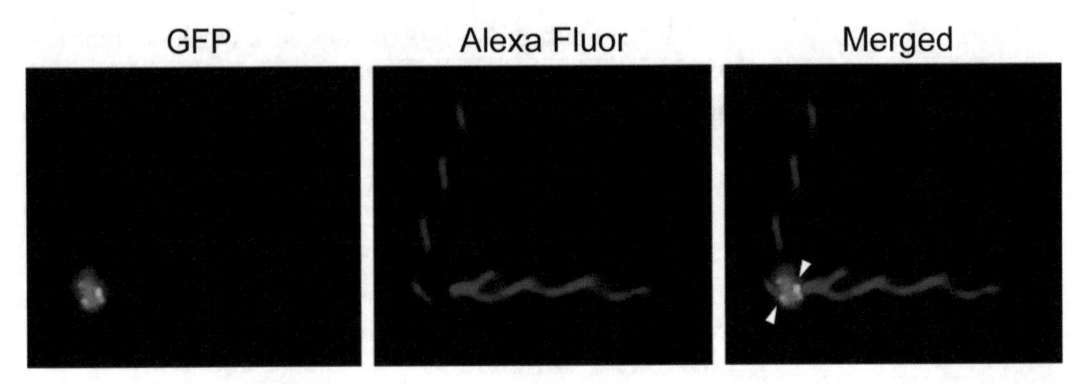

Fig. 4 Colocalization of GFP-FliG with the flagellar filament. The cells are treated with polyclonal anti-FliC antibody and Alexa Fluor 594-conjugated secondary antibody. The fluorescence images of GFP-FliG (*green*) and the filament labeled with Alexa Fluor (*red*) are merged in the right panel. *Arrowheads* indicate fluorescent spots of GFP-FliG localized to the flagellar base

8. Merge two color channels. Filament (red) and GFP spot (Green) (Fig. 4).
9. Observe the localization of GFP-MotB or GFP-FliG to the flagellar base.

3.4 Estimation of Stoichiometry by Step-Wise Photobleaching

A custom-built fluorescence microscope is set up for photobleaching experiments using a YFP-tagged flagellar component protein (e.g., YFP-FliF; YFP-MotB, FliI-YFP, FlhA-YFP) (Fig. 5) [25].

1. Split the primary beam of 514 nm gas laser into two independently attenuated paths by a polarizing beam-splitter cube to generate a separate excitation path that is used for continuous fluorescence observation, and a strong laser for photobleaching. 25% transmission ND filter is used to reduce laser power.
2. Control the on/off switching of a mechanical shutter and an EMCCD camera by an electrical stimulator and the camera control software.
3. Select a high NA objective (e.g., UApo 150XOTIRFM, NA 1.45) (*see* **Note 6**).
4. Excite YFP by a gas laser with a wavelength of 514 nm through the ND filter and detect YFP fluorescence with an emission filter of 535AF26.
5. Capture sequential fluorescence images of YFP spots by the EMCCD camera with an exposure time of 50 ms under continuous illumination until the YFP spots become disappeared (30–60 s).
6. Open the fluorescence images by the ImageJ software.
7. Apply a rectangular mask for the contribution of each YFP spot of 8 × 8 pixels to the ROI (region of interest).

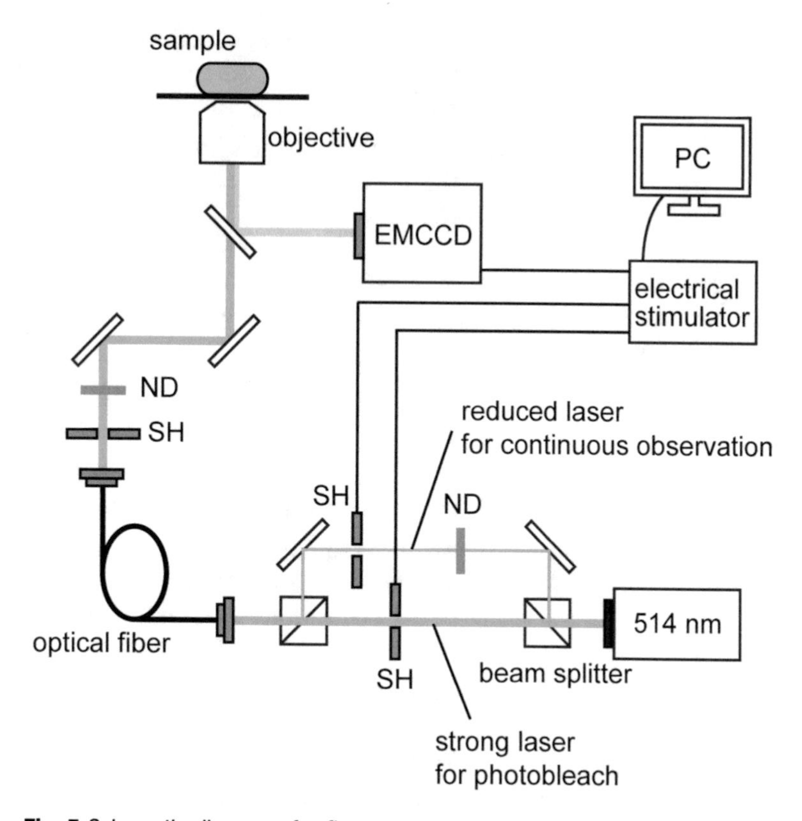

Fig. 5 Schematic diagram of a fluorescence microscope for photobleaching. An excitation laser is split into a strong laser for photobleach and a reduced laser for continuous fluorescence observation. EMCCD and mechanical shutters are controlled by an electrical stimulator. *SH* mechanical shutter, *ND* neutral density filter

8. Subtract the total background intensity from each pixel value. The instrumental background intensity is defined as the mean pixel intensity within the ROI containing no cells. The contribution to the background count due to autofluorescence of the cells and diffusing YFP molecules in the cytoplasm is calculated from the area within the same cell containing no fluorescent spots.
9. Measure the initial intensity of each fluorescent spot of the YFP-tagged flagellar component (*see* **Note 7**).
10. Detect step-wise photobleaching from the fluorescent intensity signal with continuous epi-fluorescent imaging for each fluorescent spot of YFP-tagged component. Each photobleaching intensity trace is filtered using a step finding algorithm [26, 27].
11. Calculate pairwise differences for all filtered photobleaching traces to identify the unitary step size of intensity decay [11, 28]. Fit the highest peak of the pairwise difference distribution by a

Gaussian function, and then define the peak value as a unitary step size, corresponding to the fluorescence intensity of a single YFP molecule (*see* **Note 8**).

12. Estimate the stoichiometry of the YFP-tagged component as the initial intensity of the YFP spot divided by the unitary step size.

3.5 Estimation of Turnover Rate by FRAP (Fluorescence Recovery After Photobleaching)

Fluorescence recovery of YFP after photobleaching is observed at the motor following Subheading 3.4, **step 4**.

1. Take a few sequential images of the intact YFP intensity in living *Salmonella* cells by the EMCCD camera with an exposure time of 50 ms under TIRF illumination.
2. Bleach all the YFP within the evanescent field by a strong laser excitation for 400 ms (*see* **Note 9**).
3. Acquire sequential fluorescence images of YFP within the same area by the EMCCD camera with an exposure time of 50 ms under continuous TIRF illumination for 5–10 min to observe the recovery of the YFP spots to the flagellar motor.
4. Take epi-fluorescence images of the cells expressing YFP before and after photobleaching to estimate the fraction of quenched YFP molecules after photobleaching (*see* **Note 10**).
5. Analyze the fluorescence images (*see* Subheading 3.4, **steps 6–8**).
6. Fit the increase of averaged fluorescence intensity of YFP at the flagellar base following photobleaching to an exponential curve. Exponential fit shows the recovery time constant and the apparent fraction of the fluorescence recovery. A position of the flagellar base is defined by a fluorescent spot in a pre-bleached image (*see* **Note 11**).
7. Estimate the fraction of unbleached YFP molecules after photobleaching by comparing the YFP fluorescence intensities of epi-fluorescence images of the cells expressing the YFP-tagged component before and after photobleaching (*see* **Note 10**).
8. Calculate the turnover rate from the apparent number of turnover of single YFP molecule and the fraction of unbleached YFP molecules in the whole cell.

4 Notes

1. To investigate the stoichiometry and dynamics of the flagellar motor proteins, the fluorescent protein-fused proteins had better been expressed from their original promoter on the chromosomal DNA in *Salmonella*.
2. Petrolatum (Vaseline) can be substituted for double-sided tapes as a spacer. Petrolatum can make the space narrower than double-sided tape.

3. High magnification is required to observe subcellular localization in bacteria. More than 100 nm/pixel resolution is recommended (e.g., 100× objective and 1.6× intermediate lens with iXon^{EM} + 897-BI EMCCD (pixel sixe: 16 μm × 16 μm): 100 nm/pixel).
4. To observe the colocalization of GFP-MotB with the filament stained with an Alexa Fluor 594 dye, two fluorescent images of the same area are taken separately. Although a split imaging system and two cameras are required to observe simultaneously, they are not required for observation of the flagellar motor components labeled with a fluorescent protein under a static condition.
5. Tethered cell is observed without staining filament with a fluorescent dye, because immunofluorescent labeling of the filament abolishes motility. The long helical flagellar filaments are partially sheared and then the cells are attached on the coverslip surface, allowing the cell body to be rotated. The localization of GFP-MotB and GFP-FliG to the center of rotational cell as a fluorescent spot indicates that both GFP-MotB and GFP-FliG are functional.
6. Total internal reflection fluorescence (TIRF) microscopy has two types, an objective type and a prism type. Objective type TIRF requires a high NA objective (>1.4).
7. Comparison of the initial fluorescence intensity of the YFP-tagged protein with that of FliF-YFP, of which stoichiometry is already known, is an easy method to estimate the stoichiometry. However, the standard deviation of the estimated stoichiometry with this method tends to be increased.
8. In another method, observation of a purified YFP protein as a single molecule under fluorescence microscope shows the fluorescence intensity of a single YFP molecule. However, the fluorescence of purified YFP molecules in vitro should be measured with caution, because the fluorescence intensity of YFP changes dependently of pH and chloride ion concentrations.
9. TIRF illumination excites only YFP molecules that are very close (100–200 nm) to the glass surface.
10. TIRF illumination using a strong laser bleaches only the fluorescence of YFP molecules within the evanescent field. But if YFP-tagged component protein freely diffuses over the *Salmonella* cell body within an exposure time, more YFP molecules must be photobleached by TIRF illumination. To estimate the fraction of unbleached YFP-tagged flagellar proteins, total YFP fluorescence in the cell is compared before and after photobleaching.
11. If only one or a few molecules show turnovers, counting the number of single YFP fluorescence that recovers at the flagellar base after photobleaching shows higher accuracy than a curve fitting [25].

Acknowledgments

We thank Keiichi Namba, Nobunori Kami-ike, and Masahiro Ueda for continuous support and encouragement. This research has been supported in part by JSPS KAKENHI Grant Numbers JP15K14498 and JP15H05593 (to Y.V.M.) and JP26293097 (to T.M.) and MEXT KAKENHI Grant Numbers JP15H01335 (to Y.V.M.) and JP25121718 (to T.M.).

References

1. Berg HC (2003) The rotary motor of bacterial flagella. Annu Rev Biochem 72:19–54
2. Macnab RM (2003) How bacteria assemble flagella. Annu Rev Microbiol 57:77–100
3. Minamino T, Imada K, Namba K (2008) Molecular motors of the bacterial flagella. Curr Opin Struct Biol 18:693–701
4. Sowa Y, Berry RM (2008) Bacterial flagellar motor. Q Rev Biophys 41:103–132
5. Morimoto YV, Minamino T (2014) Structure and function of the bi-directional bacterial flagellar motor. Biomolecules 4:217–234
6. Minamino T, Imada K (2015) The bacterial flagellar motor and its str diversity. Trends Microbiol 23:267–274
7. Morimoto YV, Nakamura S, Kami-ike N, Namba K, Minamino T (2010) Charged residues in the cytoplasmic loop of MotA are required for stator assembly into the bacterial flagellar motor. Mol Microbiol 78:1117–1129
8. Morimoto YV, Nakamura S, Hiraoka KD, Namba K, Minamino T (2013) Distinct roles of highly conserved charged residues at the MotA-FliG interface in bacterial flagellar motor rotation. J Bacteriol 195:474–481
9. Kojima S, Nonoyama N, Takekawa N, Fukuoka H, Homma M (2011) Mutations targeting the C-terminal domain of FliG can disrupt motor assembly in the Na^+-driven flagella of *Vibrio alginolyticus*. J Mol Biol 414:62–74
10. Takekawa N, Kojima S, Homma M (2014) Contribution of many charged residues at the stator-rotor interface of the Na^+-driven flagellar motor to torque generation in *Vibrio alginolyticus*. J Bacteriol 196:1377–1385
11. Leake MC, Chandler JH, Wadhams GH, Bai F, Berry RM, Armitage JP (2006) Stoichiometry and turnover in single, functioning membrane protein complexes. Nature 443:355–358
12. Tipping MJ, Steel BC, Delalez NJ, Berry RM, Armitage JP (2013) Quantification of flagellar motor stator dynamics through *in vivo* proton-motive force control. Mol Microbiol 87:338–347
13. Lele PP, Hosu BG, Berg HC (2013) Dynamics of mechanosensing in the bacterial flagellar motor. Proc Natl Acad Sci U S A 110:11839–11844
14. Tipping MJ, Delalez NJ, Lim R, Berry RM, Armitage JP (2013) Load-dependent assembly of the bacterial flagellar motor. mBio 24:e00551-13
15. Castillo DJ, Nakamura S, Morimoto YV, Che YS, Kamiike N, Kudo S, Minamino T, Namba K (2013) The C-terminal periplasmic domain of MotB is responsible for load-dependent control of the number of stators of the bacterial flagellar motor. Biophysics 9:173–181
16. Che YS, Nakamura S, Morimoto YV, Kami-ike N, Namba K, Minamino T (2014) Load-sensitive coupling of proton translocation and torque generation in the bacterial flagellar motor. Mol Microbiol 91:175–184
17. Fukuoka H, Inoue Y, Terasawa S, Takahashi H, Ishijima A (2010) Exchange of rotor components in functioning bacterial flagellar motor. Biochem Biophys Res Commun 394:130–135
18. Delalez NJ, Wadhams GH, Rosser G, Xue Q, Brown MT, Dobbie IM, Berry RM, Leake MC, Armitage JP (2010) Signal-dependent turnover of the bacterial flagellar switch protein FliM. Proc Natl Acad Sci U S A 107:11347–11351
19. Yuan J, Branch RW, Hosu BG, Berg HC (2012) Adaptation at the output of the chemotaxis signalling pathway. Nature 484:233–236
20. Lele PP, Branch RW, Nathan VS, Berg HC (2012) Mechanism for adaptive remodeling of the bacterial flagellar switch. Proc Natl Acad Sci U S A 109:20018–20022
21. Delalez NJ, Berry RM, Armitage JP (2014) Stoichiometry and turnover of the bacterial flagellar switch protein FliN. mBio 5:e01216-14
22. Branch RW, Sayegh MN, Shen C, Nathan VS, Berg HC (2014) Adaptive remodelling by FliN in the bacterial rotary motor. J Mol Biol 426:3314–3324

23. Yamaguchi S, Fujita H, Sugata K, Taira T, Iino T (1984) Genetic analysis of H2, the structural gene for phase-2 flagellin in *Salmonella*. J Gen Microbiol 130:255–265
24. Morimoto YV, Ito M, Hiraoka KD, Che YS, Bai F, Kami-Ike N, Namba K, Minamino T (2014) Assembly and stoichiometry of FliF and FlhA in *Salmonella* flagellar basal body. Mol Microbiol 91:1214–1226
25. Bai F, Morimoto YV, Yoshimura SD, Hara N, Kami-Ike N, Namba K, Minamino T (2014) Assembly dynamics and the roles of FliI ATPase of the bacterial flagellar export apparatus. Sci Rep 4:6528
26. Nakamura S, Kami-ike N, Yokota JP, Minamino T, Namba K (2010) Evidence for symmetry in the elementary process of bidirectional torque generation by the bacterial flagellar motor. Proc Natl Acad Sci U S A 107:17616–17620
27. Sowa Y, Rowe AD, Leake MC, Yakushi T, Homma M, Ishijima A, Berry RM (2005) Direct observation of steps in rotation of the bacterial flagellar motor. Nature 437:916–919
28. Svoboda K, Schmidt CF, Schnapp BJ, Block SM (1993) Direct observation of kinesin stepping by optical trapping interferometry. Nature 365:721–727

Chapter 17

Direct Imaging of Intracellular Signaling Molecule Responsible for the Bacterial Chemotaxis

Hajime Fukuoka

Abstract

To elucidate the mechanisms by which cells respond to extracellular stimuli, the behavior of intracellular signaling proteins in a single cell should be directly examined, while simultaneously recording the cellular response. In *Escherichia coli*, an extracellular chemotactic stimulus is thought to induce a switch in the rotational direction of the flagellar motor, elicited by the binding and dissociation of the phosphorylated form of CheY (CheY-P) to and from the motor. We recently provided direct evidence for the binding of CheY-P to a functioning flagellar motor in live cells. Here, we describe the method for simultaneously measuring the fluorescent signal of the CheY-enhanced green fluorescent protein fusion protein (CheY-EGFP) and the rotational switching of the flagellar motor. By performing fluorescence and bright-field microscopy simultaneously, the rotational switch of the flagellar motor was shown to be induced by the binding and dissociation of CheY-P, and the number of CheY-P molecules bound to the motor was estimated.

Key words Simultaneous measurements, Fluorescence imaging, *E. coli*, Chemotaxis, Signal transduction, Two-component signaling system

1 Introduction

Two-component signaling systems that sense changes in the extracellular environment are well conserved in many cell types. The two-component chemotaxis signaling system of *Escherichia coli* modulates the function of the flagella, bacterial locomotive organelles. *E. coli* cells migrate toward favorable environments in response to extracellular signals by controlling the rotational direction of the flagellar motors [1]. Chemoreceptors that sense extracellular stimuli at one of the cell poles modulate the autophosphorylation activity of a histidine protein kinase, CheA. The phosphoryl group on CheA is rapidly transferred to a response regulator, CheY [2, 3]. Phosphorylated CheY (CheY-P) functions as an activated intracellular signaling molecule; its binding to FliM and FliN subunits in the motor is believed to induce a switch in rotation from counterclockwise (CCW) to clockwise (CW) direction [4, 5]. We recently

Tohru Minamino and Keiichi Namba (eds.), *The Bacterial Flagellum: Methods and Protocols*, Methods in Molecular Biology, vol. 1593, DOI 10.1007/978-1-4939-6927-2_17, © Springer Science+Business Media LLC 2017

demonstrated the direct observation of the binding and dissociation of CheY-P to and from a functioning flagellar motor and provided evidence that the binding of CheY-P induces the change in the rotational direction of a flagellar motor [6].

To determine how *E. coli* cells respond to extracellular chemotactic stimuli, the functional interactions and dynamics of intracellular signaling proteins in a single cell should be directly measured along with the cellular response of that cell. Here, we describe a method for simultaneously analyzing the intracellular signaling molecule CheY and the rotational switching of a flagellar motor (Fig. 1). By using a tethered cell assay, a CheY-enhanced green fluorescent protein fusion protein (CheY-EGFP), and total internal reflection fluorescence (TIRF) microscopy [6], the binding and dissociation of CheY-P were visualized at a single functioning motor as the presence and loss of a fluorescent spot localized to the rotational center of the tethered cell (Fig. 2). By red light bright-field microscopy, the rotational direction of the flagellar motor was detected as the rotation of the tethered cell (Fig. 2). The binding and dissociation of CheY-EGFP and the rotational switching of the flagellar motor were measured simultaneously by capturing the fluorescent and bright-field images with electron multiplying charge-coupled device (EMCCD) and high-speed cameras, respectively. The first section (Subheading 3.1) describes the method for constructing the microscope system to simultaneously measure the fluorescence from GFP-fusion proteins and the rotation of the tethered cell. The second section (Subheading 3.2) describes the methods used for cultivating the *E. coli* cells and for preparing the samples. The third section (Subheadings 3.3–3.5) describes the method used to calculate the fluorescence intensity of the fluorescently labeled proteins localized to the rotational center of a tethered cell. The method used to estimate the number of CheY-EGFP molecules bound to the motor is also described in this section. By this microscope system and the associated algorithms, our group directly demonstrated that the rotational switching of a flagellar motor from the CCW to CW direction is induced by the binding of CheY-EGFP, and that the switch from the CW to CCW direction is induced by the dissociation of CheY-EGFP. Our group also estimated the number of CheY-P molecules bound to the motor during CW rotation [6].

2 Materials

2.1 Cell Growth Media and Sample Observation Buffer

1. Luria broth (LB): 1% bactotryptone, 0.5% yeast extract, 0.5% NaCl.
2. Tryptone broth (TB): 1% bactotryptone, 0.5% NaCl.

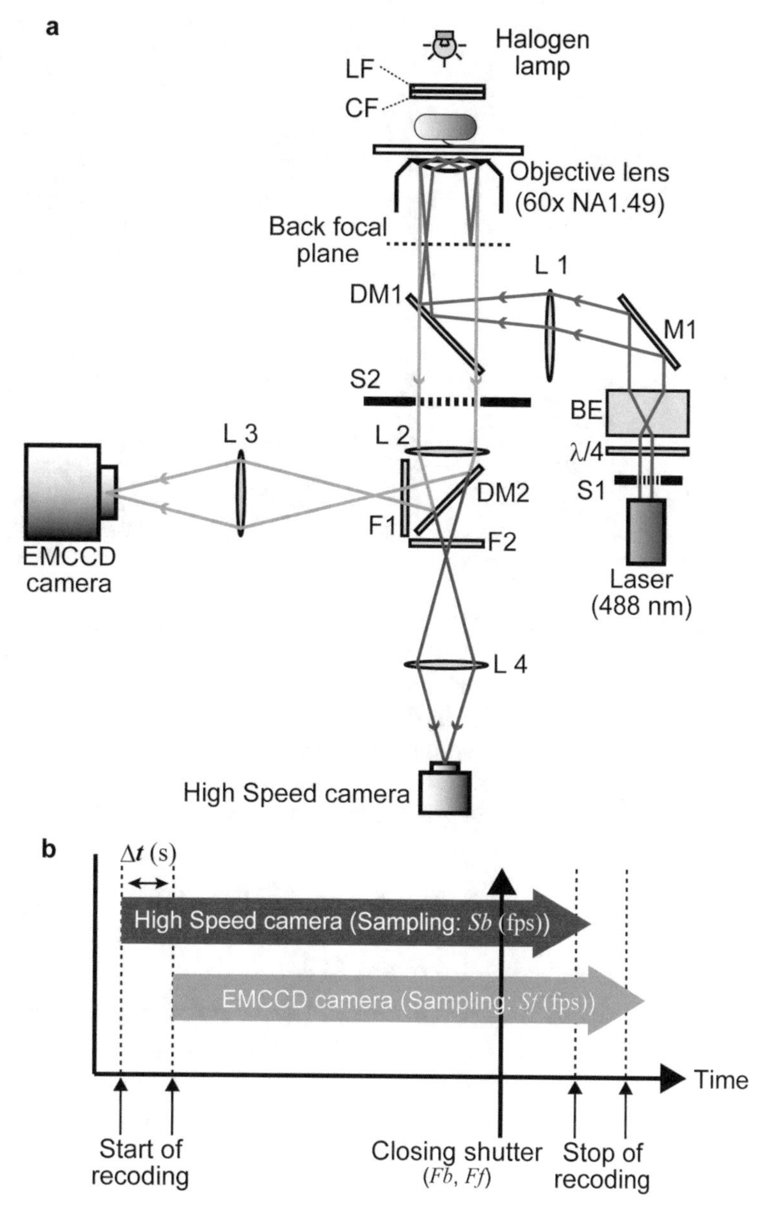

Fig. 1 Imaging system for the simultaneous observation of GFP-fusion protein localization and flagellar motor rotation. (**a**) Schematic diagram of the fluorescence and red light bright-field microscopy system. The laser (λ = 488 nm), objective lens, quartz waveplate (λ/4), mirror (M1), beam expander (BE), lenses (L1 – L4), dichroic mirrors (DM1, DM2), emission filters (F1, F2), long-pass filter (LF), cold filter (CF), and mechanical shutters (S1, S2) were positioned appropriately. The fluorescent images were magnified fivefold with L2 and L3, and focused onto the EMCCD camera. The bright-field images were magnified 1.5-fold with L2 and L4, and focused onto the high-speed CCD camera. (**b**) Schematic diagram of the recording method. Δt (s) indicates the time difference between the recording start times of the high-speed and EMCCD cameras. S_b and S_f are the sampling rates (fps) for the high-speed and EMCCD cameras, respectively. F_b and F_f are the frame numbers on which the shutter-close images were recorded by the high-speed and EMCCD cameras, respectively

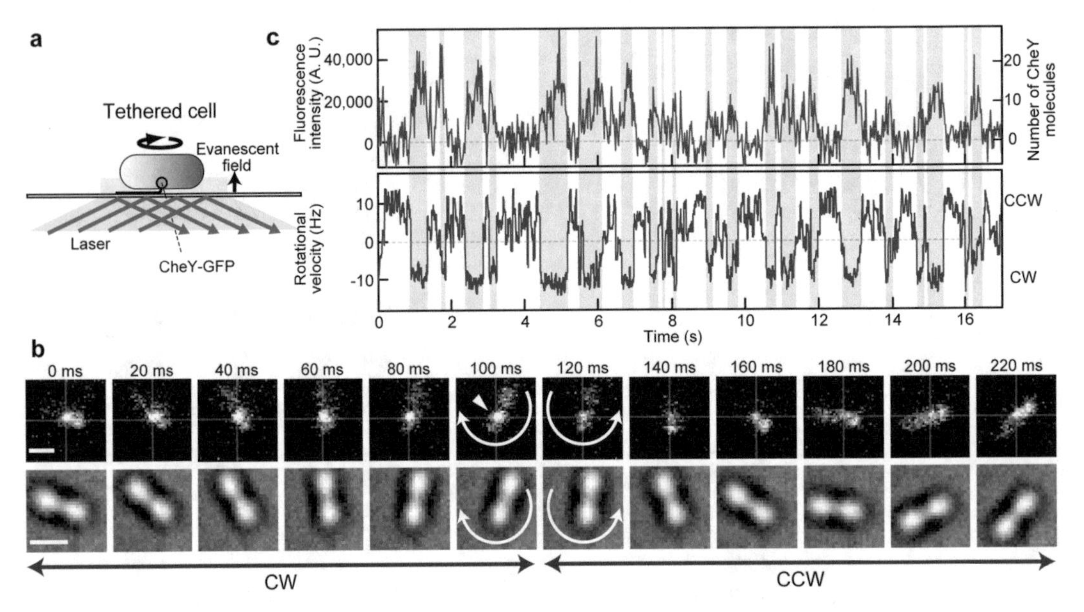

Fig. 2 Simultaneous analysis of CheY-GFP binding to a functioning motor and the rotational switching of the motor. (**a**) Schematic diagram of a tethered cell observed under TIRF microscopy. *E. coli* cells attach to the glass surface via the sticky flagellar stub, and CheY-GFP localizes to the rotational center of the tethered cell by binding to the flagellar motor. (**b**) Sequential fluorescence (*top*) and bright-field (*bottom*) images of a tethered cell expressing CheY-GFP. Arrowhead, CheY-GFP localized to a motor. Bar, 1 μm. (**c**) Fluorescence intensity at the rotational center (*top*) and rotational velocity (*bottom*) of a tethered cell. Positive and negative rotational velocity values indicate CCW and CW rotational directions, respectively. A.U., arbitrary units

3. Motility buffer (10NaMB): 10 mM potassium phosphate buffer, pH 7.0, 0.1 mM EDTA-2 K, pH 7.0, 10 mM NaCl, 75 mM KCl.
4. 10 mM isopropyl-β-d-thiogalactopyranoside (IPTG) in H_2O as a stock solution.
5. 20% arabinose in H_2O as a stock solution.
6. 25 mg/mL chloramphenicol in ethanol as a stock solution.
7. 50 mg/mL ampicillin in H_2O as a stock solution.
8. Rotary shaker.

2.2 E. coli Strains and Plasmids

1. Δ*cheY* strain (*see* **Notes 1** and **2**).
2. Δ*fliM* strain (*see* **Notes 1** and **3**).
3. Plasmid encoding the *cheY-egfp* gene (*see* **Note 4**).
4. Plasmid encoding the *fliM-egfp* gene (*see* **Note 5**).

2.3 Cell Preparation for Observation

1. Polystyrene tube.
2. Syringe.
3. Centrifuge.

4. Coverslip 18 × 18 mm.
5. Coverslip 24 × 50 mm.

2.4 Microscopy Equipment

1. Inverted microscope (IX71, Olympus, Japan).
2. Objective lens (APON 60XOTIRF NA1.49, Olympus, Japan).
3. Blue laser (λ = 488 nm, Sapphire 488-20-SV, Coherent, USA).
4. EMCCD camera (DU860D-CS0-BV, Andor Technology, UK).
5. High-speed CCD camera (ICL-B0610M-KC, Imperx, USA).
6. Laser beam expander (BE) (LBED-5, Sigma Koki, Japan).
7. Quartz waveplate (λ/4) (WPQ-4880-4 M, Sigma Koki, Japan).
8. Mirror (M1) (TFM-30C05-500, Sigma Koki, Japan).
9. Dichroic mirror (DM1) (FF495-Di02, Semrock, USA).
10. Long-pass filter (LF) (R60, Hoya, Japan).
11. Cold filter (CF) (SC751, Asahi Spectra, Japan).
12. Dichroic mirror (DM2) (FF593-Di03, Semrock, USA).
13. Emission filter (F1) (FF01-520/35, Semrock, USA).
14. Long-pass filter (F2) (BLP01-488R-25, Semrock, USA).
15. Mechanical shutter (S2) (F116, Suruga Seiki, Japan).

3 Methods

3.1 Microscope System for the Simultaneous Observation of Fluorescent and Bright-Field Images

The microscope system described here was constructed based on the IX71 inverted microscope (Olympus, Japan) and is shown in Fig. 1.

1. Use a blue laser to observe GFP-fusion proteins by objective-type TIRF microscopy [7].
2. Position a laser beam expander (BE) to magnify the blue laser beam to the appropriate diameter.
3. Position a quartz waveplate (λ/4) to convert the linearly polarized beam to a circularly polarized beam.
4. Position a mirror (M1) at the focal point of the lens (L1) and focus the blue laser beam to back focal plane of the objective lens (NA1.49).
5. Position a dichroic mirror (DM1) to reflect and direct the blue laser beam into the objective lens. (The fluorescence from the EGFP-fusion protein and the red light for the bright-field image will pass through DM1 (*see* below)).
6. Rotate M1 to adjust the incident angle of the laser beam. The incident angle should be adjusted between the critical angle (61°) and maximum angle (78.6°) to allow total internal reflection of the laser beam at the interface of the sample and glass coverslip (*see* **Note 6**).

7. Position the long-pass (LF) and cold (CF) filters in front of the halogen lamp to allow the transmission of red light for the bright-field observation and to remove heat generated by the lamp. (The LF and CF block light under 600 nm and over 750 nm, respectively).
8. Position the dichroic mirror (DM2) to separate the fluorescence from the EGFP-fusion protein and the red light. DM2 reflects the green fluorescence from the EGFP-fusion protein to the EMCCD camera and passes the red light for bright-field imaging to the high-speed camera.
9. To observe green fluorescence from the EGFP-fusion proteins, position an emission filter (F1) in front of the EMCCD camera. F1 blocks the light from the excitation laser and halogen lamp that leaks through the dichroic mirrors, and allows the green fluorescence from the EGFP-fusion proteins to pass through.
10. Position two lenses (L2 and L3) to magnify the fluorescent image fivefold and focus the fluorescent light from the EGFP-fusion proteins onto the EMCCD camera.
11. To generate a bright-field image, position a long-pass filter (F2) in front of the high-speed CCD camera. F2 blocks the excitation laser light that leaks through the dichroic mirrors.
12. Position two lenses (L2 and L4) to magnify the bright-field image 1.5-fold and focus the red light from the sample onto the high-speed CCD camera.
13. Position a mechanical shutter (S2) in front of L2 to record "shutter-close" images using both the EMCCD and high-speed cameras. The sampling time between the fluorescent and bright-field images is adjusted by calculating the delay time of the start of recoding (Δt) by the EMCCD camera to the high-speed camera from the "shutter-close" images recorded by both cameras (*see* Subheading 3.2 and Fig. 1b).
14. Adjust the laser power and sampling rates of the EMCCD and high-speed CCD cameras to allow optimum recording (*see* **Note 7**).

3.2 Cell Preparation and Analysis by Bright-Field and Fluorescence Microscopy

1. To initiate cell cultivation, collect a small aliquot of frozen cells with a sterilized toothpick and transfer them into 2 mL of LB containing 25 μg/mL chloramphenicol. Cultivate the cells overnight at 30 °C.
2. To prepare a frozen stock for measurement, add 200 μL of DMSO to 2 mL of cell culture, and freeze 50 μL aliquots in liquid nitrogen. Store the frozen aliquots at −80 °C.
3. To cultivate the CheY-EGFP-expressing cells for experiments, thaw a 50 μL frozen stock sample at room temperature, and transfer it into 5 mL of TB containing 5 μM IPTG and 25 μg/mL chloramphenicol. Cultivate the cells at 30 °C for 5.25 h with a rotary shaker at 180 rpm (*see* **Note 4**).

4. To cultivate the FliM-EGFP-expressing cells for an experiment, thaw a 50 μL frozen stock sample at room temperature, and transfer it into 5 mL of TB containing 0.002% arabinose and 50 μg/mL ampicillin. Cultivate the cells at 30 °C for 5.25 h with a rotary shaker at 180 rpm (*see* **Note 5**).
5. To remove the flagellar filaments, pass 1 mL of the cell culture through a narrow polystyrene tube connected between two syringes (*see* Subheading 2.3) at least 100 times.
6. Centrifuge the cell suspension at ~8,000 × *g* (*see* Subheading 2.3) for 1 min at room temperature and remove all of the supernatant with a pipette (*see* **Note 8**).
7. Resuspend the cell pellet in 1 mL of motility buffer (10NaMB).
8. Centrifuge the cell suspension as above and remove all of the supernatant with a pipette.
9. Resuspend the cell pellet in 350 μL of 10NaMB.
10. Load 15 μL of the cell suspension into a flow chamber constructed from 18 × 18 and 24 × 50 mm coverslips (*see* Subheading 2.3) with a spacer, and incubate for 12 min to allow the cells to attach to the coverslip (*see* **Notes 9** and **10**).
11. Remove unattached cells by gentle perfusion of the flow chamber with an additional 30 μL of 10NaMB.
12. Set the sample on the microscope, and using bright-field microscopy, select a tethered cell rotating smoothly (Fig. 2a).
13. Open the mechanical shutter (S2) in front of L2 (Fig. 1a), and start recording with the high-speed camera to obtain bright-field images of the tethered rotating cell (Figs. 1b, 2b, bottom).
14. Next, start recording with the EMCCD camera. Open the shutter (S1) to illuminate the tethered cell with the blue laser beam to obtain fluorescent images (Figs. 1b, 2b, top).
15. Close the S2, and stop recording with both cameras (Fig. 1b).
16. Obtain the time course of fluorescence intensity using the procedure described below (*see* Subheading 3.3).
17. Obtain the time course of switches in the rotational direction of the tethered cell using the procedure described in Chapter 12.
18. Adjust the timing between the fluorescent and bright-field images by estimating the time difference between the recording start times of the fluorescence and bright-field imaging (Δt (s), Fig. 1b) using the following equation:

$$\Delta t = \left(\frac{F_{\mathrm{b}}}{S_{\mathrm{b}}} - \frac{F_{\mathrm{f}}}{S_{\mathrm{f}}} \right) \quad (1)$$

where F_{b} and F_{f} are the frame numbers on which the shutter-close images were captured by the high-speed and EMCCD cameras, respectively. S_b and S_f are the sampling rates of the high-speed and EMCCD cameras, respectively.

3.3 Estimating the Fluorescence Intensity of CheY-EGFP at the Rotational Center of a Tethered Cell

1. Apply two regions of interest (ROIs, ROI 1 and ROI 2) around the fluorescent spot on the tethered cell (Fig. 3a, left) (*see* **Note 11**).
2. Estimate the fluorescence intensity of CheY-EGFP at the rotational center with the following equation:

$$F_{\text{motor}} = F_1 - C \times \frac{F_2 - F_1}{P_2 - P_1} \times P_1 \tag{2}$$

where $F1$ and $F2$ are the total fluorescence intensities of ROI 1 and ROI 2, and P_1 and P_2 are the total numbers of pixels in ROI 1 and ROI 2, respectively. C is a correction factor for the estimation of fluorescence intensity using the tethered cell method and TIRF microscopy (*see* Subheading 3.4).

3.4 Estimating the Correction Factor (C)

Because the intensity of evanescent light decays exponentially and *E. coli* cells are rod-shaped, a correction factor (C) should be estimated by the following procedure and apply it to Eq. 2.

1. Apply two ROIs (ROI 3 and ROI 4) around the rotational center of the tethered cell producing EGFP (Fig. 3a, right). The size of the ROIs should be the same as those of ROI1 and ROI2.
2. Estimate C with the following procedure. First, estimate the apparent background intensity (ABg) with the following equation:

$$ABg = \frac{F_4 - F_3}{P_4 - P_3} \times P_3 \tag{3}$$

where $F3$ and $F4$ are the total fluorescence intensities of ROI 3 and ROI 4, and $P3$ and $P4$ are the total numbers of pixels in ROI 3 and ROI 4, respectively (Fig. 3a).
3. Next, calculate an estimate for C using the following equation (Eq. 4), and apply it to Eq. 2 (*see* **Note 12**).

$$C = \frac{F_3}{ABg} \tag{4}$$

3.5 Estimating the Number of CheY-EGFP Molecules Bound to a Flagellar Motor

1. Observe a tethered cell expressing an EGFP-fusion protein of FliM (FliM-EGFP), one of the rotor components of flagellar motors [8, 9], using the same microscope system described above. FliM-EGFP should be expressed in a Δ*fliM* strain.

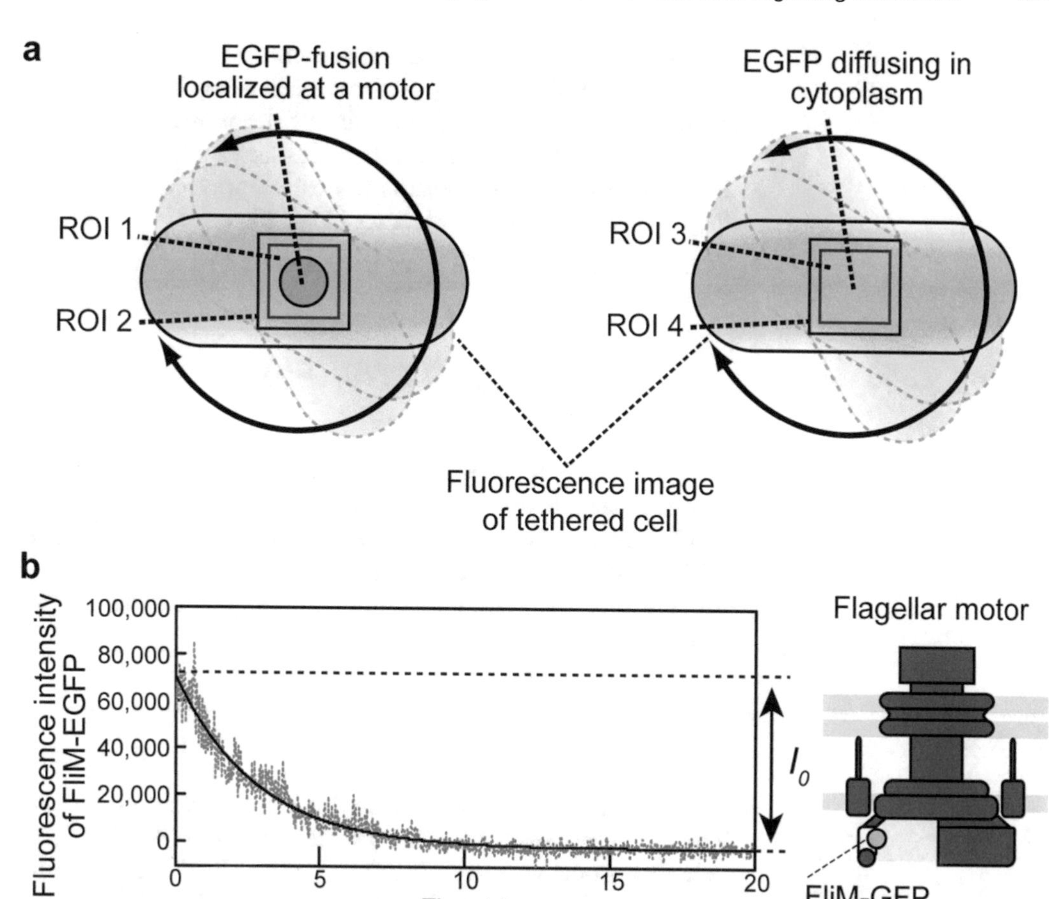

Fig. 3 Estimation of the fluorescence intensity of the GFP-fusion proteins localized to the rotational center of a tethered cell. (**a**) Scheme depicting the method used to estimate the fluorescence intensity of the GFP-fusion proteins localized to the flagellar motor. *Left*, tethered cells in which the flagellar motors were labeled with GFP-fusion proteins were observed under TIRF microscopy. Two ROIs (ROI 1 and 2) surrounding a fluorescent spot on the tethered cell were applied. *Right*, tethered cells producing EGFP were observed under TIRF microscopy. Two ROIs (ROI 3 and 4) surrounding the rotational center of the tethered cell were applied. (**b**) Estimation of the fluorescence intensity of the FliM-EGFP molecules assembled in a single flagellar motor. (*Left*) Typical photobleaching profile of a motor containing FliM-EGFP. *Gray dotted* and *black solid* lines indicate the fluorescence intensity of the FliM-EGFP-labeled motor and the fitted curve of the photobleaching profile using a single exponential function, respectively. The value of the fitted curve at time 0 (*I0*) corresponds to the intensity from all FliM-EGFP molecules (34 ± 2) incorporated into the flagellar motor. (*Right*) Illustration of a flagellar motor depicting one of the FliM-EGFP molecules

2. Estimate the time course of fluorescence intensity of FliM-EGFP at the rotational center of a cell by performing the same procedure described in Subheadings 3.3 and 3.4.
3. The time course of fluorescence intensity of FliM-EGFP was fitted by a single exponential function (Fig. 3b):

$$F_{\text{FliM}\ GFP} = I_0 \times e^{-\lambda \cdot t} \tag{5}$$

where $F_{FliM\text{-}GFP}$ is the fluorescence intensity of a motor labeled with FliM-EGFP, t is the laser exposure time, λ is a rate constant, and $I0$ is the initial intensity of the motor when $t = 0$. Since a single flagellar motor contains an assembly of 34 ± 2 FliM molecules, $I0$ represents the total fluorescence intensity from ~34 FliM-EGFP molecules [10–13].

4. The number of CheY-GFP molecules bound to a motor can be estimated by Eq. 6.

$$N_{\mathrm{CheY}\ GFP} = (34 \pm 2) \times \frac{F_{\mathrm{CheY}\ GFP}}{I_0} \tag{6}$$

where $N_{CheY\text{-}GFP}$ is the number of CheY-GFP molecules bound to a flagellar motor, $F_{CheY\text{-}GFP}$ is the fluorescence intensity of the CheY-GFP that is bound to the motor, $I0$ is the initial intensity of the motor labeled with FliM-GFP when $t = 0$, and 34 ± 2 is the number of FliM molecules in a single flagellar motor [10–13].

4 Notes

1. All *E. coli* strains are derived from the *E. coli* K12 strain, RP437, which exhibits wild-type chemotaxis function [14]. The gene deletion procedures are performed using the λ red recombinase method and tetracycline sensitivity selection [15, 16].
2. CheY-EGFP localization is analyzed in a Δ*cheY* strain in which the *fliC* gene, which encodes a flagellar filament protein, is replaced with the *fliC-sticky* gene [6]. The flagellar filament made of FliC-sticky mutant proteins is attached to glass surface spontaneously; therefore, *E. coli* cells containing these filaments are easily tethered to the glass surface [17].
3. Localization of FliM-EGFP to the flagellar motors is observed in a Δ*fliM* strain in which the *fliC* gene is replaced with the *fliC-sticky* gene [9]. This strain is used as the control strain to estimate the number of CheY-EGFP molecules associated with a single flagellar motor (*see* Subheading 3.5).
4. To express CheY-EGFP, the *egfp* gene is fused to the 3′ end of the *cheY* gene, and the fusion gene is placed under the control of an IPTG-inducible promoter on pMMB206 [6]. The plasmid-containing cells are grown in TB and induced with 5 μM IPTG to produce physiological levels of the CheY-EGFP protein. Higher expression of CheY-EGFP leads to increased diffusion of CheY-EGFP into the cytoplasm and high levels of background fluorescence that prevent visualization of the CheY-EGFP-bound flagellar motors.

5. To express FliM-EGFP, the *egfp* gene is fused to the 3′ end of the *fliM* gene, and the fusion gene is placed under the control of an arabinose-inducible promoter on pBAD24 [9]. The plasmid-containing cells are grown in TB and induced with 0.002% arabinose to produce physiological levels of the FliM-EGFP protein.
6. The incident angle of laser beam will be determined by the numerical aperture of the objective lens.
7. To generate fluorescent and bright-field images of the cells, the laser power, the sampling rate of the EMCCD camera, and the sampling rate of the high-speed camera should be optimized. The laser power and the sampling rates of the EMCCD and the high-speed cameras are set at 1.3 mW, 50 fps, and 200 fps, respectively [6].
8. The G-force and centrifugation times should be optimized to avoid damaging the cells.
9. The volumes of the cell suspension and motility buffer loaded into a flow chamber should be adjusted for the size of the chamber.
10. The incubation time for attachment of *E. coli* cells to the glass surface should be adjusted to obtain optimal numbers of attached cells for the observation. Long incubation times result in high cell densities that can interfere with the rotation of tethered cells.
11. In a paper of Fukuoka et al. [6], the small ROI (ROI 1) was 7 × 7 pixels (560 × 560 nm) and the large one (ROI 2) was 9 × 9 pixels (720 × 720 nm).
12. If the background intensity is uniformly distributed throughout the cell body, as is the case when using conventional Epi-fluorescence microscopy, *F3* and *ABg* should be the same. However, we found that *F3* is several times higher than the experimentally estimated *ABg* when observing *E. coli* cells expressing EGFP with TIRF microscopy [6], probably because of the rod shape of the *E. coli* cell and its effect on the fluorescence profile. The correction factor (*C*) can be estimated from Eq. 4 and inserted into Eq. 2. Therefore, ~1.3 was applied as correction factor (*C*) as described by Fukuoka et al. [6].

Acknowledgments

I thank A. Ishijima (Osaka University) for critical reading and useful discussions in this manuscript. This work was supported by Grants-in-Aid for Scientific Research from MEXT KAKENHI JP23115004 (to A.I.), and from JSPS KAKENHI JP26440073 (to H.F.).

References

1. Wadhams GH, Armitage JP (2004) Making sense of it all: bacterial chemotaxis. Nat Rev Mol Cell Biol 5:1024–1037
2. Stewart RC (1997) Kinetic characterization of phosphotransfer between CheA and CheY in the bacterial chemotaxis signal transduction pathway. Biochemistry 36:2030–2040
3. Sourjik V, Berg HC (2002) Binding of the *Escherichia coli* response regulator CheY to its target measured in vivo by fluorescence resonance energy transfer. Proc Natl Acad Sci U S A 99:12669–12674
4. Bren A, Eisenbach M (1998) The N terminus of the flagellar switch protein, FliM, is the binding domain for the chemotactic response regulator, CheY. J Mol Biol 278:507–514
5. Welch M, Oosawa K, Aizawa S, Eisenbach M (1993) Phosphorylation-dependent binding of a signal molecule to the flagellar switch of bacteria. Proc Natl Acad Sci U S A 90:8787–8791
6. Fukuoka H, Sagawa T, Inoue Y, Takahashi H, Ishijima A (2014) Direct imaging of intracellular signaling components that regulate bacterial chemotaxis. Sci Signal 7:ra32
7. Tokunaga M, Kitamura K, Saito K, Iwane AH, Yanagida T (1997) Single molecule imaging of fluorophores and enzymatic reactions achieved by objective-type total internal reflection fluorescence microscopy. Biochem Biophys Res Commun 235:47–53
8. Macnab R (1996) Flagella and motility. In: Neidhardt FC (ed) *Escherichia coli and Salmonella*. American Society for Microbiology, Washington, DC, pp 123–145
9. Fukuoka H, Inoue Y, Terasawa S, Takahashi H, Ishijima A (2010) Exchange of rotor components in functioning bacterial flagellar motor. Biochem Biophys Res Commun 394:130–135
10. Suzuki H, Yonekura K, Namba K (2004) Structure of the rotor of the bacterial flagellar motor revealed by electron cryomicroscopy and single-particle image analysis. J Mol Biol 337:105–113
11. Thomas DR, Francis NR, Xu C, DeRosier DJ (2006) The three-dimensional structure of the flagellar rotor from a clockwise-locked mutant of *Salmonella enterica* serovar Typhimurium. J Bacteriol 188:7039–7048
12. Delalez NJ, Wadhams GH, Rosser G, Xue Q, Brown MT, Dobbie IM, Berry RM, Leake MC, Armitage JP (2010) Signal-dependent turnover of the bacterial flagellar switch protein FliM. Proc Natl Acad Sci U S A 107:11347–11351
13. Lee SH, Shin JY, Lee A, Bustamante C (2012) Counting single photoactivatable fluorescent molecules by photoactivated localization microscopy (PALM). Proc Natl Acad Sci U S A 109:17436–17441
14. Parkinson JS, Houts SE (1982) Isolation and behavior of *Escherichia coli* deletion mutants lacking chemotaxis functions. J Bacteriol 151:106–113
15. Datsenko KA, Wanner BL (2000) One-step inactivation of chromosomal genes in *Escherichia coli* K-12 using PCR products. Proc Natl Acad Sci U S A 97:6640–6645
16. Maloy SR, Nunn WD (1981) Selection for loss of tetracycline resistance by *Escherichia coli*. J Bacteriol 145:1110–1111
17. Ryu WS, Berry RM, Berg HC (2000) Torque-generating units of the flagellar motor of *Escherichia coli* have a high duty ratio. Nature 403:444–447

Part IV

Structural Diversity of the Bacterial Flagellar Motors Derived from Different Bacterial Species

Chapter 18

In Situ Structural Analysis of the Spirochetal Flagellar Motor by Cryo-Electron Tomography

Shiwei Zhu, Zhuan Qin, Juyu Wang, Dustin R. Morado, and Jun Liu

Abstract

The bacterial flagellar motor is a large multi-component molecular machine. Structural determination of such a large complex is often challenging and requires extensive structural analysis in situ. Cryo-electron tomography (cryo-ET) has emerged as a powerful technique that enables us to visualize intact flagellar motors in cells with unprecedented details. Here, we detail the procedure beginning with sample preparation, followed by data acquisition, tomographic reconstruction, sub-tomogram analysis, and ultimately visualization of the intact spirochetal flagellar motor in *Borrelia burgdorferi*. The procedure is applicable to visualize other molecular machinery in bacteria or other organisms.

Key words Cryo-electron tomography, Periplasmic flagellum, Spirochete, Flagellar motor, Sub-tomogram averaging

1 Introduction

The bacterial flagellum is a fascinating organelle for motility [1], mechanosensing [2], colonization and pathogenicity [3]. Over 30 gene products are required for its assembly and function [4]. The flagellum is composed of three parts: the filament, the hook, and the flagellar motor [5]. The flagellar motor is a large rotary machine responsible for flagellar rotation, switching, and assembly [6]. Although structural studies have revealed stunning details of many motor components, structural determination of the entire motor complex in situ remains challenging. Recently, cryo-electron tomography (cryo-ET) has been utilized to determine the flagellar motor structures in bacteria [7–9], providing a new avenue for studying the structure and function of these molecular machines [10, 11]. Here, we describe a detailed protocol of the in situ structural analysis of the spirochetal flagellar motor in *Borrelia burgdorferi*, the causative agent of Lyme disease (Fig. 1). We particularly focus on sample preparation, tomographic reconstruction (Fig. 2), sub-tomogram analysis (Fig. 3), and 3-D visualization (Fig. 4).

Tohru Minamino and Keiichi Namba (eds.), *The Bacterial Flagellum: Methods and Protocols*, Methods in Molecular Biology, vol. 1593, DOI 10.1007/978-1-4939-6927-2_18,

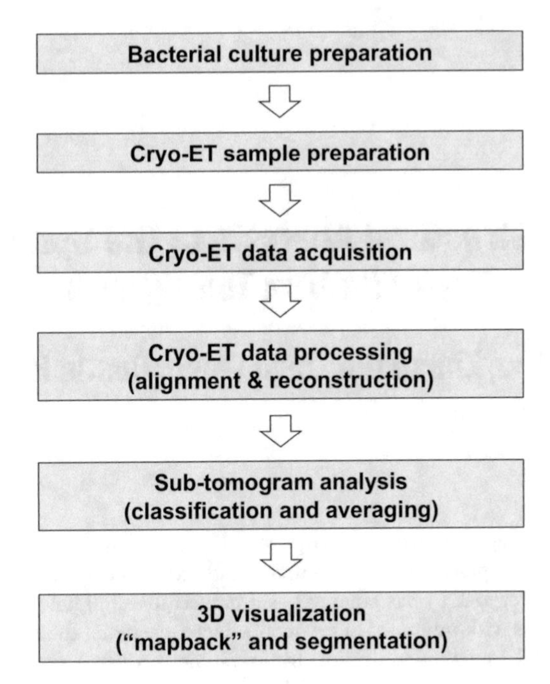

Fig. 1 Overall scheme of in situ flagellar motor structural analysis by using cryo-ET

The procedure described here should be applicable to many other large molecular complexes in bacteria or other organisms suited for cryo-ET.

2 Materials

2.1 Strains and Reagents

1. Bacterial strain: *Borrelia burgdorferi* B31 5A18NP1.
2. Culture medium: Barbour-Stoenner-Kelly (BSK-II) Liquid medium without gelatin.
3. Dilution buffer: Sterile Phosphate buffered saline (PBS) (*see* **Note 1**).
4. Fiducial markers: Aurion BSA-coated gold tracers, 10 nm (5×10^{12} particles/mL, www.emsdiasum.com).
5. Cryogen: Compressed Ethane gas (*see* **Note 2**) and liquid Nitrogen.
6. Filter paper: Whatman® qualitative filter paper, Grade 1.
7. Grids: QUANTIFOIL, R2/2, Cu 200 mesh.

2.2 Equipment and Software

1. Glow Discharger: PELCO easiGlow™ Glow Discharge Cleaning System (TED PELLA).
2. Electron microscope: FEI Polara, 300 kV, field emission gun.
3. Direct Detection Device: Gatan K2 Summit.

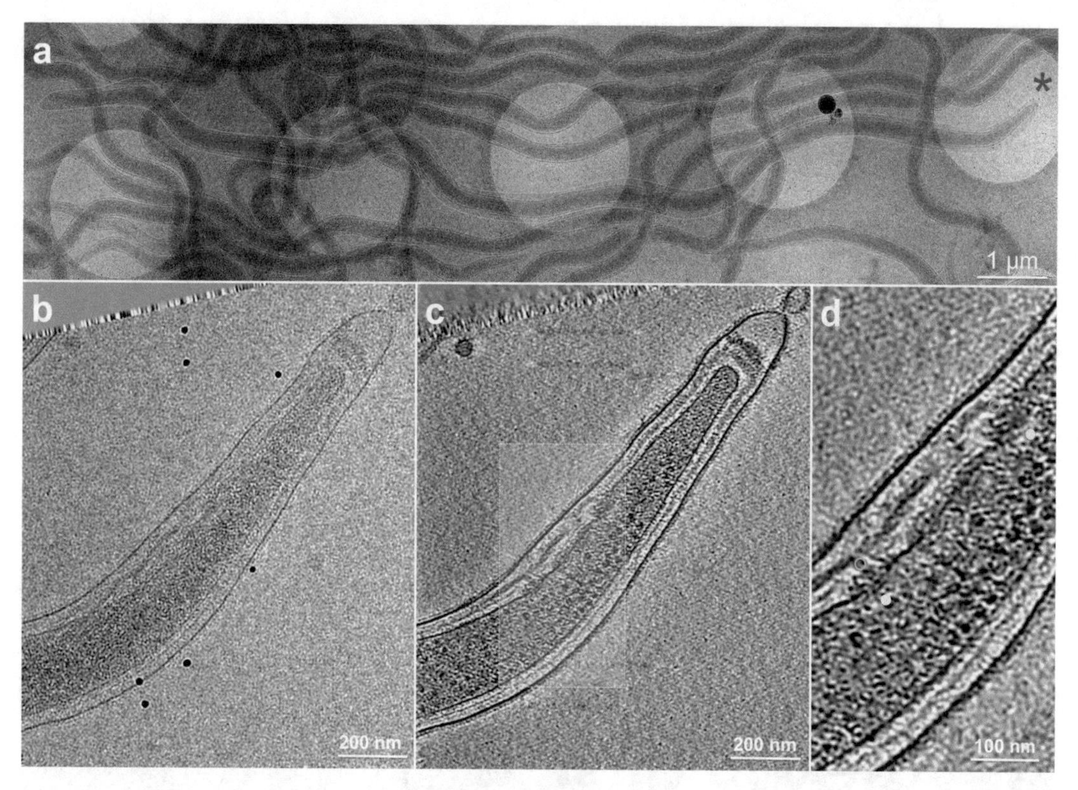

Fig. 2 Cryo-ET reveals a whole *Borrelia burgdoferi* cell and multiple flagella at the cell tip. (**a**) A snapshot of *B. burgdorferi* cells in one region of a low-magnification montage. The montage picture was obtained at 2300× magnification with a dose of 0.042 e/Å^2, and with approximately −68 μm defocus. An example target cell is depicted with a *dashed white outline* and *red fill*. The *red star* symbol indicates the region selected to collect a tilt series. (**b**) One image from the tilt series after data acquisition, drift correction, and alignment. (**c**) A central slice of the tomographic reconstruction shows several flagellar motors. We emphasize the region containing motors with a *gray rectangle*. (**d**) Enlarged view from the selected area in panel (**c**) showing two motors. For each motor, two points (*green and yellow*) have been manually selected to provide the location and orientation required for sub-tomogram analysis

4. Tweezers: EMS 26, SA (*see* **Note 3**).
5. Pipettes.
6. Automated EM data acquisition: SerialEM [12] (http://bio3d.colorado.edu/SerialEM).
7. Beam-induced motion correction: MOTIONCORR [13] (http://cryoem.ucsf.edu/software/driftcorr.html).
8. Tomographic data analysis: IMOD [14] (http://bio3d.colorado.edu/imod), tomoauto [15] (https://github.com/DustinMorado/tomoauto), tomo3d [16] (https://sites.google.com/site/3demimageprocessing/tomo3d).
9. Sub-tomogram analysis: Protomo/I3 [17, 18] (http://www.electrontomography.org).
10. Three-dimensional visualization: IMOD [19], UCSF Chimera [20] (http://www.cgl.ucsf.edu/chimera).

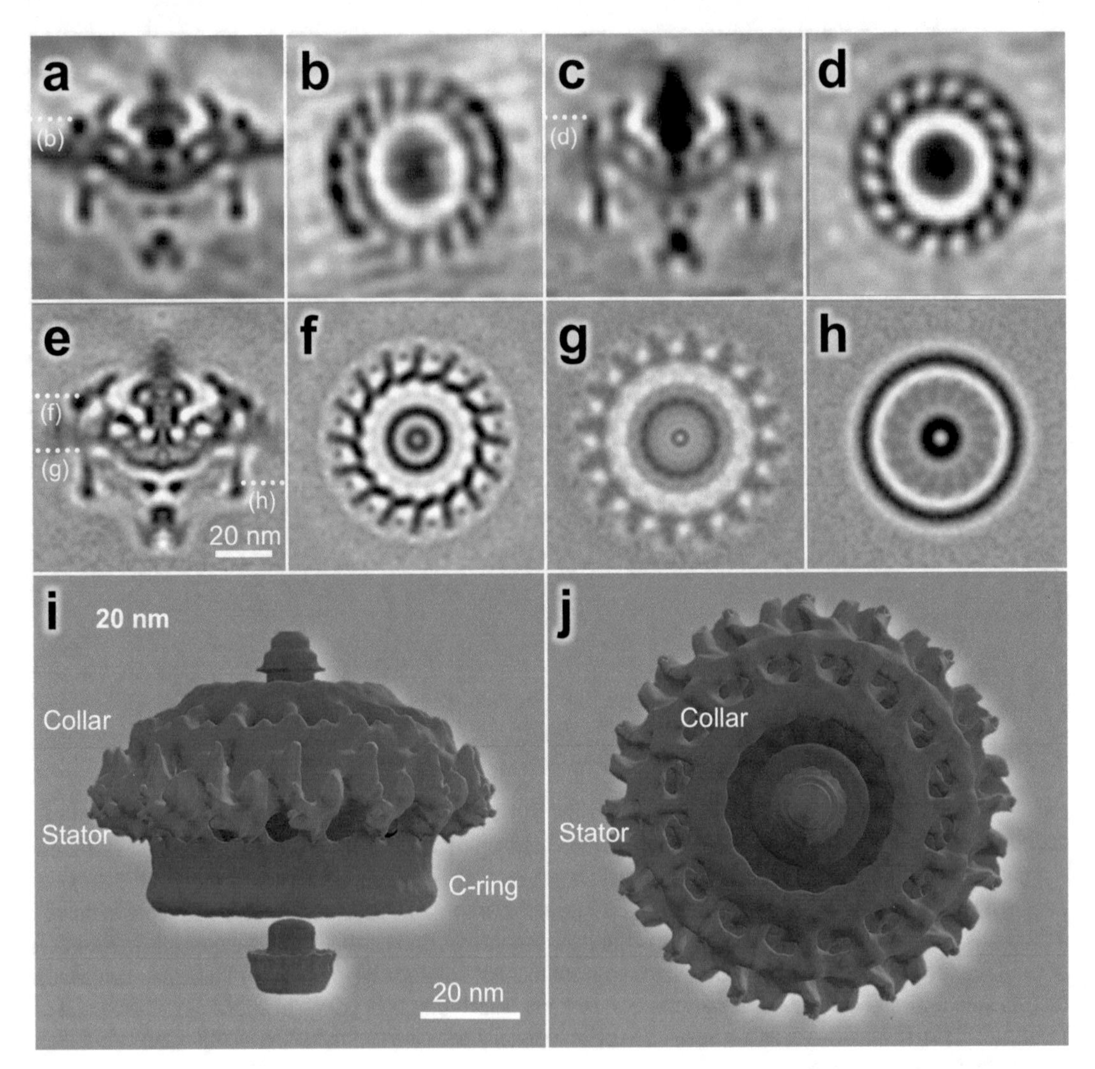

Fig. 3 Sub-tomogram analysis of in situ flagellar motor structures. (**a–d**) Snapshots of two representative class averages from SIRT reconstructions. *Side view* (**a**) and *top view* (**b**) of one class average. *Side view* (**c**) and *top view* (**d**) of another class average. Note the 16-fold symmetric features of the motor visible in panel (**d**). (**e–h**) Snapshots of the global average from WBP reconstructions with 16-fold symmetry imposed to improve the signal-to-noise ratio. The central section is shown in panel (**e**). (**f–h**) Horizontal cross sections through the average, as indicated in panel (**e**). Surface views of the global average of the motor are shown in a *side view* (**i**) and a *top view* (**j**)

3 Methods

3.1 Sample Preparation

Culture *Borrelia burgdorferi* strain B31A cells in Barbour-Stoenner-Kelly (BSK) medium and incubate at 34 °C in the presence of 3.4% carbon dioxide for 7–10 days [9].

Centrifuge the *B. burgdorferi* culture at 5000 × *g* for 5 min and suspend the resulting pellet in 20 μL PBS (*see* **Note 1**). Mix the culture with 2 μL colloidal gold solution (10 nm in diameter).

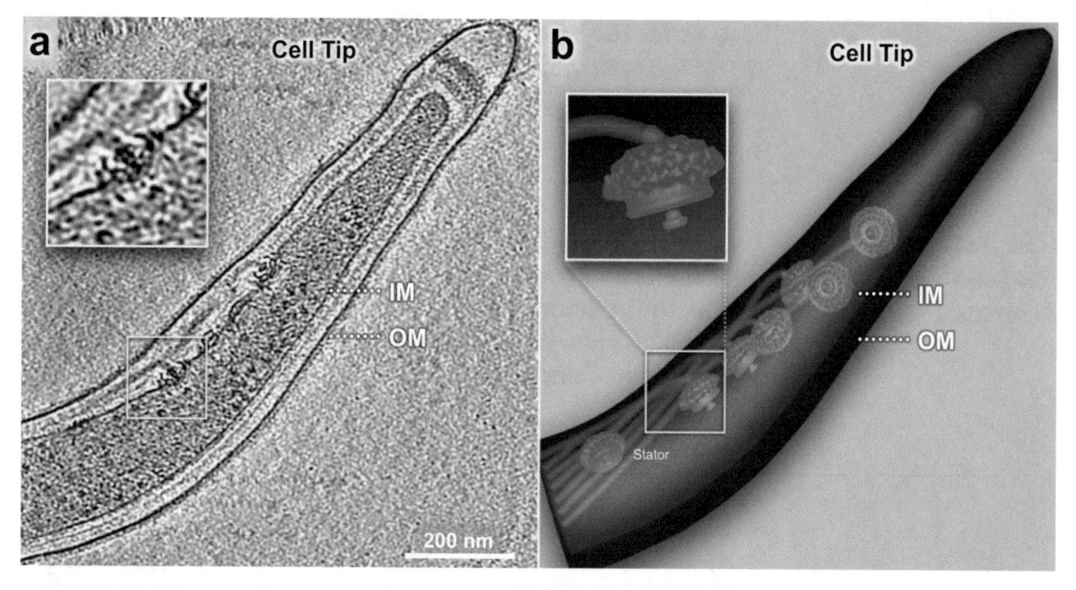

Fig. 4 Mapping the averaged motor structures into their original cellular context. (**a**) Tomogram of the original cell tip of *B. burgdoferi* where the original motors have been replaced with the average motor structure. The inset picture shows an enlarged view of a single motor. (**b**) Three-dimensional surface view of the cell tip as manually segmented using IMOD. The inset picture shows an enlarged view of the model with a motor embedded in the inner membrane (IM) and surrounded by the outer membrane (OM)

1. Use tweezers to gently pick up four grids, being careful to not bend or damage them. Place the grids on a glass microscope slide with the carbon surface, which is the duller side of the grid, facing upward.
2. To make the grids hydrophilic, place the slide with the grids into the glow discharger and treat the grids for 20 s by pressing the "AUTO" function.

3.2 Cryo-ET Sample Preparation

1. Add clean liquid nitrogen into the plastic foam chamber of a plunge-freezing apparatus. Do not add ethane until the rapid boiling of liquid nitrogen subsides completely.
2. Flow ethane gas slowly into the precooled cup causing the gas to liquefy (*see* **Note 2**).
3. Use the tweezers to pick up the glow discharged grid and clamp the tweezers using an O-ring (*see* **Note 3**).
4. Transfer 5 μL of the bacterial sample onto the carbon surface of the glow discharged grid. Wait for 1 min.
5. Install the tweezers clamped with grid into the gravity-driven plunger apparatus.
6. Use a small piece of filter paper to blot the grid for about 3 s to form a thin layer of bacterial sample on the grid. Plunge the grid immediately after the blotting into the liquid ethane in the central cup (*see* **Note 4**).

7. Transfer the grid quickly from the liquid ethane into a grid storage box, which has been precooled in liquid nitrogen (*see* **Note 5**).
8. Store the grid box at liquid nitrogen temperature until ready to transfer it to the microscope (*see* **Note 6**).

3.3 Transfer Cryo-ET Sample into Electron Microscope

1. Add clean liquid nitrogen into the cartridge-loading chamber of the cryo-transfer holder workstation (*see* **Note** 7) and add liquid nitrogen to the multi-specimen holder dewar. Wait for about 20 min for both to cool down before continuing to transfer the sample.
2. Unscrew the lock ring of a sample cartridge using a precooled (*see* **Note 8**) lock ring tool, both of which are specifically designed for the FEI Polara microscope.
3. Use precooled (*see* **Note 9**) tweezers to first move the lock ring to the slot above the cartridge, then transfer the grid into the cartridge, and finally replace the lock ring onto the cartridge.
4. Gently screw the lock ring to secure the grid in the cartridge (*see* **Note 10**).
5. Transfer the cartridge into the multi-specimen holder and subsequently into the microscope (*see* ref. 21).

3.4 Configure Low Dose Mode in Electron Microscope

1. Turn on low dose mode in SerialEM [12] to minimize beam exposure on the area you want to record by defining separate areas with different settings of magnification, spot size, beam intensity, and location for tracking and focusing during the tilt series.
2. Configure the "Record" area with a magnification of 9400×, and spot size and beam intensity such that the beam completely covers the camera and has an exposure rate of about 0.5 e/Å^2/s (*see* **Note 11**).
3. Configure the "View" area, which is centered on the same location as the record area, with a magnification of 2300×, and spot size and beam intensity such that the beam covers the camera and has an exposure rate of about 0.02 e/Å^2/s.
4. Configure the "Focus" area with the same magnification, spot size, and intensity as the record area but displaced 3 μm away along the tilt axis. Set the "Trial" area to be identical to the focus area.
5. Set a defocus offset of −60 μm for the view area in the low dose control panel to increase the contrast of low-magnification images.
6. Correct any image shift between record and view areas by setting the image shift offset for view in the low dose control panel necessary to bring a view image into alignment with a record image (*see* **Note 12**).

3.5 Cryo-ET Data Acquisition

1. Use the "View" mode to screen the grid and identify areas with bacteria embedded in thin vitreous ice (*see* **Note 13**).
2. Set the eucentric height of an area by tilting the holder to 30° and then adjusting the stage Z position using the stage controller to minimize image movement during tilting.
3. Open a new navigator window using the "Navigator" menu in SerialEM and in the navigator window click "Add Stage Position."
4. Set up a new montage using the "File" menu in SerialEM and configure the montage as 8 × 8 pieces in *X* and *Y* using the "View" parameters to collect montage images and moving the stage as opposed to applying image shifts.
5. Toggle the "Acquire" checkbox in the navigator window for the stage position and then run "Acquire at Points" in the "Navigator" menu in SerialEM. Select "Rough eucentricity" as the only prior action and "Acquire map image" as the primary task to generate a low-magnification montage.
6. Click "Load map" in the navigator window to show the low-magnification montage (Fig. 2a).
7. Click "Add Points" in the navigator window and select the areas to be collected in the montage. The flagellar motors are located at the cell tips in *B. burgdorferi*, so select cell tips that lay within grid holes (Fig. 2a). Unclick "Add Points" when finished.
8. Toggle the "Tilt series" checkbox in the navigator window for the first selected point. In the Tilt Series Setup Dialog set the tilt angle range as −51° to +51° with a fixed increment of 3° to be run in two directions starting at 0° with a target defocus of −8 μm (*see* **Note 14**). Toggle the "Tilt series" checkbox for the remaining selected points.
9. Set the parameters for the recorded images under the "Camera" menu in SerialEM. Set the Record parameter to collect images using dose-fractionation mode and to save the subframes. Set the folder for the subframes to be the same as the tilt series and set the exposure time such that the total cumulative dose of the tilt series is about 50 $e^-/Å^2$.
10. Run "Acquire at Points" in the "Navigator" menu in SerialEM and select "Rough Eucentric," "Realign to Item," "Autofocus" as prior actions and select "Acquire Tilt Series" as the primary task.

3.6 Tomographic Data Processing

3.6.1 Drift Correction

1. During data acquisition, the raw tilt image subframes are saved as time stamped files with the suffix ".mrc," and the original tilt series are saved as a file with the suffix ".st." SerialEM tilt series log file is also saved with the same name as the tilt series but with the suffix ".log" (*see* **Note 15**).

2. Generate beam-induced motion corrected sums of the subframes using the *dosefgpu_driftcorr* command from MOTIONCORR [13]. Assemble the corrected sums into a corrected tilt series using the *dose_fractioned_to_stack* command from tomoauto [15].
3. Create a copy of the corrected tilt series, binned by a factor of 4 using the *newstack* command from IMOD [14]. Use the *3dmod* command also from IMOD [18], to visually check the tilt series.
4. Trim images that are obscured by grid bars, or are too thick to show any meaningful subject contrast from the tilt series using *newstack* (*see* **Note 16**).

3.6.2 Tilt Series Alignment

1. Align tilt series deemed suitable for further processing using either the *tomoauto* command from tomoauto or the *batchruntomo* command from IMOD with automated processing set to stop after the generation of the aligned tilt series with the suffix ".ali" (*see* **Note 17**).
2. Visually assess the quality of the alignment using *3dmod*, the quality of the alignment can also be measured by the mean residual error that is output in the log file from the automated alignment procedure (*see* **Note 18**).
3. If the tilt series cannot be aligned automatically, use the *etomo* command from IMOD to do the alignment step by step (*see* ref. 22).

3.6.3 Tomographic Reconstruction

1. Create two copies of the aligned tilt series, one binned by a factor of 2, the other by a factor of 4 using *newstack*.
2. Use the command *tomo3d* [16] to generate a Simultaneous Iterative Reconstruction Technique (SIRT) reconstruction from the binned by four aligned tilt series (*see* **Note 19**).
3. Generate a Weighted Back Projection (WBP) reconstruction from the binned by two aligned tilt series using *tomo3d*.
4. The output tomograms are originally oriented with slices perpendicular to the untilted tilt image, but sub-tomogram picking is preferably done on slices parallel to the untilted tilt image. To do this rotate the tomogram by –90° along the *X*-axis using the command *clip* from IMOD.

3.7 Sub-tomogram Analysis

3.7.1 Particle Picking in Tomogram

1. Select particles manually using the command *i3display* from i3 [17, 18]. To select a single-motor particle, left click the mouse while holding the Shift key at two points to establish the particle position and initial orientation within the tomogram. Specifically, select the first point, shown in yellow in Fig. 2d, at the base of the basal body and the second point at the hook, shown in green in Fig. 2d. Repeat until all motors in the tomo-

gram are selected. The motor positions are stored as *x*, *y*, and *z* coordinates relative to the tomogram in text files with the suffix ".pos" (*see* **Note 20**).

2. Repeat **step 1** until all tomograms have been processed. In total, 758 motor particles were identified from 102 tomograms.
3. Calculate the approximate center and orientation of each flagellar motor based on the *x*, *y*, and *z* coordinates of the two locations in the ".pos" file (*see* **Note 21**). Store the information regarding the motor location and orientation in the tomogram in a new file with the suffix ".trf."

3.7.2 Sub-tomogram Alignment and Averaging

1. Analyze the sub-tomograms using the tomographic package i3, which is based on the "alignment by classification" method with missing wedge compensation.
2. To speed up the processing, first align the sub-tomograms alignment using the binned by four tomograms reconstructed by the SIRT method. As the pixel size of the binned tomogram is 1.8 nm and the flagellar motor is about 80 nm in size, use a sub-tomogram size of 64 × 64 × 64 pixels for initial alignment.
3. Sub-tomogram averaging within i3 is processed in relation to a parameter shell script named "mraparam.sh" and from the transforms calculated above. The alignment proceeds iteratively with each iteration consisting of three parts in which references and classification masks are generated, sub-tomograms are aligned and classified, and finally class averages are aligned to each other (*see* **Note 22**).
4. In the first iteration generate a global average by averaging all the sub-tomograms with their predefined orientation and then align all sub-tomograms to a template (*see* **Note 23**). Divide the sub-tomograms into multiple classes using Multivariable Statistical Analysis and Hierarchical Ascendant Classification within i3 (*see* **Note 24**).
5. Align the class averages to each other, which consequently updates the geometry parameters of each sub-tomogram. Run six iterations of translational and rotational alignment to refine the geometry parameters of the sub-tomograms.
6. Test the possible symmetries calculated by classification by applying the proposed symmetry using the command *i3sym* from i3 to the global average so far and use that as a reference for several iterations of alignment (*see* **Note 25**). The quality of this alignment should suggest a best-determined symmetry (*see* **Note 26**).
7. Replace the sub-tomograms extracted from the SIRT tomograms with sub-tomograms extracted from the WBP tomograms. Remember to extract the doubled size of

128 × 128 × 128 pixels and also double the *x*, *y*, and *z* coordinates in the ".trf" files. Generate the global average and apply the determined symmetry to create an updated reference.

8. Refine the geometry parameters of sub-tomograms by translational, polar, and spin alignment through a few more iterations till convergence, when the mean value of the cross correlation coefficient of all sub-tomograms compared to the reference no longer improve, and the global average no longer improves (*see* **Note 27**).
9. Calculate the Fourier Shell Correlation between global averages generated even and odd half-sets of all sub-tomograms to estimate the resolution.
10. Symmetrize the final global average as shown in Fig. 3e. Take snapshots of central slices from the side and top views of the motor using *3dmod* that exhibit several interesting features of the flagellar motor like the collar structure (Fig. 3f), stator ring (Fig. 3g), and C-ring (Fig. 3h).
11. Use UCSF Chimera [20] to visualize a three-dimensional (3-D) surface-rendered map of the final structure (Fig. 3i, j).

3.8 Mapping the Averaged Motor Structures into Their Cellular Context

1. Use the refined geometry parameters of all the motors in each tomogram or cell to map the average structure back into the original *B. burgdoferi* tomogram creating a new tomogram with the original motors replaced with the final average as shown in Fig. 4a.
2. Create a new model on the new tomogram using 3dmod to manually segment the inner membrane, flagellar filaments, and outer membrane.
3. Use an open contour object to model the inner membrane, as shown in green in Fig. 4b. Draw points along the center of the cross-sectional area of the inner membrane along the major axis of the cell, adjusting the size of each point to roughly fit the cross-sectional area. Once finish drawing the contour, mesh the object by using the meshing panel in the object edit dialog from the "Edit" menu in the model view window.
4. Model the outer membrane, shown in blue in Fig. 4b, as a new object in the model using the same method as for the inner membrane.
5. Segment the flagellar filaments, shown in purple in Fig. 4b, all as one object with a separate contour for each filament. Draw fixed size points along each flagellar filament. After drawing contours for all flagella, mesh the object as for the membranes and adjust the "Diam" option in the meshing panel, to match the filament diameter.

6. Add the motor structures, also shown in purple in Fig. 4b, to the model as a single-isosurface object setting the threshold in the isosurface view dialog such that just the motors are rendered.
7. Display all objects together in the model view window to visualize a representation of the motors and their distribution in native cellular environment (Fig. 4b).

4 Notes

1. PBS buffer is commonly used to dilute the bacterial culture. However, some bacteria such as *Vibrio alginolyticus* are unstable in PBS [23].
2. Caution needs to be taken when working with ethane because liquid ethane can burn skin on contact.
3. Make sure to avoid capillary effects of the tweezers. You can purchase special anti-capillary tweezers, but we have found that simply inserting the tips of open tweezers into a wax candle also works.
4. We find that we get the best results, in terms of ice thickness, by carefully looking at the sample drop absorbing to the filter paper while blotting and looking for when the meniscus between the grid and the filter paper begins to separate, which creates a visible lightening of the absorbing drop. Once the meniscus is completely separated, we immediately plunge the sample. Plunging too early will make ice that is too thick, while too late will cause the remaining sample dehydrated.
5. The exposure of the grid to water vapor has to be avoided to minimize ice contamination on the sample. Be careful to not breathe directly into liquid nitrogen containers, or lift the tweezers holding the frozen grid too high out of the container.
6. Grids boxes can be stored in smaller portable dewars for a few hours, or kept long term in large storage dewars.
7. Again use clean liquid nitrogen to avoid contamination. Fill a dry dewar with liquid nitrogen and use as soon as possible. Avoid breathing directly into the workstation chamber.
8. Always precool tools used for grid transfer by first making sure the tool is completely dry, using a blow-dryer if necessary to quickly dry a tool that has been recently used, and placing the tool in the chamber and waiting for the bubbling of liquid nitrogen to subside.

9. Precooled tweezers are very cold and caution should be taken to avoid burning the skin. Gloves can be worn or plastic coated tweezers can be purchased.
10. The lock ring is very fragile and it is very easy to damage the threads. Do not force the tool. We also find that lock rings "match" to their cartridge, so care should be taken to not mix lock rings to other cartridges.
11. We calculate the approximate dose rate by using the mean value of a 1 s exposure of an empty area of the grid and divide the value by the square of the object pixel size.
12. We do this using a noticeable feature found in a view image.
13. Select areas in which the vitreous ice layer is thin, such that the grid holes are easily distinguished, and in which the bacterial cell concentration is reasonable.
14. There are many more parameters in tilt series setup, but most of the defaults are reasonable and the most important variables are the ones mentioned.
15. Note that SerialEM and IMOD use descriptive suffixes to differentiate the data throughout processing, ".mrc," ".st," ".ali," ".rec," etc., but all data are stored as MRC format files.
16. We usually limit the trimming to three sections at each end of the tilt range and if more sections should be trimmed the tilt series is moved to a separate directory and discarded from further processing.
17. The basic principal for alignment is that the tilt series is first coarsely aligned by cross-correlation and then a subset of the gold trackers are used as fiducial markers and modeled in each tilt image to produce a 3-D reference against which the tilt series is brought into alignment.
18. Visual inspection of the aligned series should always be performed, as there can be cases where a low residual error is reported based on a very poorly model of the gold markers. The residual error is better used as a quantitative measure of comparing the more precise quality of a good alignment.
19. We find that SIRT reconstructions have higher contrast that makes them better for particle picking and early iterations of alignment, but WBP reconstructions are better for obtaining higher resolutions. To enhance contrast further use the *mtffilter* command from IMOD to apply low-pass filter.
20. Remember to keep the picking sequence consistent for all particles, and that exactly two points must exist per particle to define a position and initial orientation. Incorrectly selected points can be deleted by left clicking the mouse while holding Control.

21. The code used to generate this transform file varies greatly based on the project and how the particle picking is done, but the algorithm is not difficult to implement.
22. The sub-tomogram processing parameters for each iteration are stored in a bash shell script named "*mraparam.sh*" as environmental variables. The initial iteration is executed by four commends: *i3mraintial.sh*, *i3mramsacls.sh*, *i3cp.sh*, and *i3mraselect.sh*. For the second iteration and thereafter, the following commands are used: *i3mranext.sh*, *i3mramsacls.sh*, *i3cp.sh*, and *i3mraselect.sh*.
23. The choice of a template is important to provide a reliable start to alignment, while preventing over fitting and over interpreting the data. Initial projects can use the first global average as the template, while projects where an existing similar structure exists, a conservatively low-pass filtered version can be used as well.
24. The number of class averages generally depends on the total numbers of sub-tomograms. Four classes are reasonable for early alignment.
25. In the *Borrelia* cells, we chose to search for the symmetry of the collar structure due to its stationary position compared to the high-revolution speed of other motor components. Masks are generated manually using the *i3mask* command from the i3 package.
26. One feature to use in measuring the success of an alignment on a symmetrized reference is the resolvability of individual subunits in a component. A structure with blurry or circularized implies that the alignment is not successful.
27. We excluded sub-tomograms that have a cross-correlation coefficient value more than 2 standard deviations below the mean value. Thus, 740 subvolumes were used to generate the final averaged structure of the flagellar motor.

Acknowledgment

This work was supported by grants from National Institute of Allergy and Infectious Diseases (NIAID) (R01AI087946, R21AI113014), and Welch Foundation (AU-1714).

References

1. Terashima H, Kojima S, Homma M (2008) Flagellar motility in bacteria structure and function of flagellar motor. Int Rev Cell Mol Biol 270:39–85
2. Lele PP, Hosu BG, Berg HC (2013) Dynamics of mechanosensing in the bacterial flagellar motor. Proc Natl Acad Sci U S A 110:11839–11844

3. Zhu S, Kojima S, Homma M (2013) Structure, gene regulation and environmental response of flagella in *Vibrio*. Front Microbiol 4:410
4. Macnab RM (2003) How bacteria assemble flagella. Annu Rev Microbiol 57:77–100
5. Minamino T, Imada K (2015) The bacterial flagellar motor and its structural diversity. Trends Microbiol 23:267–274
6. Zhou J, Lloyd SA, Blair DF (1998) Electrostatic interactions between rotor and stator in the bacterial flagellar motor. Proc Natl Acad Sci U S A 95:6436–6441
7. Murphy GE, Leadbetter JR, Jensen GJ (2006) In situ structure of the complete *Treponema primitia* flagellar motor. Nature 442:1062–1064
8. Chen S, Beeby M, Murphy GE et al (2011) Structural diversity of bacterial flagellar motors. EMBO J 30:2972–2981
9. Zhao X, Zhang K, Boquoi T et al (2013) Cryoelectron tomography reveals the sequential assembly of bacterial flagella in *Borrelia burgdorferi*. Proc Natl Acad Sci U S A 110:14390–14395
10. Zhao X, Norris SJ, Liu J (2014) Molecular architecture of the bacterial flagellar motor in cells. Biochemistry 53:4323–4333
11. Beeby M, Ribardo DA, Brennan CA et al (2016) Diverse high-torque bacterial flagellar motors assemble wider stator rings using a conserved protein scaffold. Proc Natl Acad Sci U S A 113:E1917–E1926
12. Mastronarde DN (2005) Automated electron microscope tomography using robust prediction of specimen movements. J Struct Biol 152:36–51
13. Li X, Mooney P, Zheng S et al (2013) Electron counting and beam-induced motion correction enable near-atomic-resolution single-particle cryo-EM. Nat Methods 10:584–590
14. Mastronarde DN (2007) Fiducial marker and hybrid alignment methods for single- and double-axis tomography. In: Frank J (ed) Electron tomography, 2nd edn. Springer, New York
15. Morado DR, Hu B, Liu J (2016) Using tomoauto: a protocol for high-throughput automated cryo-electron tomography. J Vis Exp 107:e53608
16. Agulleiro J-I, Fernandez J-J (2015) Tomo3D 2.0-exploitation of advanced vector extensions (AVX) for 3D reconstruction. J Struct Biol 189:147–152
17. Winkler H (2007) 3D reconstruction and processing of volumetric data in cryo-electron tomography. J Struct Biol 157:126–137
18. Winkler H, Zhu P, Liu J, Ye F, Roux KH, Taylor KA (2009) Tomographic subvolume alignment and subvolume classification applied to myosin V and SIV envelope spikes. J Struct Biol 165:64–77
19. Kremer JR, Mastronarde DN, McIntosh JR (1996) Computer visualization of three-dimensional image data using IMOD. J Struct Biol 116:71–76
20. Pettersen EF, Goddard TD, Huang CC, Couch GS, Greenblatt DM, Meng EC, Ferrin TE (2004) UCSF Chimera—a visualization system for exploratory research and analysis. J Comput Chem 25:1605–1612
21. Grassucci RA, Taylor D, Frank J (2008) Visualization of macromolecular complexes using cryo-electron microscopy with FEI Tecnai transmission electron microscopes. Nat Protoc 3:330–339
22. Mastronarde D (2006) Tomographic reconstruction with the IMOD software package. Microsc Microanal 12(S02):178–179
23. Zhu S, Takao M, Li N, Sakuma M, Nishino Y, Homma M, Kojima S, Imada K (2014) Conformational change in the periplamic region of the flagellar stator coupled with the assembly around the rotor. Proc Natl Acad Sci U S A 111:13523–13528

Chapter 19

Motility of Spirochetes

Shuichi Nakamura and Md. Shafiqul Islam

Abstract

Spirochetes are bacteria distinguished by an undulate or helical cell body and intracellular flagellar called periplasmic flagella or endoflagella. Spirochetes translate by rotating the cell body. In this chapter, we show a method for simultaneous measurement of the cell body rotation and swimming speed in individual spirochete cells. We also describe a simple chemotaxis assay capable of observing the response of spirochete in real time under a microscope and quantitatively evaluating the response magnitude to attractants and repellents.

Key words Spirochetes, One-sided dark-field microscope, Microscopic agar-drop assay, Chemotaxis

1 Introduction

The spirochete is a group of motile bacteria having a flat-wave or spiral-shaped cell body. Several spirochete species are causative agents of clinically important diseases; e.g., *Treponema pallidum* causing a sexually transmitted disease syphilis [1], *Borrelia burgdorferi* causing the Lyme disease [2], *Leptospira* spp. causing zoonosis leptospirosis [3], and *Brachyspira hyodysenteriae* causing swine dysentery [4]. The spirochete flagella, called periplasmic flagella or endoflagella, are present beneath the outer membrane, and their rotation within the periplasmic space is responsible for propagating the cell-body wave backward, propelling the cell (Fig. 1) [5]. Spirochetes move more actively in high viscosity environments such as gel-like polymer solutions [6, 7]. Quantitative measurements of kinematic parameters in spirochete movements are required for understanding their motility mechanism, but it is difficult to determine the fast cell-body rotation by conventional optical microscopy.

Here, we show "one-sided dark-field illumination" technique to measure the rotation of helical or sinusoidal cell body and the swimming speed simultaneously [8]. For observation of spirochetes, a dark-field microscope is generally used to clearly visualize the thin cell body of the bacteria (about 200 nm in diameter) (Fig. 2a).

Tohru Minamino and Keiichi Namba (eds.), *The Bacterial Flagellum: Methods and Protocols*, Methods in Molecular Biology, vol. 1593, DOI 10.1007/978-1-4939-6927-2_19, © Springer Science+Business Media LLC 2017

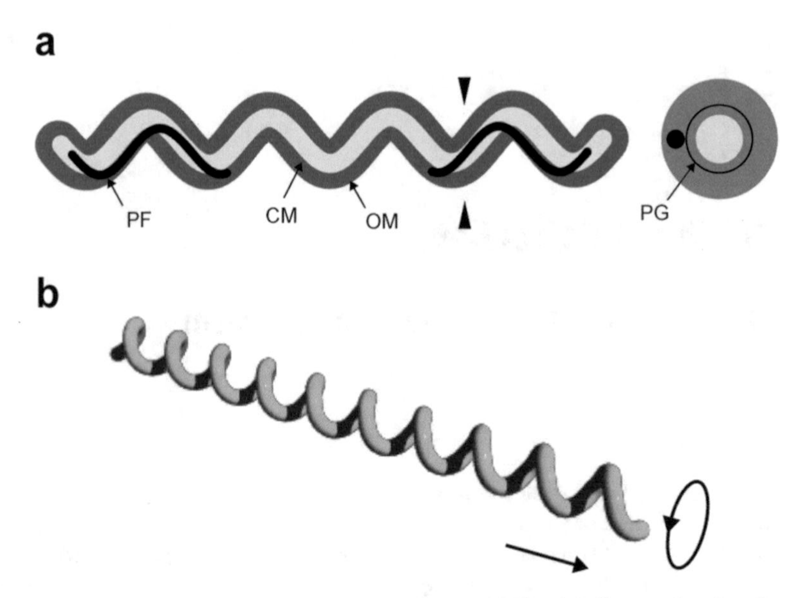

Fig. 1 Structure and motile form of spirochetes. (**a**) The *left* diagram is a longitudinal representation of a typical spirochete, and the *right* one is a cross section at the position indicated by *triangles* in the *left*. OM, CM, PF, and PG denote outer membrane, cytoplasmic membrane, periplasmic flagellum, and peptidoglycan layer, respectively. PG is omitted in the *left*. The PFs of *Borrelia*, *Brachyspira*, *Spirochaeta*, and most of *Treponema* species overlap in the cell center, but those of *Leptospira* and *Treponema phagedenis* are too short for overlapping as illustrated in this diagram. (**b**) Spirochetes translate by rotating the spiral cell body. The cell shapes of some species, e.g., *Borrelia*, are sinusoidal, in which wave propagation is responsible for propulsion

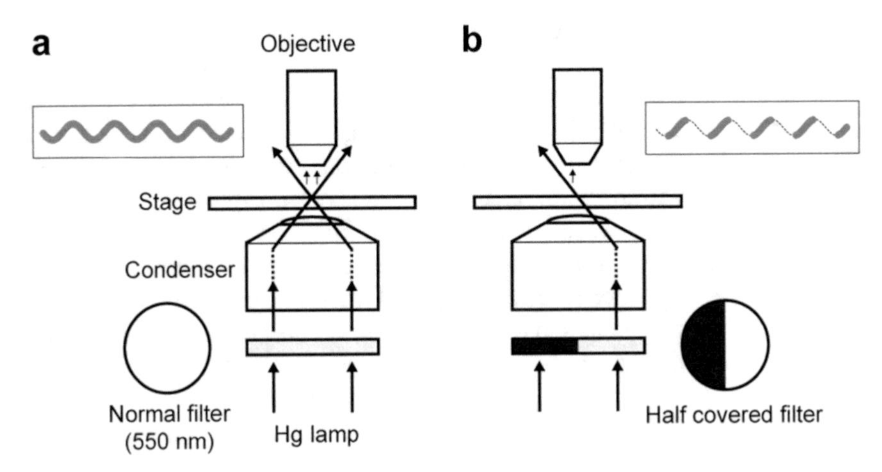

Fig. 2 Diagram of dark-field illumination. (**a**) Conventional dark-field illumination. (**b**) One-sided dark-field illumination. The optical path is simplified, e.g., one indicated by a *dashed line* within the condenser

A unidirectionally illuminated spirochete cell is observed as a series of bright spots with intervals corresponding to the helix pitch of the cell body (Fig. 2b). The cell-body rotation rate is determined from the bright spots moving backward against swimming direction, and the swimming speed is measured by analyzing changes in the centroid coordinates of the cell. In this chapter, we describe the application of this technique to motility assays of *Leptospira*.

We also show a simple chemotaxis assay, designated as "microscopic agar-drop assay" (MAA). In this method, a droplet of agar-containing attractants or repellents is placed at the center of a flow chamber, and bacteria detect chemical compounds diffusing from the agar drop and migrate to or go away from the agar drop. This technique was developed for a chemotaxis study of *Leptospira* in ref. 9. Chemotaxis studies by conventional capillary assay [10] can also be performed by a simple setup, but requiring a time for cell growth (e.g., several weeks for observing colonies of *Leptospira*). MAA allows us to evaluate the magnitude of chemotactic responses in real time by counting the cell number near and/or far from the agar drop, completing less than 20 min for *Leptospira* [9]. Because individual cells are observed under a microscope, parameters characterizing motility and chemotaxis (swimming trajectories, swimming speeds, reorientation frequencies and so on) can be analyzed in the presence of chemical concentration gradients.

2 Materials

2.1 Korthof's Liquid Medium

1. Korthof's basement solution (×10 stock solution): Dissolve 8 g of Peptone, 14 g of NaCl, 0.2 g of $NaHCO_3$, 0.4 g of KCl, 2.4 g of KH_2PO_4, and 8.8 g of Na_2HPO_4 in 900 mL of distilled water and then adjust pH to 7.4. Sterilize the solution at 121 °C for 20 min and cool down to room temperature (Solution 1). Dissolve 0.4 g of $CaCl_2$ in 100 mL of distilled water and sterilize using a filter with 0.22 μm pore size. Add the $CaCl_2$ solution to Solution 1 under the germfree condition. Store at 4 °C.
2. Rabbit serum: Inactivate rabbit serum at 56 °C for 30 min.
3. Mix 100 mL of ×10 basement solution, 100 mL of inactivated rabbit serum, and 800 mL of sterilized distilled water under the germfree condition. Add 5-fluorouracil to the medium to be 100 μg/mL for preventing contamination if necessary.

2.2 Ellinghausen–McCullough–Johnson–Harris (EMJH) Liquid Medium

1. Stock solutions: Prepare all stock solutions (Sol. 1–8) at a volume of 100 mL. Weights of chemicals dissolved in stock solutions are as follows: 0.4 g $ZnSO_4{\cdot}7H_2O$ for Sol. 1, 1.5 g $MgCl_2{\cdot}6H_2O$ for Sol. 2, 1.5 g $CaCl_2{\cdot}2H_2O$ for Sol. 3, 0.5 g $FeSO_4{\cdot}7H_2O$ for Sol. 4 (*see* **Note 1**), 10 g sodium pyruvate for Sol. 5, 10 g glycerol for Sol. 6, 10 g Tween 80 for Sol. 7, and 0.02 g cyanocobalamin for Sol. 8. Sterilize Sols. 2, 3, 6, and 7 at 121 °C for 20 min, and sterilize Sols. 1 and 5 using a filter with 0.22 μm pore size. Store at 4 °C.

2. Albumin solution: Dissolve 10 g of Bovine serum albumin (Probumin Universal Grade (K): Millipore, code 81-003-7) in 50 mL of distilled water. Mix 1 mL of Sol. 1, 1 mL of Sol. 2, 1 mL of Sol. 3, 10 mL of Sol. 4, 12.5 mL of Sol. 7, and 1 mL of Sol. 8 with the BSA solution. Sterilize the mixture by filtration and store at 4 °C.
3. EMJH basement solution: Mix 2.3 g of Difco™ *Leptospira* medium base EMJH (BD), 1 mL of Sol. 5, 1 mL of Sol. 6, and 900 mL of distilled water, and adjust pH to 7.4. Sterilize at 121 °C for 20 min and store at 4 °C.
4. Mix the albumin solution and the EMJH basement solution under the germfree condition. Store at 4 °C.

2.3 Motility Buffer

Because various kinds of buffer are available for motility assay of *Leptospira*, select appropriate buffers depending on experiments (*see* **Note 2**). Convenient buffers are listed below.

1. 20 mM sodium phosphate buffer, pH 7.4.
2. 20 mM Tris–HCl, pH 7.4.
3. Buffer shown in ref. 11: Dissolve Na_2HPO_4, 0.3 g KH_2PO_4, 1 g NaCl, 0.25 g NH_4Cl, 5 mg thiamine hydrochloride, and 5 g BSA into 1 L distilled water. Adjust pH to 7.4 and sterilize by filtration. Store at 4 °C.

2.4 Microscope Setup

1. Dark-field microscope: An oil condenser (NA 1.2–1.4), objectives (ca. 100× for motility assay, ca. 40× for chemotaxis assay, *see* **Note 3**), a mercury lamp, and a half covered filter. Set a relay lens if necessary. Figure 2b shows a schematic of optical setup.
2. Video camera: Recording at 30 frames per second (normal video frame rate) is enough for chemotaxis assay. For rotation assay, set a high-speed camera capable of recording at >500 frames per second (*see* **Note 4**).
3. Flow chamber: Attach a coverslip with a glass slide via a double-sided tape.

3 Methods

3.1 Measurement of Cell Rotation Rate and Swimming Speed

1. Grow *Leptospira* cells aerobically in Korthof's liquid medium or EMJH liquid medium at 30 °C for 4 days.
2. Dilute the cell culture into fresh Korthof's media or fresh EMJH medium at a 1:5 ratio and infuse the cell suspension to a flow chamber (*see* **Note 5**).
3. Observe the cells using the dark-field microscope with a 100× objective and one-sided illumination. Find cells translating

unidirectionally and adjust the direction of illumination by rotating the half-covered filter to make sure of visualizing the cell as a series of bright spots.

4. Record microscopic images of translating cells on a computer with a high-speed digital camera at 500 frames per second.
5. Read the movie file of cell movements into the image analysis software "ImageJ" (*see* **Note 6**).
6. Duplicate the movie into two and smooth one of them to connect adjacent spots for swimming speed analysis described next (Fig. 3a, right). The unprocessed images (Fig. 3a, left) are used for cell-body rotation analysis described in **step 4**.

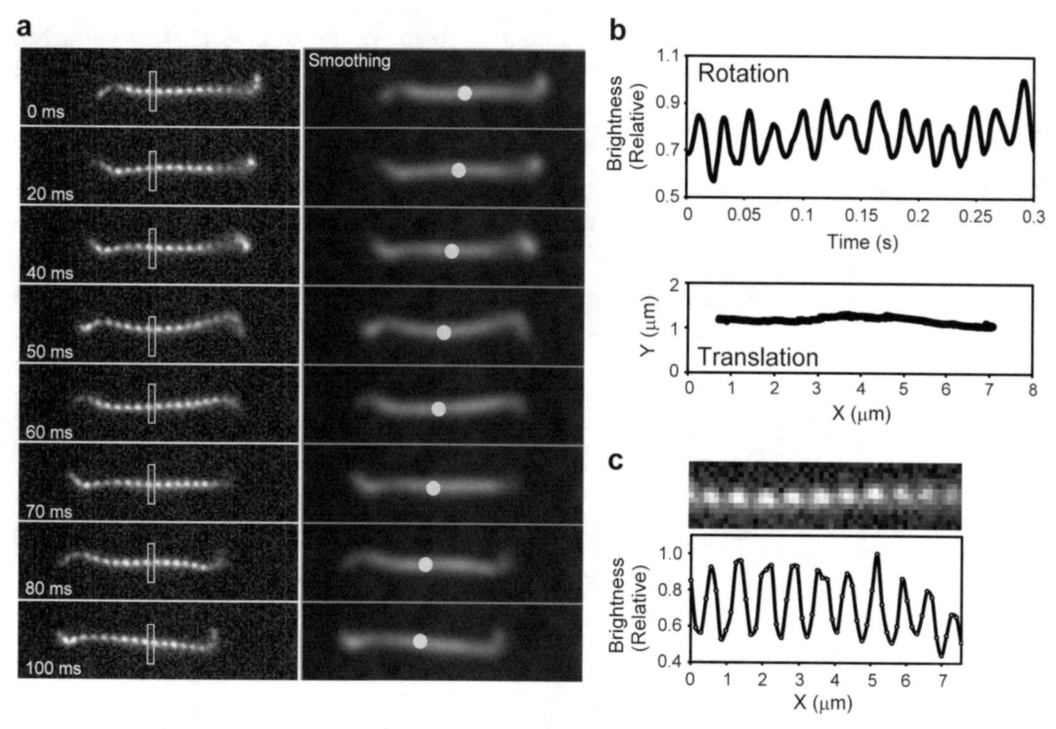

Fig. 3 Image analysis in one-sided dark-field microscopy. (**a**) Time-sequence images of swimming *Leptospira*. The *left* column is a raw data. Set a square-type ROI near the center of the image (a *yellow square*) so that the cell crosses the ROI. Because the cell body separately visualized by one-sided illumination was recognized as multiple cells by software, perform the smoothing process to the original images so that the software determines the centroid of entire cell body (*yellow dots*). (**b**) Brightness changes obtained by analyzing bright spots (*upper panel*), indicating cell body rotation, and swimming trajectory for 0.3 s (*lower panel*). These data indicate that the rotation rate is 46 rps and swimming speed is 25 μm/s. However, note that the rotation speed contains the effect of cell migration (*see* Fig. 4); correction using the values of swimming speed and helix pitch of the cell body is required. (**c**) The helix pitch is determined by brightness analysis of the cell body. The pitch of this cell is 0.75 μm; thus, the actual cell-body rotation rate is 83 rps (use eq. 1)

7. Determine the time-sequence centroid change in the cell body by using the smoothed images (Fig. 3a, right) and calculate its swimming speed: the migration distance (μm) per unit time (second).
8. Set a thin rectangle ROI (region of interest) as a virtual slit in the unprocessed movie (Fig. 3a, left). Determine the change in brightness at the virtual slit. The forward and backward movements of the bright spots, resulting from the translational migration and the cell body rotation respectively, are mixed in the periodic brightness change (Fig. 4). Therefore, calculate the actual rotation rate by removing the effect of cell migration as follows:

$$f = f' + \frac{v}{p} \tag{1}$$

where f, f', v, and p are the actual cell rotation rate, apparent rotation rate obtained from the brightness change (Fig. 3b, upper panel), the swimming speed (Fig. 3b, lower panel), and the helix pitch of the cell body (Fig. 3c).

3.2 Microscopic Agar Drop Assay

1. Add 0.2 g of powder agar into 20 mL of distilled water in a beaker and mix well using a stirrer bar. Add attractant or repellent chemicals if necessary.
2. Heat the agar solution in a microwave oven for 20–30 s and then mix the agar solution using a stirrer.

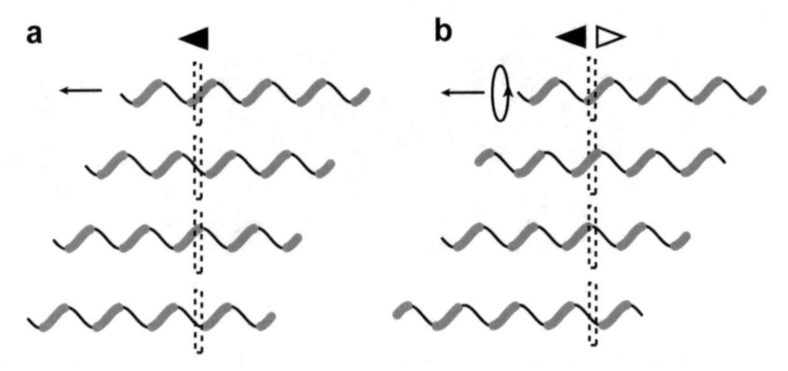

Fig. 4 Explanation for correction of the rotation rate. (**a**) If a cell is pulled somehow in the direction indicated by an *arrow* without rotating, *dots* along the cell body move in the direction indicated by a filled *triangle*. The frequency of the brightness change of ROI (*dashed squares*) (f) corresponds to the number of dots crossing ROI per unit time, which is obtained by dividing the migrating speed with the helix pitch (= v/p in eq. 1). (**b**) A cell translating with rotating contains forward and backward (an *open triangle*) movements of dots due to translation and rotation, respectively; therefore, the raw data must be corrected as shown in eq. 1

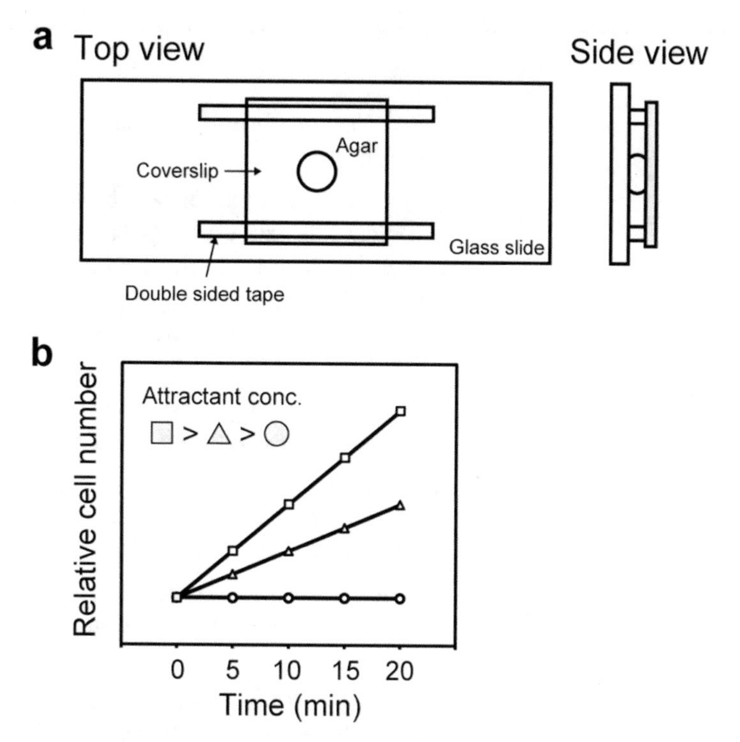

Fig. 5 Schematic representation of MAA. (**a**) A flow chamber. (**b**) Idealized results when examining chemotactic responses to different concentrations of attractant. The area about 1 mm away from the edge of the agar drop is convenient for measuring the number of cells. When the agar drop contains glucose, significant cell accumulation is observed within 20 min. Because the diffusion area of sugar molecules (~0.5 nm radius) is estimated to be ca. 1.5 mm square in 20 min at room temperature, the area 3–4 mm away from the agar drop can be analyzed as a negative control

3. Repeat **step 2** until the agar is completely dissolved (*see* **Note** 7).
4. Place one drop of the agar solution around the center of a glass slide, and then immediately attach a coverslip with the glass slide using double-sided tapes (Fig. 5a).
5. Infuse bacterial suspension into the flow chamber and seal both sides of the chamber using silicon oil.
6. Place the flow chamber on a microscope stage and observe the cell behavior.
7. Count the number of cells (Fig. 5b) or analyze the motility parameters (swimming speeds, reversal frequency and so on) near the agar drop. Record over the experiment as necessary (*see* **Notes 8** and **9**).

4 Notes

1. Solution 4 should be prepared just before making EMJH medium.
2. Growth media can usually be used for motility assay. However, for chemotaxis assay, potassium phosphate buffer is recommended to reduce the background responses to components of media. Although *Leptospira* are motile over a pH range from 6 to 9 [12], neutral pH (about pH 7.5) is recommended for conventional observation.
3. NA of 100× objective must be smaller than that of the condenser to ensure dark-field illumination.
4. For recording using a high-speed video camera, Hg lamp is essential to acquire high brightness images. A black paper can be used for covering half of the filter.
5. When replacing growth media with some buffer is required, the cell suspension can be gently centrifuged at ca. 1000 × *g* for 10 min.
6. ImageJ software and VBA macros developed originally were used for image analysis in refs. [7, 8]; however, any software will do if it can deal with a movie file and acquire the values of brightness and centroid.
7. The stirrer bar should be spun slowly in the agar solution for preventing the mixing of air bubble.
8. MAA can be used for other bacteria besides spirochetes, taking just 5 min for analysis of *Salmonella*.
9. A capillary assay in combination with real-time PCR (developed in ref. [11]) is also useful for chemotaxis studies of bacteria. This method requires equipment for genetic experiments and takes several hours, but it could be quantitative and highly sensitive.

Acknowledgment

We thank N. Koizumi and K. Takabe for technical support. This work was partly supported by the Grant-in-Aid for Scientific Research on Innovative Areas "Harmonized Supramolecular Motility Machinery and Its Diversity" to S.N. (JP25117501), Research Foundation for Opto-Science and Technology to S.N.

References

1. Norris SJ (1993) Polypeptides of *Treponema pallidum*: progress toward understanding their structural, functional, and immunologic roles. Treponema pallidum Polypeptide Research Group. Microbiol Rev 57:750–779
2. Radolf JD, Caimano MJ, Stevenson B, Hu LT (2012) Of ticks, mice and men: understanding the dual-host lifestyle of Lyme disease spirochetes. Nat Rev Microbiol 10:87–99
3. Adler B, Moctezuma AP (2010) *Leptospira* and leptospirosis. Vet Microbiol 140:287–296
4. Duhamel GE (2001) Comparative pathology and pathogenesis of naturally acquired and experimentally induced colonic spirochetosis. Anim Health Res Rev 2:3–17
5. Charon NW, Cockburn A, Li C et al (2012) The unique paradigm of spirochete motility and chemotaxis. Annu Rev Microbiol 66:349–370
6. Kaiser GE, Doetsch RN (1975) Enhanced translational motion of *Leptospira* in viscous environments. Nature 225:656–657
7. Nakamura S, Adachi Y, Goto T, Magariyama Y (2006) Improvement in motion efficiency of the spirochete *Brachyspira pilosicoli* in viscous environments. Biophys J 90:3019–3026
8. Nakamura S, Leshansky A, Magariyama Y, Namba K, Kudo S (2014) Direct measurement of helical cell motion of the spirochete *Leptospira*. Biophys J 106:47–54
9. Islam MS, Takabe K, Kudo S, Nakamura S (2014) Analysis of the chemotactic behaviour of *Leptospira* using microscopic agar drop assay. FEMS Microbiol Lett 356:39–44
10. Adler J (1973) A method for measuring chemotaxis and use of the method to determine optimum conditions for chemotaxis by *Escherichia coli*. J Gen Microbiol 74:77–91
11. Lambert A, Takahashi N, Charon NW, Picardeau M (2012) Chemotactic behavior of pathogenic and non-pathogenic *Leptospira* species. Appl Environ Microbiol 78:8467–8469
12. Islam MS, Morimoto YV, Kudo S, Nakamura S (2015) H^+ and Na^+ are involved in flagellar rotation of the spirochete *Leptospira*. Biochem Biophys Res Commun 466:196–200

Chapter 20

Structure of the Sodium-Driven Flagellar Motor in Marine *Vibrio*

Yasuhiro Onoue and Michio Homma

Abstract

Most bacteria can swim by rotating the flagellum. The basal body of the flagellum is an essential part for this motor function. Recent comprehensive analysis of the flagellar basal body structures across bacteria by cryo-electron tomography has revealed that they all share core structures, the rod, and rings: the C ring, M ring, S ring, L ring, and P ring. Furthermore, it also has uncovered that in some bacteria, there are extra ring structures in the periplasmic space and outer-membrane. Here, we describe a protocol to isolate the basal body of the flagellar basal body from a marine bacterium, *Vibrio alginolyticus*, for structural analysis of additional ring structures, the T ring and H ring.

Key words Bacterial flagellum, Basal body, *Vibrio*, T ring, H ring, MotX, MotY, FlgT

1 Introduction

1.1 The Flagellar Motor

The bacterial flagellum is a rotary motor that is responsible for bacterial motility. It is composed of three parts: the filament, hook, and basal body [1, 2]. The filament is shaped like helix and can produce the propulsive force when it is rotated. The basal body is located in bacterial membranes and working as a rotary motor. The hook is a universal joint that connects these two parts and transmits rotary motion efficiently from the basal body to the filament.

Salmonella and *Escherichi coli* have been a model bacterium for the structural analysis of the flagellar basal body [3, 4]. The basal body consists of a rod and five ring structures: the C ring, M ring, S ring, L ring, and P ring. Recently, the intact structure of the basal body of the flagellar motor throughout bacteria has been studied by cryo-electron tomography [5]. This observation suggests that the structure observed in *Salmonella* is common as the core structure throughout the bacteria. In addition, it has been found that in some bacteria the basal body has additional ring structure(s), especially in the periplasmic apace and outer membrane (Fig. 1).

Tohru Minamino and Keiichi Namba (eds.), *The Bacterial Flagellum: Methods and Protocols*, Methods in Molecular Biology, vol. 1593, DOI 10.1007/978-1-4939-6927-2_20, © Springer Science+Business Media LLC 2017

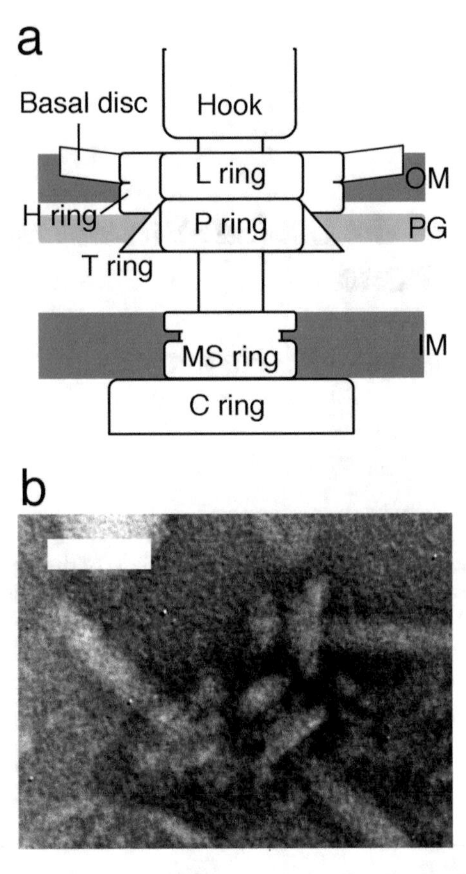

Fig. 1 The structure of the Na+-driven bacterial flagellar motor of *Vibrio*. (**a**) Schematic representation of the basal body of *Vibrio* embedded in the cellular membranes. The C ring, M ring, S ring, P ring, and L ring surround the rod. They are the common structure of the basal body throughout bacteria. In addition, *Vibrio* basal body has additional ring-like structures: the T ring, H ring, and basal disk. OM, PG, and IM represent outer membrane, peptidoglycan layer, and inner membrane, respectively. (**b**) Electron micrograph of basal bodies of *Vibrio* purified by CHAPS. The C ring and the basal disk cannot be observed because they are dissociated during the purification. Scale bar, 50 nm

1.2 The Basal Body of the Flagellar Motor from Vibrio

The basal body of the Na+-driven flagellar motor from *Vibrio* has the common core and additional ring structures: the T ring, H ring, and basal disk [6–8] (Fig. 1). MotX and MotY are the components of the T ring, and FlgT is one of the components of the H ring. FlgP forms the basal disk. Moreover, other unknown components should be present in these structures. These structures are suggested to be important for higher rotational speed and torque generation by the *Vibrio* motor than those of *Salmonella* and *Escherichia coli*.

1.3 Overview of the Methods

Cryo-electron tomography can obtain the intact structure of the flagellar motor but for this analysis a specialized electron microscope is required. On the other hand, observation of isolated basal body by negative stain electron microscopy is more accessible to many researchers and technically simpler. Therefore, in this book chapter, we described the method to purify the basal body of the flagellar motor from a marine bacterium, *Vibrio alginolyticus*, for the analysis of additional ring structures, the T ring and the H ring.

Isolation of the basal body from *Vibrio* follows almost the same protocol developed by Aizawa [9]. Cells are treated with lysozyme in the presence of EDTA to prepare spheroplast. The spheroplast is lysed by TritonX-100, and DNA released from the bacterial cytoplasm is digested by DNase I. Then PEG fractionation and differential centrifugation are used for further purification. This purified fraction contains the complex of the basal body and the filament, which obstructs the observation of the basal body clearly. Therefore, this sample is incubated in a low pH solution to depolymerize the filament. After centrifugation the purified basal body is used for structural observation by negative stain electron microscopy.

2 Materials

2.1 Growth of Vibrio

2.1.1 Strains

1. KK148 [VIO5 *flhG* (multi-Pof^{+})] (*see* **Notes 1** and **2**).

2.1.2 Growth Media

1. VC medium: 0.5% (w/v) Bacto tryptone, 0.5% (w/v) yeast extract, 0.4% (w/v) K_2HPO_4, 3% (w/v) NaCl, 0.2% (w/v) glucose.
2. VPG500 medium: 1% (w/v) Bacto tryptone, 0.4% (w/v) K_2HPO_4, 500 mM NaCl, 0.5% (w/v) glycerol (*see* **Note 3**).

2.2 Isolation of Flagellar Basal Body

1. Sucrose solution: 0.5 M sucrose, 50 mM Tris–HCl, pH 8.0.
2. 10 mg/mL lysozyme dissolved in sucrose solution.
3. 0.1 M EDTA-Na, pH 8.0.
4. 20% (w/v) Triton X-100.
5. 1 M $MgSO_4$.
6. 10 mg/mL DNase I.
7. 50% PEG6000.
8. 2 M NaCl.
9. TET buffer: 10 mM Tris–HCl, pH 8.0, 5 mM EDTA, 0.1% (w/v) Triton X-100.

2.3 Depolymerization of Flagellar Filament

1. 50 mM glycine-HCl, pH 3.5, 0.1% (w/v) Triton X-100.

2.4 Electron Microscopy

1. 2% (w/v) phosphotungstic acid, pH 7.4.
2. 3% (w/v) uranyl acetate.

3 Methods

3.1 Growth of Vibrio

1. Inoculate 5 mL of overnight culture (grown in VC medium) into VPG500 medium (500 mL) at a 100-fold dilution.
2. Culture at 30 °C for 3.5 h.

3.2 Isolation of Flagellar Basal Body

1. Harvest cultures cells and suspend cell pellet in 20 mL of sucrose solution.
2. To convert them into spheroplasts, add lysozyme and EDTA at final concentrations of 0.1 mg/mL and 5 mM, respectively.
3. Gently stir the suspension for 30–40 min on ice.
4. Add Triton X-100 at a final concentration of 1% (w/v) to lyse the spheroplasts (*see* **Note 4**).
5. Gently stir the suspension for 1 h on ice.
6. To reduce the viscosity of the lysed samples, add $MgSO_4$ and DNase I at final concentrations of 10 mM, and 0.1 mg/mL, respectively.
7. Gently stir the suspension for 1 h on ice (*see* **Note 5**).
8. Add EDTA at a final concentration of 5 mM.
9. Remove unlysed cells and cellular debris by centrifugation at 17,000 × *g* for 20 min at 4 °C.
10. Add PEG 6000 and NaCl to the supernatant at final concentrations of 2% and 100 mM, respectively.
11. Incubate the sample at 4 °C for 60 min.
12. Centrifuge the suspension at 27,000 × *g* for 30 min at 4 °C.
13. Resuspend the pellet in 6 mL of TET buffer.
14. To remove cellular debris, centrifuge the suspension at 1000 × *g* for 15 min at 4 °C.
15. Centrifuge the supernatant at 100,000 × *g* for 30 min at 4 °C.
16. Resuspend the pellet in 200 μL of TET buffer.

3.3 Depolymerization of Flagellar Filament

1. Dilute the suspension to 30-fold in 50 mM glycine-HCl, pH 3.5, 0.1% (w/v) Triton X-100.
2. Shake the suspension for 60 min at room temperature.
3. Centrifuge at 1000 × *g* for 15 min at 4 °C.
4. Centrifuge the supernatant at 150,000 × *g* for 40 min at 4 °C.
5. Resuspend the pellet in 100 μL of TET buffer (*see* **Notes 6** and **7**).

3.4 Electron Microscopy

Stain the isolated flagellar basal body negatively with phosphotungstic acid or uranyl acetate. Observe this sample with a JEM-2010 electron microscope (JEOL).

4 Notes

1. We usually use the *Vibrio alginolyticus* KK148 strain, which has multiple polar flagella [10], to improve the yield of purification of the basal body as compared to a wild-type VIO5 strain, which has a single polar flagellum.
2. Construction of the deletion strain is essential for the identification of the component of the new ring structure [6, 8], but it is difficult to predict the candidates. We usually try to find out new components by performing SDS-PAGE of the purified basal body and identifying the new bands of the gel by N-terminal amino acid sequencing by Edman degradation. We have succeeded in finding that FlgT is the component of the H ring by this strategy [7].
3. We usually use polypeptone instead of Bacto tryptone for culturing *Vibrio.*
4. CHAPS is also useful as a detergent for isolation of the basal body [11].
5. If viscosity of the suspension is still high at this step, add more DNase I to the suspension.
6. In this protocol, the C ring and the basal disk are detached from the core structure of the basal body presumably due to the effect of detergent. Therefore, it is difficult to analyze these rings by the method described here.
7. If the sample has many impurities, we further purify it by sucrose density gradient centrifugation. To prepare stepwise sucrose gradient, 500 μL of TET buffer containing 60, 50, 40, 30, or 20% (w/w) of sucrose is successively layered from the bottom to the top of a centrifugation tube. After centrifugation at 72,000 × *g* for 90 min at 4 °C, the sample is fractionated into 20 factions. The fraction of the basal body is detected by SDS-PAGE/Western blotting. If necessary, remove sucrose by centrifugation.

Acknowledgments

This work was supported by JSPS KAKENHI Grants (JP23247024 and JP24117004 to M.H.).

References

1. Berg HC (2003) The rotary motor of bacterial flagella. Annu Rev Biochem 72:19–54
2. Sowa Y, Berry RM (2008) Bacterial flagellar motor. Q Rev Biophys 41:103–132
3. DePamphilis ML, Adler J (1971) Fine structure and isolation of the hook-basal body complex of flagella from *Escherichia coli* and *Bacillus subtilis*. J Bacteriol 105:384–395
4. Thomas DR, Francis NR, Xu C, DeRosier DJ (2006) The three-dimensional structure of the flagellar rotor from a clockwise-locked mutant of *Salmonella enterica* Serovar Typhimurium. J Bacteriol 188:7039–7048
5. Chen S, Beeby M, Murphy GE, Leadbetter JR, Hendrixson DR, Briegel A, Li Z, Shi J, Tocheva EI, Müller A, Dobro MJ, Jensen GJ (2011) Structural diversity of bacterial flagellar motors. EMBO J 30:2972–2981
6. Terashima H, Fukuoka H, Yakushi T, Kojima S, Homma M (2006) The *Vibrio* motor proteins, MotX and MotY, are associated with the basal body of Na-driven flagella and required for stator formation. Mol Microbiol 62:1170–1180
7. Terashima H, Koike M, Kojima S, Homma M (2010) The flagellar basal body-associated protein FlgT is essential for a novel ring structure in the sodium-driven *Vibrio* motor. J Bacteriol 192:5609–5615
8. Beeby M, Ribardo DA, Brennan CA, Ruby EG, Jensen GJ, Hendrixson DR (2016) Diverse high-torque bacterial flagellar motors assemble wider stator rings using a conserved protein scaffold. Proc Natl Acad Sci USA 113:E1917–E1926
9. Aizawa S-I, Dean GE, Jones CJ, Macnab RM, Yamaguchi S (1985) Purification and characterization of the flagellar hook-basal body complex of *Salmonella typhimurium*. J Bacteriol 161:836–849
10. Kusumoto A, Kamisaka K, Yakushi T, Terashima H, Shinohara A, Homma M (2006) Regulation of polar flagellar number by the *flhF* and *flhG* genes in *Vibrio alginolyticus*. J Biochem 139:113–121
11. Koike M, Terashima H, Kojima S, Homma M (2010) Isolation of basal bodies with C-ring components from the Na^+-driven flagellar motor of *Vibrio alginolyticus*. J Bacteriol 192:375–378

Chapter 21

Chemotactic Behaviors of *Vibrio cholerae* Cells

Ikuro Kawagishi and So-ichiro Nishiyama

Abstract

Vibrio cholerae, the causative agent of cholera, swims in aqueous environments with a single polar flagellum. In a spatial gradient of a chemical, the bacterium can migrate in "favorable" directions, a property that is termed chemotaxis. The chemotaxis of *V. cholerae* is not only critical for survival in various environments and but also is implicated in pathogenicity. In this chapter, we describe how to characterize the chemotactic behaviors of *V. cholerae*: these methods include swarm assay, temporal stimulation assay, capillary assay, and receptor methylation assay.

Key words *Vibrio cholerae*, Chemotaxis, Swarm assay, Temporal stimulation assay, Capillary assay, Receptor methylation assay

1 Introduction

1.1 Background

Vibrio cholerae, the causative agent of cholera, is a Gram-negative highly motile bacterium with a single-polar flagellum, the rotation of which is driven by sodium-motive force. Since the bacterium inhabits nutrient-poor aquatic environments such as rivers and estuaries, as well as the lumen of the human gastrointestinal tract [1], chemotaxis of *V. cholerae* toward favorable locations is proposed to play critical roles in its survival [2] and pathogenicity [3–14].

The genome sequence of *V. cholerae* (O1 serogroup, El Tor biotype) predicts that the bacterium has three sets of chemotactic signaling proteins, or Che, and 45 MCP-like proteins (referred to as MLPs) [15]. Each set of *che* genes are linked together in the genome to form a cluster except that the *cheR2* (VC2201) gene is in the *fla* gene cluster located just next to the *che* cluster II (Fig. 1). The products of the genes in a particular cluster are known or hypothesized to constitute a coherent signaling system (note that CheR2 is assigned to system II). Out of the three Che systems, only system II is involved in taxis [16, 17]. Among 45 MLPs, only a few have been proven to mediate tactic responses (and hence to

Tohru Minamino and Keiichi Namba (eds.), *The Bacterial Flagellum: Methods and Protocols*, Methods in Molecular Biology, vol. 1593, DOI 10.1007/978-1-4939-6927-2_21, © Springer Science+Business Media LLC 2017

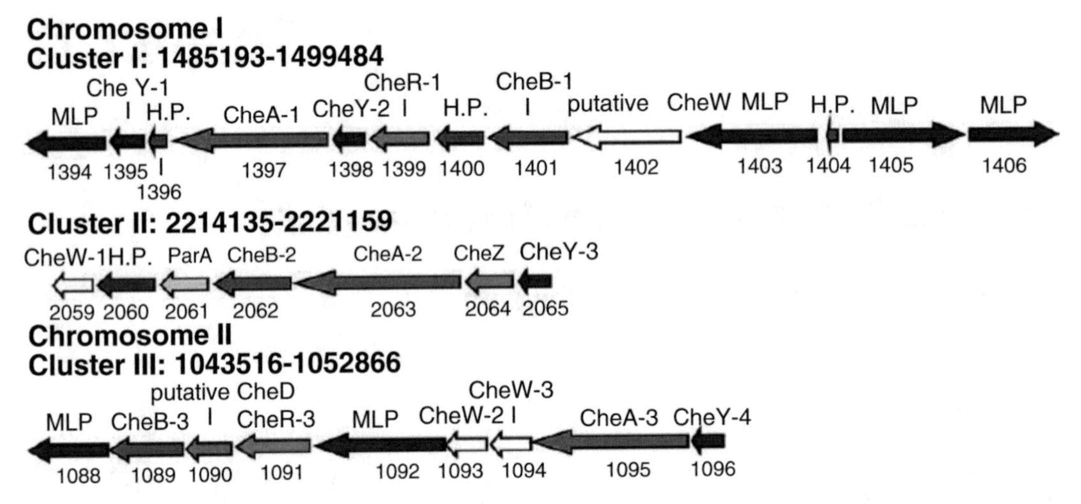

Fig. 1 The *che* gene clusters of *V. cholerae*. The genes located in the three *che* gene clusters of the genome of *V. cholerae* (El Tor). Cited from the literature [2]

belong to Che system II): Mlp24 for taxis to amino acids [18], Mlp32 for redox taxis [19], and Mlp37 [20] for taxis to taurine and amino acids.

In this chapter, we present protocols of capillary assay and receptor methylation assay to characterize chemotactic responses of *V. cholerae*, in conjunction with those of potentially useful swarm assay and temporal stimulation assay.

1.2 Overview of Methods

Swarm assay is, in principle, easier than other chemotaxis assays. Wild-type *V. cholerae* cells form chemotactic rings when inoculated onto a tryptone swarm plate (0.25–0.3% agar), whereas a mutant strain lacking *cheA2* does not spread [14], demonstrating that spreading in semisolid agar depends not only on motility but also on chemo/aerotaxis. Swarm assay is also well suited for screening of mutants defective in motility and chemotaxis (for instance, *see* [21]). In *E. coli*, a strain lacking all of the chemoreceptors is often used as a standard strain to examine functions of a chemoreceptor of interest by expressing it from a plasmid. No such strain is available for *V. cholerae* due to high numbers of chemoreceptors (MLPs). Although a mutant strain lacking both *mlp24* and *mlp37* that encode major chemoreceptors for amino acid taxis shows significantly weaker responses to amino acids in capillary assay [18, 20], this defect is eluded in swarm assay using tryptone semisolid agar plates (Fig. 2). Interestingly, swarm assay for the chemotaxis of *Pseudomonas aeruginosa* became possible when the *aer* gene, encoding the sensor for redox taxis, was deleted [22]. We therefore examined a triple mutant strain of *V. cholerae* lacking the *aer* homolog (*mlp32*) as well as *mlp24* and *mlp37* with swarm assay (R. Iwazaki, T. Nakagawa, S.N. & I.K., unpublished): this strain

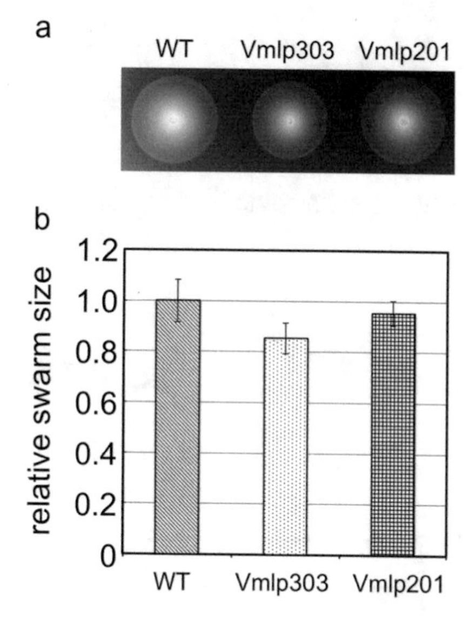

Fig. 2 Swarm assay of *V. cholerae* cells. (**a**) The wild-type (O395N1, WT), the triple mutant (Vmlp303: Δ*mlp24* Δ*mlp32* Δ*mlp37*), and the double mutant (Vmlp201: Δ*mlp24* Δ*mlp37*) strains were subjected to tryptone swarm assay (37 °C, 5.5 h). (**b**) Swarm sizes of four different colonies were measured for each strain strain: error bars denote standard deviations

swarmed slower in tryptone semisolid agar than the Δ*mlp24* Δ*mlp37* double mutant (Fig. 2), possibly providing a useful tool for the in vivo characterization of a chemoreceptor of interest.

Temporal stimulation assay has proved useful in the study of the chemotaxis of enterobacteria with peritrichous flagella [23]. A typical protocol of temporal stimulation assay to obtain a dose-response curve for a particular attractant is as follows: cells are made tumbling by applying a repellent such as a high concentration of glycerol, and then are exposed to various concentrations of an attractant. This protocol may seem not much suitable for a bacterial species with a single-polar flagella, such as *Vibrio* spp., because a repellent would only cause a backward motion of a cell which otherwise swims forward. Homma et al. [24] reported, however, that a marine bacterium *V. alginolyticus* with a single-polar flagellum (Pof) undergoes a frequent back-and-forth movement upon exposure to a millimolar concentration of phenol. The reversal frequency was about 1 s^{-1} without chemoeffectors. When phenol was added at 5 mM, the reversal frequency was increased to 3–4 s^{-1} with little adaptation. After phenol addition, L-serine was added at various concentrations (Fig. 3). At 1 mM, L-serine elicited an attractant response and its effect was saturated at 5–10 mM. Essentially, the same results were obtained for *V. cholerae* (Y. Miura, S. N., and I. K., unpublished).

Capillary assay was originally developed for *Escherichia coli* by Adler and coworkers [25–27]. In brief, this assay involves insertion of a glass capillary containing an attractant into a suspension of

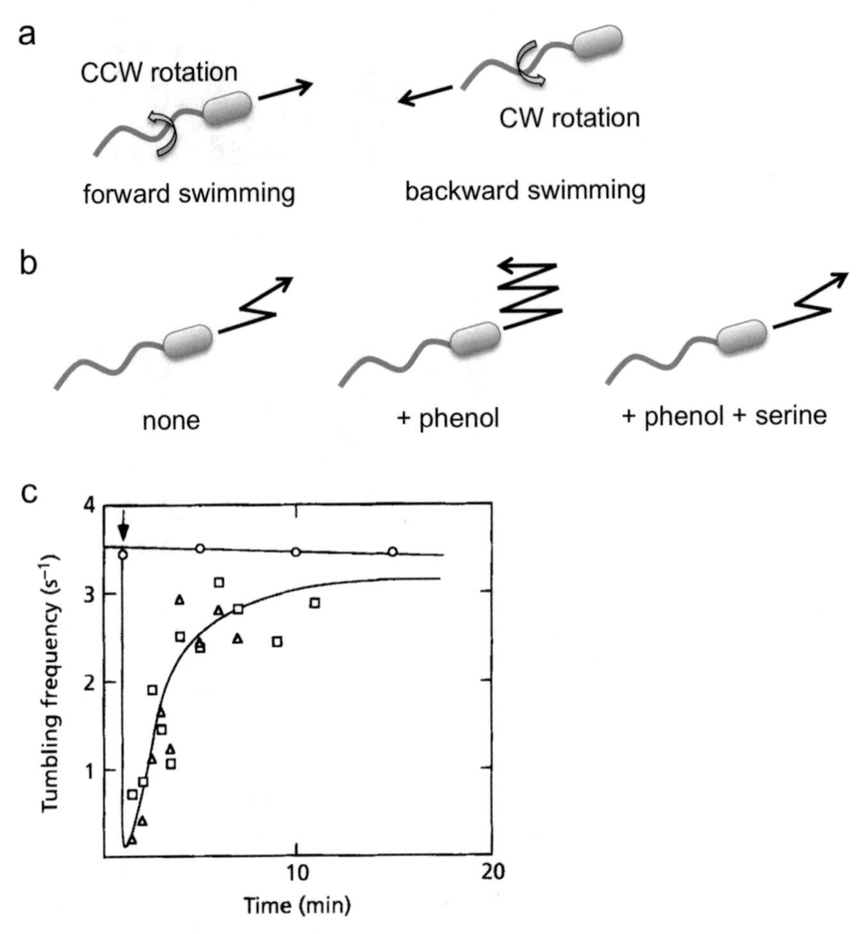

Fig. 3 Temporal stimulation assay of *Vibrio* spp. cells. (**a**) Flagellar rotational senses and swimming patterns. (**b**) Swimming behaviors of polar-flagellated cells of *V. alginolyticus* [24] and *V. cholerae* (Y. Miura, S.N. and I.K., unpublished) in the presence of a repellent and an attractant. (**c**) Phenol and serine responses of *V. alginolyticus*. Cited from the literature [24]

cells and counting of the number of cells in the capillary (Fig. 4a, b). We have modified the original protocol for *V. cholerae* [18, 20] to characterize amino acid chemotaxis of the bacterium (Fig. 4c). Although we have only tested attractant responses, it is possible in principle to assay repellent responses using a repellent added to cell suspensions with capillaries filled with buffer, a procedure named the chemical-in-pond method.

Receptor methylation assay has also been used to examine responses to particular chemoeffectors. Receptor methylation and demethylation are central to adaptation in chemotaxis and are catalyzed by CheR and CheB (Fig. 5a): upon binding of an attractant or a repellent, the relevant chemoreceptor (or MLP) becomes more or less methylated to attenuate signal output. A change in the methylation level of the relevant receptor can therefore be used as

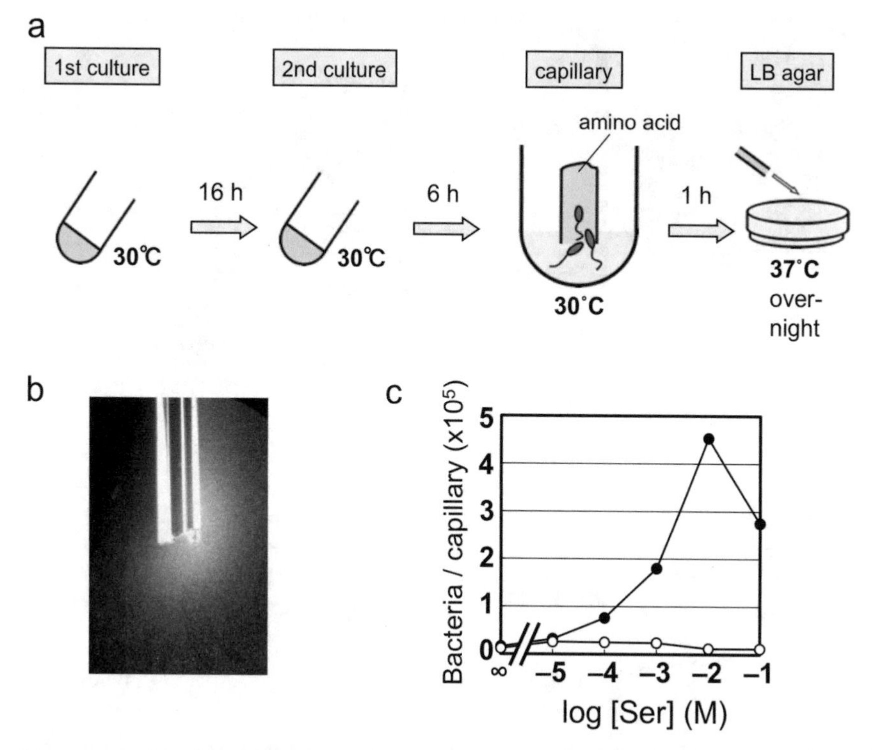

Fig. 4 Capillary assay of *V. cholerae* cells. (**a**) concept (**b**) *E. coli* cells swarming toward a mouth of capillary filled with an attractant (L-serine). (**c**) A typical result of capillary assay. Responses of strain O395N1 (*closed circles*) and its derivative VcheA2 (Δ*cheA2*, open circles) to L-serine. Cited from the literature [18]

a measure of a response to a chemoeffector. Methylation of a chemoreceptor increases its mobility upon SDS-PAGE [28–31] and can be detected by immunoblotting [32] (Fig. 5b). To detect methylation of a specific MLP in *V. cholerae*, the targeted MLP is fused to a tag sequence. With this setup, attractant-induced changes in methylation levels of Mlp24 or Mlp37 have been observed [18, 20] (Fig. 5c).

Other assays for chemotaxis of *V. cholerae* remain to be developed. In principle, bead assay [33] (*see* Chapter 14) and FRET analysis [34] can also be used as quantitative measurements of behavioral responses though *V. cholerae* has not been analyzed by these methods. To monitor ligand binding in vitro, radioactive ligands [35] have been used conventionally, but isothermal calorimetry (ITC) has now become much popular as a reliable and handy method [18, 20, 36–38]. Furthermore, a fluorescently labeled ligand has recently been used to visualize ligand binding to an amino acid chemoreceptor of *V. cholerae* [20], a new methodology that would be applicable for in vivo analyses of ligand-binding kinetics to chemoreceptors.

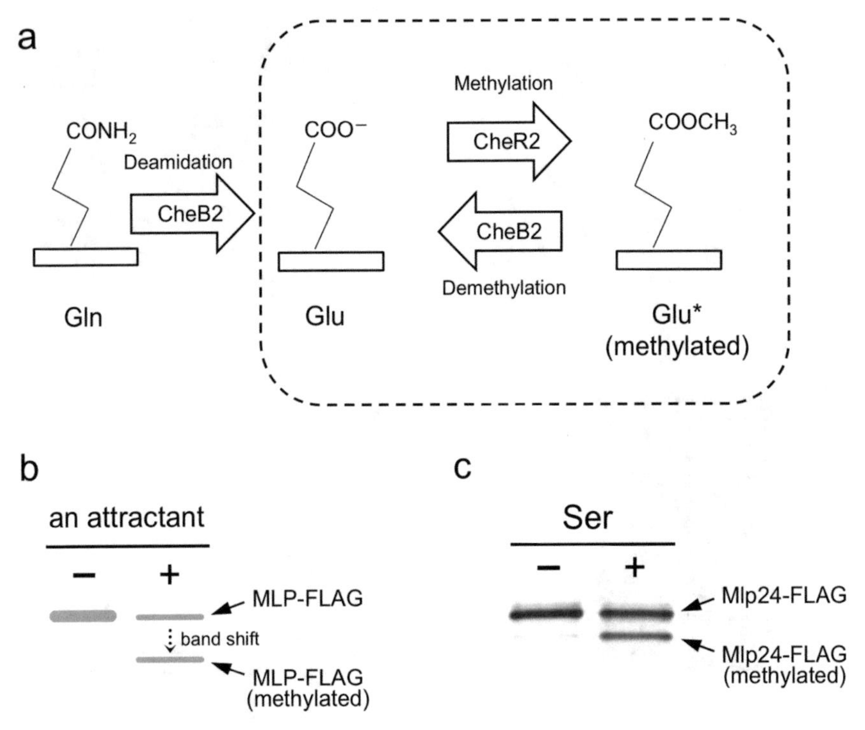

Fig. 5 Methylation assay of *V. cholerae* MLPs. (**a**) Receptor methylation and demethylation. (**b**) Effect of methylation of an MCP on its mobility in SDS-PAGE. (**c**) A typical result of methylation assay. O395N1 cells (classical) expressing Mlp24-FLAG were incubated with (+) or without (−) L-serine (Ser), and Mlp24-FLAG was detected by immunoblotting with anti-FLAG antibody. Cited from the literature [18]

2 Materials

2.1 Culture Media of V. cholerae to Examine Its Chemotaxis

1. LB agar plate: 1% tryptone, 0.5% yeast extract, 0.5% NaCl, 1.5% agar.
2. TG broth: 1% tryptone, 0.5% NaCl, 0.5% glycerol.
3. TMN buffer: 50 mM Tris–HCl, pH 7.4, 5 mM glucose, 100 mM NaCl, 5 mM $MgCl_2$ (*see* **Note 1**).

2.2 Swarm Assay Media for V. cholerae Chemotaxis

1. Tryptone semisolid agar plate: 1% tryptone, 0.5% NaCl, 0.25–0.3% agar.

2.3 Temporal Stimulation Assay for V. cholerae Chemotaxis

1. Dark-field microscope.
2. 20× objective lens.
3. Video camera with a computer installed with the camera control software.
4. Glass test tubes (15 mm diameter, 90 mm tall).

5. TMN buffer (*see* Subheading 2.1).
6. Stock solutions of phenol and an attractant (100-fold of the working concentrations).

2.4 Capillary Assay for V. cholerae Chemotaxis

1. Glass test tubes (15 mm in diameter, 90 mm tall).
2. Capillaries (2 μL, Drummond).
3. Beakers (10 mL).
4. TMN buffer (*see* Subheading 2.1).
5. TG broth (*see* Subheading 2.1).
6. Sterilized distilled water.
7. Solutions of the attractant of interest.
8. Cell suspension in TMN buffer.

2.5 Assay for Receptor Methylation of V. cholerae Chemoreceptors

1. TMN buffer (*see* Subheading 2.1).
2. Cell suspension in TMN buffer.
3. 100-fold concentrated solutions of attractants of interest, e.g., L-serine and taurine (aminoethane sulfonate).
4. SDS loading buffer: 67 mM Tris–HCl, pH 6.8, 8% glycerol, 1% SDS, 0.003% bromophenol blue, 7.7% 2-mercaptoethanol.
5. Reagents and equipment for SDS-PAGE. Acrylamide stock solution: 33% acrylamide, 0.225% N, N′-methylene bis-acrylamide (final 15.1% and 5% acrylamide for the separation gel and the stacking gel, respectively). L buffer for separation gel: 1.5 M Tris–HCl, pH 8.8, 0.4% SDS (×4 dilution for use). M buffer for stacking gel: 0.5 M Tris–HCl, pH 6.8, 0.4% SDS. Q buffer for electrophoresis: 0.1 M Tris–HCl, pH 6.8, 192 mM glycine, 0.1% SDS (×4 dilution for use). APS stock solution: 14% ammonium persulfate (×100 dilution for use). N, N, N′, N′-Tetramethylethylenediamine (0.083% and 0.15% for the separation gel and the stacking gel, respectively). A standard power supply (for example AE-8135, ATTO, Japan) and a mini gel electrophoresis apparatus (for example, BE-240, BIO CRAFT, Japan) are used.
6. Reagents, antibodies, and equipment for immunoblotting. Blotting buffer: 20 mM Trizma base (Sigma-Aldrich, USA), 150 mM glycine, 20% methanol. TBS: 20 mM Tris–HCl, pH 7.0, 0.5 M NaCl. Alkaline phosphatase (AP) chromogenic buffer: 0.1 M Tris–HCl, pH 9.5, 1 M NaCl, 5 mM $MgCl_2$. NBT stock solution: 5% nitro blue tetrazolium chloride in 70% dimethylformamide (final 0.033% in AP buffer). BCIP stock solution: 50 mg/mL 5-bromo-4-chloro-3-indlyl phosphate disodium salt 1.5-hydrate (final 0.165 mg/mL in AP buffer). Skim milk powder (3% for blocking and 1% for antibody reaction) in TBS buffer.

Anti-FLAG M2 antibody (Sigma-Aldrich, USA) and alkaline phosphatase-labeled anti-mouse antibody (KPL, USA). Polyvinylidene difluoride (PVDF) membrane (Immobilon-P, Merck Millipore, USA). A standard power supply (for example AE-8135, ATTO, Japan) and a membrane transfer apparatus (for example, BE-300, BIO CRAFT, Japan) are used.

3 Methods

3.1 Culture of *V. cholerae* to Examine Its Chemotaxis

Procedures to prepare *V. cholerae* cells common to the following assays are described in this section.

1. Culture cells overnight in TG with shaking at 30 °C.
2. Dilute the overnight culture 1:30 in TG and further incubate cells with shaking at 30 °C for 5–6 h.
3. Centrifuge cells in a plastic microtube at 7700 × *g* for 3 min at room temperature.
4. Wash the cells twice with TM buffer and resuspend the cells in TMN buffer unless otherwise noted (*see* **Note 2**).

3.2 Swarm Assay for *V. cholerae* Chemotaxis

1. Prepare an overnight culture (3 mL of TG medium, at 30 °C) or fresh colonies on LB agar.
2. Inoculate an aliquot (e.g., 2 μL) of the overnight culture or a toothpick-picked fresh colony onto a tryptone semisolid agar plate (*see* **Note 3**).
3. Incubate the plate at 37 °C for 5.5 h (*see* **Note 4**).
4. Take photographs of the swarm plates and measure diameters of the swarm rings.

3.3 Temporal Stimulation Assay for *V. cholerae* Chemotaxis

1. Dilute cell suspension about 1:100 with TMN buffer and incubate the resulting cell suspension in a glass test tube (*see* **Note 5**).
2. Without adding any chemoeffector, observe cells under a dark-field microscope with a 20× objective lens at room temperature and record the behavior of the cells with the video camera.
3. Add phenol at an appropriate concentration (e.g., 5 mM).
4. Immediately after the addition of phenol, observe cells and record their behavior as **step 2**.
5. Add phenol (at the concentration used at **step 3**) to a new cell suspension and then add an attractant at an appropriate concentration (up to 10 mM) or add the two chemoeffectors simultaneously.
6. Immediately after the addition of the attractant or the mixture, observe cells and record their behavior as **step 2**.

7. Analyze the images with an appropriate software, such as ImageJ or LabVIEW, to measure turning frequencies of cells under the conditions tested.

3.4 Capillary Assay for *V. cholerae* Chemotaxis

1. During the culture, prepare appropriate volumes (e.g., 1 mL each) of TG in plastic microtubes and keep them at 4 °C (*see* **Note 6**).
2. Adjust the OD_{600} values of the cell suspension to 0.1 with TMN buffer.
3. Put 0.2 mL each of the diluted cell suspension to glass test tubes and incubate them at 30 °C for 1 h (*see* **Note 7**).
4. Prepare capillaries during the incubation times (*see* **Note 8**).
   (a) Pour aliquots (2 mL) of various concentrations of the attractant of interest or buffer into small beakers.
   (b) Melt-shut one end of each capillary with Bunsen gas flame. Prepare all capillaries used in a single assay at this step.
   (c) Mildly heat the capillary up by putting it into gas flame (less than 1 s) for five to six times (*see* **Note 9**).
   (d) Put the capillary into the attractant solution or buffer in the beaker (*see* **Note 10**). The liquid will be sucked into the capillary while it is cooling down.
   (e) Repeat **steps 4(c)** and **4(d)** for all attractant and control solutions to be tested (*see* **Note 11**).
5. Drop each capillary into the cell suspension in a test tube (every 30 s–1 min) (*see* **Note 12**).
6. Incubate them at 30 °C for 1 h.
7. Recover cells from the capillaries.
   (a) Pick up each capillary (every 30 s–1 min).
   (b) Wash out the outer wall of the capillary with sterile distilled water to remove attached cells.
   (c) Wipe remaining water off the capillary wall quickly by filter paper or Kimwipe.
   (d) Break up the closed end of the capillary.
   (e) Blow the capillary content with a microcap into TG broth in a microtube (*see* **Note 13**).
8. Dilute the cell suspension (typically 1:1000) with TG broth, spread onto LB plates containing 50 μg/mL of streptomycin (*see* **Note 14**).
9. Incubate the plates overnight at 37 °C.
10. Count the number of colonies on each plate.

3.5 Assay for Receptor Methylation of *V. cholerae* Chemoreceptors (See Note 15)

1. Adjust OD_{600} of the cell suspension to 0.5 with TMN buffer.
2. Take into aliquots (490 μL each) (*see* **Note 16**).
3. Add 10 μL of H_2O or 0.5 M attractant solution.
4. Incubate cell suspensions for 30 min at 30 °C in a water bath.
5. Harvest cells by centrifugation for 3 min at top speed (21,500 × *g*).
6. Remove supernatant completely (*see* **Note 17**).
7. Resuspend the pellets with 250 μL of 1× SDS buffer, boil for 5 min.
8. Apply the samples (5–10 μL, *see* **Note 18**) to SDS-PAGE (*see* **Note 19**) followed by immunoblotting (*see* **Note 20**).

4 Notes

1. Stock solution of 1 M $MgCl_2$ should be prepared separately, filtered and added to the buffer after autoclave.
2. Cells must be resuspended gently to avoid shearing of flagella.
3. Make triplicates or more as swarm sizes substantially vary.
4. The temperature and the duration can be modified.
5. For cell incubation, avoid plastic tubes to maintain their motility and responsiveness.
6. The volume of TG should be determined in accordance with dilution ratio at **step 8**.
7. For cell incubation, glassware (for instance, glass test tubes (15 mm diameter, 90 mm tall), IWAKI Glass Co., Tokyo, Japan) is recommended to maintain their motility and responsiveness.
8. Handle capillaries with precision tweezers (SS. S., ideal-tek S.A., Switzerland).
9. Use the low-temperature, reducing flame.
10. Hold the capillary until it cools down to avoid bumping or explosive boil, which would make it jumping out from the beaker.
11. The volume of the solution to be sucked is about two thirds of the content of a capillary (i.e., 2/3 solution and 1/3 air). The volumes must be equal among all capillaries with minimal variations. Heating time at **step 4(c)** is critical for it.
12. Make sure that the ends of all capillaries are certainly reached to the bottom of test tubes.
13. Place the microtube stands on ice to prevent cell proliferation, which would affect the accuracy of the assay.

14. *V. cholerae* is naturally resistant to streptomycin. This antibiotic prevents growth of other bacteria, such as *Staphylococcus* spp., which would otherwise contaminate the samples.
15. Here, we describe the protocol to detect methylation of a specific MLP that is fused to a FLAG tag with anti-FLAG antibody. Unlike the major chemoreceptors (Tsr and Tar) of *E. coli*, each of which yields multiple bands upon methylation, thus far we observed only single methylated band for Mlp24 or Mlp37 [18, 20] (Fig. 5c), suggesting that these MLPs might have less methylation sites than *E. coli* Tsr and Tar.
16. Glass test tubes are highly recommended for efficient thermal conductivity and to keep cells healthy.
17. The pellets can be stored at −30 °C if necessary.
18. Sample volumes to be applied to SDS-PAGE should be varied by the expression levels of the target MLPs.
19. For SDS-PAGE, 15% (w/v) of acrylamide mini gel is used. Conditions for the electrophoresis are 25 mA, 2.5 h. The run time is optimized for Mlp24 or Mlp37 (molecular weight is approximately 67,000). It should be varied by the molecular weights of the targeted chemoreceptors.
20. Dilution rates for the first and secondary antibody are both 1:4000.

Acknowledgments

We thank Drs. M. Nishikawa and Y. Sowa, our colleagues at Hosei University, for critically the reading manuscript and R. Iwazaki, Y. Miura and T. Nakagawa, our present or former students, for preliminary examination of some of the assay protocols.

References

1. Reidl J, Klose KE (2002) *Vibrio cholerae* and cholera: out of the water and into the host. FEMS Microbiol Rev 26:125–139
2. Boin MA, Austin MJ, Häse CC (2004) Chemotaxis in *Vibrio cholerae*. FEMS Microbiol Lett 239:1–8
3. Alm RA, Manning PA (1990) Characterization of the *hlyB* gene and its role in the production of the El Tor haemolysin of *Vibrio cholerae* O1. Mol Microbiol 4:413–425
4. Banerjee R, Das S, Mukhopadhyay K, Nag S, Chakrabortty A, Chaudhuri K (2002) Involvement of *in vivo* induced *cheY-4* gene of *Vibrio cholerae* in motility, early adherence to intestinal epithelial cells and regulation of virulence factors. FEBS Lett 532:221–226
5. Everiss KD, Hughes KJ, Kovach ME, Peterson KM (1994) The *Vibrio cholerae acfB* colonization determinant encodes an inner membrane protein that is related to a family of signal-transducing proteins. Infect Immun 62:3289–3298
6. Everiss KD, Hughes KJ, Peterson KM (1994) The accessory colonization factor and toxin-coregulated pilus gene clusters are physically linked on the *Vibrio cholerae* O395 chromosome. DNA Seq 5:51–55
7. Butler SM, Camilli A (2004) Both chemotaxis and net motility greatly influence the infectivity of *Vibrio cholerae*. Proc Natl Acad Sci U S A 101:5018–5023
8. Freter R, O'Brien PC (1981) Role of chemotaxis in the association of motile bacteria with

intestinal mucosa: fitness and virulence of non-chemotactic *Vibrio cholerae* mutants in infant mice. Infect Immun 34:222–233

9. Freter R, Allweiss B, O'Brien PC, Halstead SA, Macsai MS (1981) Role of chemotaxis in the association of motile bacteria with intestinal mucosa: *in vitro* studies. Infect Immun 34:241–249
10. Freter R, O'Brien PC, Macsai MS (1981) Role of chemotaxis in the association of motile bacteria with intestinal mucosa: *in vivo* studies. Infect Immun 34:234–240
11. Gupta S, Chowdhury R (1997) Bile affects production of virulence factors and motility of *Vibrio cholerae*. Infect Immun 65:1131–1134
12. Hang L, John M, Asaduzzaman M, Bridges EA, Vanderspurt C, Kirn TJ, Taylor RK, Hillman JD, Progulske-Fox A, Handfield M, Ryan ET, Calderwood SB (2003) Use of *in vivo*-induced antigen technology (IVIAT) to identify genes uniquely expressed during human infection with *Vibrio cholerae*. Proc Natl Acad Sci U S A 100:8508–8513
13. Krukonis ES, DiRita VJ (2003) From motility to virulence: sensing and responding to environmental signals in *Vibrio cholerae*. Curr Opin Microbiol 6:186–190
14. Lee SH, Butler SM, Camilli A (2001) Selection for *in vivo* regulators of bacterial virulence. Proc Natl Acad Sci U S A 98:6889–6894
15. Heidelberg JF, Eisen JA, Nelson WC, Clayton RA, Gwinn ML, Dodson RJ, Haft DH, Hickey EK, Peterson JD, Umayam L, Gill SR, Nelson KE, Read TD, Tettelin H, Richardson D, Ermolaeva MD, Vamathevan J, Bass S, Qin H, Dragoi I, Sellers P, McDonald L, Utterback T, Fleishmann RD, Nierman WC, White O, Salzberg SL, Smith HO, Colwell RR, Mekalanos JJ, Venter JC, Fraser CM (2000) DNA sequence of both chromosomes of the cholera pathogen *Vibrio cholerae*. Nature 406:477–483
16. Gosink KK, Kobayashi R, Kawagishi I, Häse CC (2002) Analyses of the roles of the three *cheA* homologs in chemotaxis of *Vibrio cholerae*. J Bacteriol 184:1767–1771
17. Hyakutake A, Homma M, Austin MJ, Boin MA, Häse CC, Kawagishi I (2005) Only one of the five CheY homologs in *Vibrio cholerae* directly switches flagellar rotation. J Bacteriol 187:8403–8410
18. Nishiyama S, Suzuki D, Itoh Y, Suzuki K, Tajima H, Hyakutake A, Homma M, Butler-Wu SM, Camilli A, Kawagishi I (2012) Mlp24 (McpX) of *Vibrio cholerae* implicated in pathogenicity functions as a chemoreceptor for multiple amino acids. Infect Immun 80:3170–3178
19. Boin MA, Häse CC (2007) Characterization of *Vibrio cholerae* aerotaxis. FEMS Microbiol Lett 276:193–201
20. Nishiyama S, Takahashi Y, Yamamoto K, Suzuki D, Itoh Y, Sumita K, Uchida Y, Homma M, Imada K, Kawagishi I (2016) Identification of a *Vibrio cholerae* chemoreceptor that senses taurine and amino acids as attractants. Sci Rep 6:20866
21. Wolfe AJ, Berg HC (1989) Migration of bacteria in semisolid agar. Proc Natl Acad Sci U S A 86:6973–6977
22. Alvarez-Ortega C, Harwood CS (2007) Identification of a malate chemoreceptor in *Pseudomonas aeruginosa* by screening for chemotaxis defects in an energy taxis-deficient mutant. Appl Environ Microbiol 73: 7793–7795
23. Macnab RM, Koshland DE Jr (1972) The gradient-sensing mechanism in bacterial chemotaxis. Proc Natl Acad Sci U S A 69:2509–2512
24. Homma M, Oota H, Kojima S, Kawagishi I, Imae Y (1996) Chemotactic responses to an attractant and a repellent by the polar and lateral flagellar systems of *Vibrio alginolyticus*. Microbiology 142:2777–2783
25. Adler J (1973) A method for measuring chemotaxis and use of the method to determine optimum conditions for chemotaxis by *Escherichia coli*. J Gen Microbiol 74:77–91
26. Adler J, Dahl MM (1967) A method for measuring the motility of bacteria and for comparing random and non-random motility. J Gen Microbiol 46:161–173
27. Mesibov R, Adler J (1972) Chemotaxis toward amino acids in *Escherichia coli*. J Bacteriol 112:315–326
28. Boyd A, Simon MI (1980) Multiple electrophoretic forms of methyl-accepting chemotaxis proteins generated by stimulus-elicited methylation in *Escherichia coli*. J Bacteriol 143:809–815
29. Chelsky D, Dahlquist FW (1980) Structural studies of methyl-accepting chemotaxis proteins of *Escherichia coli*: evidence for multiple methylation sites. Proc Natl Acad Sci U S A 77:2434–2438
30. Dunten P, Koshland DE Jr (1991) Tuning the responsiveness of a sensory receptor via covalent modification. J Biol Chem 266:1491–1496
31. Engström P, Hazelbauer GL (1980) Multiple methylation of methyl-accepting chemotaxis proteins during adaptation of *E. coli* to chemical stimuli. Cell 20:165–171
32. Okumura H, Nishiyama S, Sasaki A, Homma M, Kawagishi I (1998) Chemotactic adaptation

is altered by changes in the carboxy-terminal sequence conserved among the major methyl-accepting chemoreceptors. J Bacteriol 180:1862–1868

33. Sowa Y, Hotta H, Homma M, Ishijima A (2003) Torque-speed relationship of the Na^+-driven flagellar motor of *Vibrio alginolyticus*. J Mol Biol 327:1043–1051
34. Sourjik V, Vaknin A, Shimizu TS, Berg HC (2007) *In vivo* measurement by FRET of pathway activity in bacterial chemotaxis. Methods Enzymol 423:365–391
35. Clarke S, Koshland DE Jr (1979) Membrane receptors for aspartate and serine in bacterial chemotaxis. J Biol Chem 254:9695–9702
36. Glekas GD, Foster RM, Cates JR, Estrella JA, Wawrzyniak MJ, Rao CV, Ordal GW (2010) A PAS domain binds asparagine in the chemotaxis receptor McpB in *Bacillus subtilis*. J Biol Chem 285:1870–1878
37. Lin LN, Li J, Brandts JF, Weis RM (1994) The serine receptor of bacterial chemotaxis exhibits half-site saturation for serine binding. Biochemistry 33:6564–6570
38. Tajima H, Imada K, Sakuma M, Hattori F, Nara T, Kamo N, Homma M, Kawagishi I (2011) Ligand specificity determined by differentially arranged common ligand-binding residues in bacterial amino acid chemoreceptors Tsr and Tar. J Biol Chem 286:42200–42210

Chapter 22

Purification of Fla2 Flagella of *Rhodobacter sphaeroides*

Javier de la Mora, Laura Camarena, and Georges Dreyfus

Abstract

The photosynthetic bacterium *R. sphaeroides* expresses two flagellar systems that are encoded by two complete gene clusters that have distinct phylogenetic origins. The isolation and purification of the Filament-Hook Basal Body (F-HBB) or the Hook Basal Body (HBB) structure is a troublesome task given the complexity of this nano-machine that is composed of multiple loosely bound substructures that can be lost during the isolation and purification procedure. A successful procedure requires adjustments to the standard method established for *Salmonella*. In this chapter, we describe a detailed protocol to isolate and purify the Fla2 F-HBB and HBB from *R. sphaeroides* a photosynthetic bacterium that has a complex intracellular membrane system that frequently interferes with isolation of high-quality samples.

Key words *Rhodobacter sphaeroides*, Motility organelle, Bacterial flagella, HBB purification

1 Introduction

Motility is a fundamental survival trait in microorganisms, therefore the widespread presence of dedicated organelles in many bacterial species. The bacterial flagellum is a complex and efficient organelle capable of propelling bacteria through liquid and viscous environments. Furthermore, this structure plays an important role in adhesion, biofilm formation, and virulence and it is composed by more than 30 different proteins with copy numbers that run from a few to thousands [1–5]. The structure can be divided into three parts, a long helical filament, a connecting structure known as the hook, and a basal body that spans the bacterial cell envelope [6]. The filament is a cylindrical structure helical in shape that works as an Archimedes screw or propeller [7–9]. The hook is also a cylindrical structure constructed in a similar fashion to the filament that functions as a universal joint. In bacteria with multiple flagella the hooks enable filaments to work as a bundle [10, 11]. The basal body consists of an integral membrane ring called the MS ring, a rod that traverses the periplasmic space, a periplasmic P ring, and an outer membrane L ring [12, 13]. In the cytoplasm

Tohru Minamino and Keiichi Namba (eds.), *The Bacterial Flagellum: Methods and Protocols*, Methods in Molecular Biology, vol. 1593, DOI 10.1007/978-1-4939-6927-2_22,

the basal body forms a bell-like structure known as the C ring that contains the export apparatus [14–18]. The C ring is also the input for signals that control the direction of rotation and consequently cell movement [19]. MotA and MotB form the stator and the MotA/B complex contains the proton channel that harnesses the energy provided by the electrochemical potential [20–22]. *R. sphaeroides* is a highly versatile photosynthetic microorganism that belongs to the α-subgroup of *Proteobacteria* [23]. This bacterium contains two sets of flagellar genes that produce two distinct flagella, a medially located flagellum and polar flagella. The presence of a second set of flagellar genes was discovered when the genomic sequence was released [24]; nevertheless, its functionality was in doubt given that it was known that motility was absolutely dependent on the *fla1* system. This question was solved when it was reported that the second set of flagellar genes (*fla2*) is not expressed under laboratory conditions in the wild-type strain WS8N [25]. We were able to isolate a spontaneous mutant that expressed the *fla2* genes [25, 26]. This mutant assembles several functional polar flagella that allow *R. sphaeroides* to swim in liquid media not on surfaces. The *fla2* set of genes constitutes the native flagellar genes of *R. sphaeroides*, whereas the genes of the *fla1* cluster were acquired by horizontal gene transfer from an ancestral γ-proteobacterium [25].

Recently, electron cryotomography has been used to observe the flagellar structure of different microorganisms revealing an astonishing diversity of features that never before had been fully appreciated [15, 17, 27–29]. From these studies a great interest to understand the structural diversity of the flagellar structures has recently arisen. In this regard, we developed a protocol to isolate the Fla2 F-HBB and HBB from the photosynthetic bacterium *R. sphaeroides*.

2 Materials

2.1 Reagents

1. Triton X-100.
2. Sterile and lyophilized DNase I from bovine pancreas.
3. Trizma base ≥99.9%, crystalline.
4. EDTA.
5. Lysozyme.
6. Sucrose.
7. CsCl.
8. HCl.
9. Casamino acids, casein enzymatic hydrolysate from bovine milk.

2.2 Buffers, Growth Media, and Solutions

Prepare all solutions with deionized water with a conductivity of 10 MΩ-cm at 25 °C.

1. Sucrose buffer
   Add 42.78 g of sucrose and 4.54 g of Trizma base in a 500 mL beaker. Dissolve in deionized H_2O and bring the volume to 250 mL. **Do not adjust pH** (pH should be approximately 10.1). The molar concentration of the reagents in this solution is 0.5 M sucrose and 0.15 M Trizma base. Store at 4 °C.
2. Lysozyme stock solution
   Lysozyme (10 mg) is dissolved in 1 mL of deionized H_2O. It should be freshly prepared each time (*see* **Note 1**).
3. EDTA stock solution
   Add 3.72 g of EDTA in a 250 mL beaker, dissolve in deionized H_2O, and bring to a final volume of 100 mL. The molar concentration of this solution is 0.1 M. Store at room temperature, the final pH value is 4.7.
4. $MgSO_4$ stock solution
   Add 2.46 g of $MgSO_4$ $7H_2O$ in a 250 mL beaker, dissolve in deionized H_2O, and bring volume to 100 mL. The concentration of this solution is 0.1 M. Store at 4 °C.
5. Triton X-100 stock solution
   10 mL of Triton X-100 is dissolved in deionized H_2O. Bring to a final volume of 100 mL. The concentration of this solution is 10% (v/v). Store at room temperature (*see* **Note 2**).
6. TET buffer
   Add 121.14 mg of Trizma base to a 100 mL beaker and dissolve in deionized H_2O, add 1 mL EDTA from stock solution and 1 mL Triton X-100 from stock solution. Bring final volume to 100 mL. Adjust pH to 8 with concentrated HCl. The final concentration of the reagents in this solution is 0.01 M Trizma base, 0.001 M EDTA, and 0.1% Triton X-100. Store at 4 °C (*see* **Note 3**).
7. A 34 or 40% CsCl solution in TET buffer should be freshly made for each preparation of HBBs (*see* **Note 13**).
8. 10X Sistrom's medium [30].
   Dissolve 34.8 g of K_2HPO_4 or 27.2 g of KH_2PO_4, 5.0 g of $(NH_4)_2$ SO_4 or 1.95 g of NH_4Cl, 40.0 g of succinic acid, 1.0 g of l-glutamic acid, 0.4 g of l-aspartic acid, 5.0 g of NaCl, 2.0 g of nitrilotriacetic acid, 3.0 g of $MgSO_4$ $7H_2O$ or 2.44 g of $MgCl_2$ $6H_20$, 0.334 g of $CaCl_2$ $2H_2O$, 0.020 g of $FeSO_4$ $7H_2O$, 0.002 g of $(NH_4)_6$ Mo_7O_{24} $4H_2O$ or 0.1 mL of a 2% solution of $(NH_4)_6Mo_7O_{24}$ $4H_2O$ and 1.0 mL of Trace elements solution (*see below*) in deionized water and take to 1 L. Store 100 mL aliquots at −20 °C.
9. Trace elements solution

Dissolve 1.765 g of EDTA Disodium salt dihydrate 99% min, 10.95 g of $ZnSO_4$ $7H_2O$, 5.00 g of $FeSO_4$ $7H_20$, 1.54 g of $MnSO_4$ H_2O, 0.392 g of $CuSO_4$ $5H_2O$, 0.248 g of Co $(NO_3)_2$ $6H_2O$, and 0.114 g of H_3BO_3 in deionized water and take to 100 mL, store at 4 °C.

10. 10,000× Vitamin solution

    Dissolve 1.0 g of Nicotinic acid, 0.5 g of Thiamine-HCl, and 0.010 g of d-Biotin in deionized water and take to 100 mL. Sterilize by filtration (0.22 μm sieve) and store at 4 °C.

11. 1× Sistrom's medium

    (a) Add 100 mL of 10× Sistrom's medium to a beaker containing 900 mL of deionized water.

    (b) Add while stirring 2 g of casamino acids.

    (c) Once dissolved adjust the pH to 7.0 with a 6 M solution of KOH (JT Baker, catalog number 3140-01).

    (d) Autoclave for 20 min at 120 °C. Importantly, cool it down to room temperature and then add 100 μL of the 10,000× vitamin solution.

12. Required strains to isolate and purify Fla2 flagella

    R. sphaeroides has two flagellar gene systems that allow the biogenesis of two different types of flagella. The Fla1 flagellum is a single structure located medially or at a near-polar location [31, 32], and the *fla2* system produces on average 5–6 polar flagella [33]. It has been shown that the expression of the genes encoding for the Fla2 flagella is dependent on the activation of the two component system CckA-ChpT-CtrA [34]. Despite the fact that there are some environmental conditions that favor activation of this two-component system, so far it remains to be elucidated how to turn on this flagellar system efficiently. Therefore, the best condition to isolate the Fla2 flagella is using a strain carrying a mutation that prevents the expression of the *fla1* genes, and carries a gain of function mutation in *cckA*. Several single mutations in this gene have been isolated that enable the expression of the *fla2* genes when expressed from a plasmid even in the presence of the wild-type *cckA* gene [34].

    Using this type of strain, that is phenotypically $Fla1^-$ $Fla2^+$, the isolation of the Fla2 structures can be carried out properly [33]. The following procedure was done using the strain AM1 (CckA $_{L391F}$) or a derivative of the AM1 strain carrying a mutation in the flagellin gene *flaA* (RSflaA strain, *flaA::aadA*) [26, 33, 34].

3 Methods

3.1 Growth Conditions

1. Photosynthetic bacteria represent a challenge to obtain isolated flagella with a reduced amount of membranous material from the photosynthetic apparatus. Cell cultures are performed under conditions that render a lower amount of photosynthetic pigments. For this, the anoxygenic purple bacterium *R. sphaeroides* is grown heterotrophically with shaking to provide plenty aeration and thus reduce the presence of the photosynthetic apparatus. Clumpy cultures should be avoided to prevent uneven cell lysis. For frequent isolation and purification of flagella, *R. sphaeroides* is maintained on agar plates of Sistrom's growth medium (Sis) supplemented with casamino acids, kept at 4 °C for 4 weeks, after this time a new plate should be streaked from a glycerol stock.
2. To start an overnight culture, use an assay tube (18 × 150 mm) with a metallic cap containing 5 mL of Sistrom's minimal medium that is supplemented with 0.2% casamino acids. Inoculate with *R. sphaeroides* AM1 or RSflaA strain from a stock plate and incubate at 30 °C with orbital shaking (230 rpm), for 16 h.
3. Inoculate a 1000 mL flask containing 500 mL of Sistrom's culture medium, without succinic acid and supplemented with 0.2% casamino acids, with 5 mL of the overnight culture. The culture is incubated at 30 °C with orbital shaking (230 rpm) for 24 h. At this time, the culture normally reaches an OD_{600} of 0.45–0.50 (*see* **Note 4**).
4. Before harvesting the AM1 cells, swimming is verified by high-intensity dark-field microscopy. Start the procedure only if at least 50% of the bacterial population is actively swimming.

3.2 Isolation of F-HBBs and HBBs

1. The culture is transferred to two 500 mL centrifuge bottles and centrifuged at 3909 × *g* for 20 min at 4 °C (*see* **Note 5**).
2. The supernatant is carefully discarded, and the walls of the bottles are carefully dried with a kimwipe tissue to remove all traces of culture medium. Each pellet is soaked in 3 mL of a cold sucrose solution for a few seconds before gently dissolving on the tube wall with a spatula (spoonula from Fisher) and making sure that the pellet is constantly soaked until cells are completely resuspended. The same procedure should be carried out with the second bottle. The cell suspension is transferred to a 50 mL glass beaker and kept on ice. Use 4 mL of cold sucrose buffer to recover the remaining cells from the two bottles and add the whole volume to the cold beaker. The beaker with the cell suspension immersed in an ice bath is stirred gently (50–60 rpm) with a small magnetic bar (length 2.5–3.5 cm) to effectively mix the solutions and reagents indicated below (2–10).

3. Gradually add 100 μL of the lysozyme stock solution, at a speed of approximately 100 μL in 5 min. This is done using a 200 μL pipette containing 100 μL of the lysozyme stock solution. The plastic tip is submerged in the cell suspension and the thumbwheel slowly turned from 100 to 0 μL in 5 min.
4. 1 mL of 20 mM EDTA (a 5× dilution of a 100 mM stock solution) is gradually added (approximately 1000 μL in 5 min); this is done as described in **step 2** using a 1000 μL automatic pipette.
5. The cell suspension containing lysozyme and EDTA is stirred for 5 min at 4 °C.
6. The cell suspension is removed from the ice bath and further stirred for 50–60 min at room temperature (23–24 °C). This allows the formation of spheroplasts (*see* **Note 6**).
7. 1.0 mL of a Triton X-100 stock solution is added quickly to achieve cell lysis. Stir at room temperature for 5–10 min. The suspension will shift from cloudy to transparent and it will become very viscous due to the liberation of chromosomal DNA (*see* **Note 7**).
8. 600 μL of a $MgSO_4$ stock solution (0.1 M) is added to the beaker. Immediately after DNase I powder (0.5 mg) is added while stirring continues at room temperature.
9. Maintain gentle stirring until the suspension is no longer viscous. This should happen approximately 5 min after DNase I was added (*see* **Note 8**).
10. Add 600 μL of a stock EDTA solution.
11. Add 10 mL of a cold sucrose buffer.
12. At this point the suspension is transferred to a 50 ml polycarbonate tube and centrifuged at 9790 × *g* for 10 min at 4 °C (*see* **Note 9**).
13. The supernatant (approximately 23 mL) is carefully transferred to a 30 mL polycarbonate tube for the 50.2 Ti rotor of a Beckman ultracentrifuge (*see* **Note 9**).
14. The sample is centrifuged at 56,805 × *g* for 35 min at 4 °C using maximum acceleration and deceleration (*see* **Note 10**).
15. The supernatant is carefully discarded and the walls of the tube cleaned with a kimwipe tissue to remove traces of the sucrose solution.
16. The pellet containing F-HBB or HBB structures is resuspended overnight in 400 μL of TET buffer (*see* **Note 11** and **Note 12**).

3.3 Purification of Hook Basal Bodies

1. A 13.2 mL ultraclear tube (Beckman) intended for use with the SW40 Ti rotor (Beckman) is filled with 11.0 mL of a 40% or 34% CsCl solution and the final volume (13.2 mL) is completed

with a batch of F-HBB or HBB samples from **step 15** respectively (*see* **Note 13**). We recommend the use of two preparations for this step to end up having enough material. Mix the tube gently to ensure proper mixing.

2. The sample is centrifuged for 21 h at 20,000 rpm (55,689 × *g*) at 20 °C using maximum acceleration and free deceleration (no brake applied).
3. The two kinds of preparations, i.e., F-HBBs or HBBs, form an opalescent band approximately half way down the tube (Arrow Fig. 1a). The material above the band is withdrawn with a 1 mL automatic pipette after which the sharp band is carefully withdrawn (*see* **Note 14**).
4. The collected sample (sharp opalescent band) is diluted in 20 mL of TET buffer and centrifuged at 25,000 rpm (56,805 × *g*) for 35 min at 4 °C using maximum acceleration and deceleration (*see* **Note 10**). The supernatant is carefully discarded and the walls of the tube cleaned with a kimwipe tissue to remove traces of the CsCl solution.

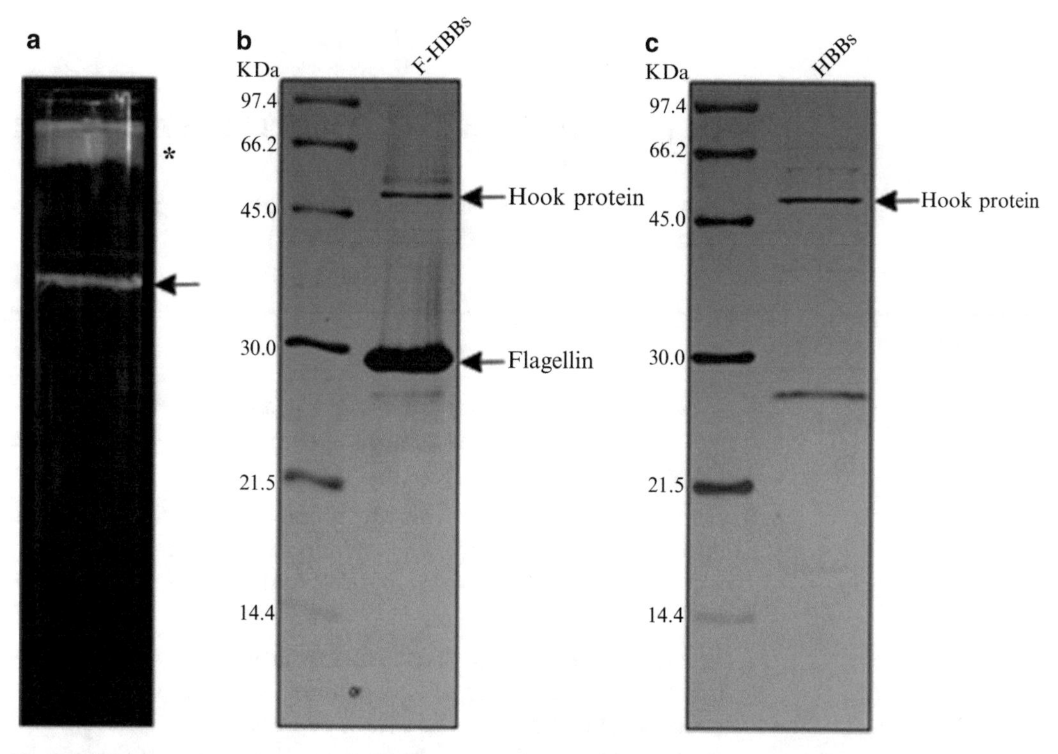

Fig. 1 Purification of Fla2 flagella. (**a**) A CsCl gradient showing three opalescent bands. At the top is the membranous material marked by an *asterisk*. The *arrow* indicates the band that contains the HBB sample. A band that always appears just above the sharp band indicated by the *arrow* contains membranes and cell debris. (**b**) SDS-PAGE (15%) showing a representative F-HBB preparation obtained from the 40% CsCl gradient. (**c**) SDS-PAGE (15%) showing a representative HBB preparation obtained from the 34% CsCl gradient. The SDS-PAGE gels were silver stained as described previously [35]

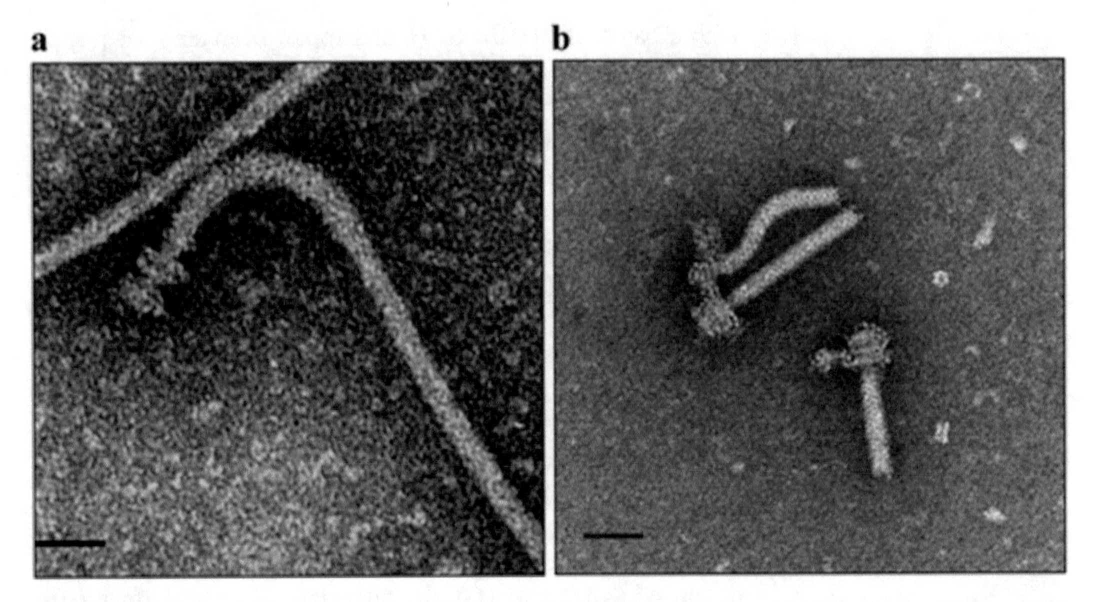

Fig. 2 Electron micrographs showing in (**a**) F-HBBs and (**b**) HBBs as they are obtained after the CsCl gradient. Samples were stained with 2% phosphotungstic acid and observed with a JEM-1200EXII electron microscope (JEOL, Tokyo Japan). Micrographs were taken at an accelerating voltage of 80 or 100 kV. Bars correspond to 50 nm

5. The pellet is resuspended in 200 μL of TET buffer and stored at 4 °C (*see* **Note 12**).
6. The sample can be analyzed by SDS-PAGE (Fig. 1b, c) and electron microscopy (Fig. 2a, b).

4 Notes

1. It is recommended that the lysozyme solution is freshly prepared for each procedure. Do not store the lysozyme solution more than 1 week at 4 °C.
2. Add Triton X-100 very slowly to 60 mL of distilled water under constant stirring. This avoids the formation of aggregates that become insoluble. Once the detergent is in solution add water to 100 mL.
3. It is recommended to adjust pH with 1 N HCl to better control the change in the pH. It can also be done with concentrated HCl (12 N) but the risk of error is high.
4. The growth procedure for the purification of HBBs from nonmotile mutants (Fla^- or Mot^-) is exactly the same as for the motile AM1 strain.

5. The centrifuge bottles should be cleaned previously with 70% ethanol and dried with a kimwipe tissue. This prevents contamination with biological material that could be attached to the inner wall of the bottle.
6. Special care should be taken to keep the temperature above 23 °C to ensure that lysozyme activity is optimal and spheroplasts form efficiently.
7. If the suspension is not completely transparent after 10 min, it is possible that the incubation time with lysozyme was insufficient, or that the enzyme was not working appropriately. It is recommended to stop the procedure at this point.
8. The fluidity of the suspension should be tested with a pipette in order to verify that it is no longer viscous.
9. The centrifuge tubes should be cleaned with 70% ethanol and dried using a kimwipe tissue to avoid contamination with biological material.
10. When isolating Fla-HBBs from the RSflaA strain, the centrifugation speed and time should be increased to 81,799 × *g* and 1 h respectively.
11. At this stage the F-HBBs and HBBs can be observed by transmission electron microscopy.
12. The pellet can be left overnight covered with 200–400 μL of TET buffer at 4 °C. The tube should be left inclined to ensure that the pellet is covered. The next day the pellet is resuspended by gentle mixing. DO NOT RESUSPEND WITH A PIPETTE.
13. To purify HBBs without filament a concentration of CsCl of 34% should be used.
14. At the bottom of the gradient it is possible to obtain hooks and various ring structures.

Acknowledgments

We thank Teresa Ballado and Aurora Osorio for technical assistance. We also thank Shin-Ichi Aizawa for many helpful and lively discussions. Financial support was from CONACyT (grant 235996) and DGAPA/UNAM (grant IN204614).

References

1. Morimoto YV, Minamino T (2014) Structure and function of the bi-directional bacterial flagellar motor. Biomolecules 4:217–234
2. Belas R (2014) Biofilms, flagella, and mechanosensing of surfaces by bacteria. Trends Microbiol 22:517–527

3. Chaban B, Hughes HV, Beeby M (2015) The flagellum in bacterial pathogens: for motility and a whole lot more. Semin Cell Dev Biol 46:91–103
4. Guttenplan SB, Kearns DB (2013) Regulation of flagellar motility during biofilm formation. FEMS Microbiol Rev 37:849–871
5. Rossez Y, Wolfson EB, Holmes A, Gally DL, Holden NJ (2015) Bacterial flagella: twist and stick, or dodge across the kingdoms. PLoS Pathog 11:e1004483
6. Macnab RM (2003) How bacteria assemble flagella. Annu Rev Microbiol 57:77–100
7. Trachtenberg S, Fishelov D, Ben-Artzi M (2003) Bacterial flagellar microhydrodynamics: laminar flow over complex flagellar filaments, analog archimedean screws and cylinders, and its perturbations. Biophys J 85:1345–1357
8. Berg HC, Anderson RA (1973) Bacteria swim by rotating their flagellar filaments. Nature 245:380–382
9. Silverman M, Simon M (1974) Flagellar rotation and the mechanism of bacterial motility. Nature 249:73–74
10. Block SM, Blair DF, Berg HC (1991) Compliance of bacterial polyhooks measured with optical tweezers. Cytometry 12:492–496
11. Samatey FA, Imada K, Nagashima S, Vonderviszt F, Kumasaka T, Yamamoto M, Namba K (2001) Structure of the bacterial flagellar protofilament and implications for a switch for supercoiling. Nature 410:331–337
12. Aizawa SI, Dean GE, Jones CJ, Macnab RM, Yamaguchi S (1985) Purification and characterization of the flagellar hook-basal body complex of *Salmonella typhimurium*. J Bacteriol 161:836–849
13. DePamphilis ML, Adler J (1971) Fine structure and isolation of the hook-basal body complex of flagella from *Escherichia coli* and *Bacillus subtilis*. J Bacteriol 105:384–395
14. Francis NR, Sosinsky GE, Thomas D, DeRosier DJ (1994) Isolation, characterization and structure of bacterial flagellar motors containing the switch complex. J Mol Biol 235:1261–1270
15. Murphy GE, Leadbetter JR, Jensen GJ (2006) In situ structure of the complete *Treponema primitia* flagellar motor. Nature 442:1062–1064
16. Minamino T, Imada K, Namba K (2008) Molecular motors of the bacterial flagella. Curr Opin Struct Biol 18:693–701
17. Chen S, Beeby M, Murphy GE, Leadbetter JR, Hendrixson DR, Briegel A, Li Z, Shi J, Tocheva EI, Muller A et al (2011) Structural diversity of bacterial flagellar motors. EMBO J 30:2972–2981
18. Fan F, Ohnishi K, Francis NR, Macnab RM (1997) The FliP and FliR proteins of *Salmonella typhimurium*, putative components of the type III flagellar export apparatus, are located in the flagellar basal body. Mol Microbiol 26:1035–1046
19. Yuan J, Berg HC (2013) Ultrasensitivity of an adaptive bacterial motor. J Mol Biol 425:1760–1764
20. Sowa Y, Berry RM (2008) Bacterial flagellar motor. Q Rev Biophys 41(2):103–132
21. Minamino T, Imada K (2015) The bacterial flagellar motor and its structural diversity. Trends Microbiol 23:267–274
22. Kojima S (2015) Dynamism and regulation of the stator, the energy conversion complex of the bacterial flagellar motor. Curr Opin Microbiol 28:66–71
23. Imhoff JF (2006) The phototrophic alpha proteobacteria. In: Dworkin M, Falkow S, Rosenberg E, Schleifer K-H, Stackebrandt E (eds) The prokaryotes, 3rd edn. Vol 5: Proteobacteria: alpha and beta subclasses. Springer, New York, pp 41–64
24. Mackenzie C, Choudhary M, Larimer FW, Predki PF, Stilwagen S, Armitage JP, Barber RD, Donohue TJ, Hosler JP, Newman JE et al (2001) The home stretch, a first analysis of the nearly completed genome of *Rhodobacter sphaeroides* 2.4.1. Photosynth Res 70:19–41
25. Poggio S, Abreu-Goodger C, Fabela S, Osorio A, Dreyfus G, Vinuesa P, Camarena L (2007) A complete set of flagellar genes acquired by horizontal transfer coexists with the endogenous flagellar system in *Rhodobacter sphaeroides*. J Bacteriol 189:3208–3216
26. del Campo AM, Ballado T, de la Mora J, Poggio S, Camarena L, Dreyfus G (2007) Chemotactic control of the two flagellar systems of *Rhodobacter sphaeroides* is mediated by different sets of CheY and FliM proteins. J Bacteriol 189:8397–8401
27. Liu J, Howell JK, Bradley SD, Zheng Y, Zhou ZH, Norris SJ (2010) Cellular architecture of *Treponema pallidum*: novel flagellum, periplasmic cone, and cell envelope as revealed by cryo electron tomography. J Mol Biol 403:546–561
28. Zhao X, Zhang K, Boquoi T, Hu B, Motaleb MA, Miller KA, James ME, Charon NW, Manson MD, Norris SJ et al (2013) Cryoelectron tomography reveals the sequential assembly of bacterial flagella in *Borrelia burgdorferi*. Proc Natl Acad Sci U S A 110:14390–14395
29. Beeby M, Ribardo DA, Brennan CA, Ruby EG, Jensen GJ, Hendrixson DR (2016) Diverse high-torque bacterial flagellar motors assemble wider stator rings using a conserved protein scaffold. Proc Natl Acad Sci U S A 113:E1917–E1926

30. Sistrom WR (1962) The kinetics of the synthesis of photopigments in *Rhodopseudomonas sphaeroides.* J Gen Microbiol 28:607–616
31. Armitage JP, Macnab RM (1987) Unidirectional, intermittent rotation of the flagellum of *Rhodobacter sphaeroides.* J Bacteriol 169:514–518
32. Fujii M, Shibata S, Aizawa S (2008) Polar, peritrichous, and lateral flagella belong to three distinguishable flagellar families. J Mol Biol 379:273–283
33. de la Mora J, Uchida K, del Campo AM, Camarena L, Aizawa S, Dreyfus G (2015) Structural characterization of the Fla2 flagellum of *Rhodobacter sphaeroides.* J Bacteriol 197:2859–2866
34. Vega-Baray B, Domenzain C, Rivera A, Alfaro-Lopez R, Gómez-Cesar E, Poggio S, Dreyfus G, Camarena L (2015) The flagellar set Fla2 in *Rhodobacter sphaeroides* is controlled by the CckA pathway and is repressed by organic acids and the expression of Fla1. J Bacteriol 197:833–847
35. Morrissey JH (1981) Silver stain for proteins in polyacrylamide gels: a modified procedure with enhanced uniform sensitivity. Anal Biochem 117:307–310

Chapter 23

Dynamics in the Dual Fuel Flagellar Motor of *Shewanella oneidensis* MR-1

Susanne Brenzinger and Kai M. Thormann

Abstract

The stator is an eminent component of the flagellar motor and determines a number of the motor's properties, such as the rotation-energizing coupling ion (H^+ or Na^+) or the torque that can be generated. The stator consists of several units located in the cytoplasmic membrane surrounding the flagellar drive shaft. Studies on flagellar motors of several bacterial species have provided evidence that the number as well as the retention time of stators coupled to the motor is highly dynamic and depends on the environmental conditions. Notably, numerous species possess more than a single distinct set of stators. It is likely that the presence of different stator units enables these bacteria to adjust the flagellar motor properties and function to meet the environmental requirements. One of these species is *Shewanella oneidensis* MR-1 that is equipped with a single polar flagellum and two stator units, the Na^+-dependent PomAB and the H^+-dependent MotAB. Here, we describe a method to determine stator dynamics by fluorescence microscopy, demonstrating how bacteria can change the composition of an intricate molecular machine according to environmental conditions.

Key words FRAP, Flagellum, Motor, Stator, Dynamics

1 Introduction

1.1 Background

The bacterial flagellar filament is rotated at its base by an intricate nanomachine, the flagellar motor (for reviews *see* [1–4]). Although the individual motors of different species may vary with respect to composition and structure, all flagellar motors consist of a rotor and a stator. The rotor disc is located within the cytoplasmic membrane and the cytoplasm and is connected via a drive shaft to the outside structures. The stator units are located in the cytoplasmic membrane in a ring-like fashion surrounding the drive shaft. The maximal number of torque-generating stator units depends on the bacterial species [5], in *Escherichia coli* up to 11 stators may simultaneously contribute to motor rotation [5–7]. Each stator consists of four A- and two B-subunits, often referred to as MotAB in H^+-dependent and PomAB in Na^+-dependent motors (reviewed in [8]).

Tohru Minamino and Keiichi Namba (eds.), *The Bacterial Flagellum: Methods and Protocols*, Methods in Molecular Biology, vol. 1593, DOI 10.1007/978-1-4939-6927-2_23, © Springer Science+Business Media LLC 2017

The A-subunits have four transmembrane regions and are thought to electrostatically interact in the cytoplasm with the rotor component FliG via a cytoplasmic loop connecting the transmembrane regions 2 and 3. The B-subunits have a single transmembrane domain and extend through the periplasmic space to bind to the peptidoglycan of the cell wall, which enables the protein complex to act as the static motor element in torque generation.

A number of studies on several bacterial motors have shown that the stator ring is highly dynamic. The stator units are thought to be produced as nonactive complexes that are diffusing in the cytoplasmic membrane. Only when engaging with the flagellar motor, binding to the peptidoglycan and ion channeling is activated. In actively running motors, the stator units are constantly exchanged with units from the pool of inactive stators in the cytoplasmic membrane; in *E. coli*, the exchange occurs over a rate of about 30s for each stator unit [6]. In addition, the number of active stators within the flagellar motors has been shown to directly depend on environmental conditions such as the corresponding ion motive force (*imf*) and the load acting on the flagellar motor [9–16]. Thus, full occupancy of stators in the motor only occurs under conditions of sufficient *imf* and high load, allowing the bacteria to adjust the motor performance accordingly. The mechanism underlying stator recruitment and activation is still mostly unclear.

While species such as *E. coli* or *Vibrio* sp. possess a single distinct set of stators to drive flagellar motor rotation, numerous bacterial species such as *Shewanella oneidensis*, *Bacillus subtilis*, or *Pseudomonas aeruginosa* harbor two or even more stator sets (reviewed in [17]). In *S. oneidensis* MR-1, two stators, the Na^+-dependent PomAB and the H^+-dependent MotAB, were demonstrated to be solely able to drive flagellar rotation [16]. Here, we describe an approach in which we employed fluorescence microscopy to determine number and exchange rate of PomAB and MotAB stator units within the flagellar motor. In addition, we determined the contribution of the two stators to torque generation by swimming speed assays. The results demonstrated that both PomAB and MotAB are synchronously recruited into the flagellar motor. Number and exchange rate of the stators were dependent on the concentration of Na^+ in the medium and were mainly governed by the enhanced recruitment and lower turnover of MotAB under conditions of low Na^+. Speed measurements strongly indicated that MotAB significantly contributed to swimming speed of wild-type cells when the Na^+ concentration was low. Thus, *S. oneidensis* MR-1 may run a "hybrid motor" synchronously using Na^+- and H^+-gradients to power the flagellar motor whose motor-stator configuration is directly governed by external Na^+ levels [15, 16].

1.2 Overview of the Methods

The approach of fluorescence microscopy requires suitable stable and active fusions of the protein of choice to an appropriate fluorophore. In case of the flagellar stator, this was particularly challenging. In previous approaches, GFP derivatives were fused to the N-terminus of the B-subunit which resides in the cytoplasm. However, since this region localizes to the rotor-stator interface within the cytoplasm, the presence of GFP was accompanied by a significant decrease in, or even absence of, activity [6, 14], which may also affect stator recruitment (and number) as well as the exchange rate. In our hands, any fluorescent fusions to the PomA or MotA subunits of *S. oneidensis* MR-1 did not result in active motors. An N-terminal fusion of MotB to superfolder GFP (sfGFP) similar to that used in *E. coli* [6] was severely affected in activity. Best results were obtained with a C-terminal fusion of PomB and MotB to mCherry, which is fluorescent after export into the periplasm. To obtain similar levels of protein, the hybrid genes *pomB-mCherry* and *motB-mCherry* were integrated into the chromosome to replace the corresponding native gene. The construction of these strains by conjugation is described in the first part.

In the second part of this chapter, we describe how we determined the activity of the produced fusion proteins. Standard Western Immunoblotting using antibodies raised against the stator proteins (or, alternatively, antibodies against dsRED) was applied to verify levels and stability of the proteins. Both PomB-mCherry and MotB-mCherry were stably produced at similar levels as in the wild type. In more detail, we describe the swimming assays that were carried out to monitor the functionality and activity of PomB-mCherry and MotB-mCherry by soft-agar assays and cell tracking. Although not reaching wild-type activity, both fusion proteins conferred robust motility to the corresponding strains.

Finally, we further describe fluorescent microscopy approaches we used to determine the turnover of the *S. oneidensis* MR-1 motor composition, which we will describe in more detail. However, determination of stator numbers was carried out in a cooperating lab using their custom-made equipment and scripts, which we will therefore not further elaborate on here. Measurements were carried out in strains in which one of the two systems was fluorescently marked. By this, we were able to show that stator turnover of PomAB appears to be constant while that of MotAB was strongly depending on the Na^+-concentration in the medium. Thus, we concluded that stator dynamics and stator-motor configuration in the *S. oneidensis* MR-1 flagellar motor is mainly governed by MotAB [15, 16].

2 Materials

2.1 Construction of Tagged Stators

1. *E. coli* WM3064 (W. Metcalf, University of Illinois, Urbana-Champaign) carrying the insertion plasmid (donor strain).
2. *S. oneidensis* MR-1 (recipient strain).
3. Lysogeny Broth (LB) agar and LB medium.
4. Diaminopimelic acid (DAP) stock solution (60 mM in ddH2O), filter sterilized.
5. Kanamycin stock solution (50 mg/mL in ddH2O), filter sterilized.
6. Sucrose stock (80% (w/v) in ddH2O), filter sterilized.
7. Two orbital shakers at 30 °C and 37 °C, respectively.
8. Petri dishes.
9. 100 mL flasks without beaker.
10. Tubes.
11. Centrifuge.
12. Toothpicks.

2.2 Motility Assays

1. *S. oneidensis* MR-1 (with tagged stator).
2. Lysogeny broth (LB).
3. LM100 broth: 10 mM HEPES, pH 7.3, 100 mM NaCl, 100 mM KCl, 0.02% yeast extract, 0.01% peptone, 15 mM lactate.
4. Agar.
5. Petri dishes.
6. 100 mL flasks without beaker.
7. Photometer and cuvettes to measure optical density.
8. Toothpicks.
9. Silicone grease (for example Baysilone from GE Bayer Silicones).
10. Microscopic slides.
11. 20 × 20 mm microscope coverslips.
12. Inverse DMI 6000 B microscope (Leica) using a HC PL APO 63×/1.40-0.6. Oil DIC (Differential Interference Contrast) objective or a HCX PL APO 100×/1.40–0.70 Oil DIC objective and a pco.edge 5.5 sCMOS camera (PCO).
13. ImageJ (Fiji package) [18, 19].

2.3 FRAP

1. *S. oneidensis* MR-1 (with tagged stator).
2. LB broth.

3. LM100 broth: 10 mM HEPES, pH 7.3, 100 mM NaCl, 100 mM KCl, 0.02% yeast extract; 0.01% peptone, 15 mM lactate.
4. Chloramphenicol stock solution (30 mg/mL in 100% EtOH).
5. 100 mL flasks without beaker.
6. Photometer and cuvettes to measure optical density.
7. Microscopic slides.
8. 20 × 20 mm microscope coverslips.
9. 24 × 60 mm microscope coverslips.
10. Scalpel (optional).
11. Agarose.
12. Axio Imager.M1 microscope (Zeiss) using a Zeiss Plan Apochromat 100×/1.40 Oil DIC objective and a Cascade:1 K CCD camera (Photometrics) equipped with a 488 nm solid-state laser and a 2D–VisiFRAP Galvo System multi-point FRAP module (Visitron Systems, Germany).
13. VALAP (Lanolin, Paraffin, Vaseline, in a 1:1:1 ratio).
14. ImageJ (Fiji package) [18, 19] to acquire raw fluorescence values.
15. Excel (Microsoft).
16. OriginPro (OriginLab Corporation).

3 Methods

3.1 Construction of Tagged Stators

1. Grow *S. oneidensis* MR-1 in 10 mL of LB broth in a 100 mL flask at 30 °C with orbital shaking at 200 rpm to stationary phase.
2. At the same time, grow *E. coli* WM3064 carrying the insertion plasmid (*see* **Notes 1** and **2**) in 10 mL LB broth supplemented with 50 μg/mL kanamycin and 300 μM DAP in a 100 mL flask at 37 °C with orbital shaking at 200 rpm to stationary phase.
3. Harvest 1 mL of each culture by centrifugation and wash three times with LB broth + DAP (300 μM).
4. Pool both cultures in ~200 μL LB broth + DAP (300 μM) and place all cells in one spot onto an LB + DAP (300 μM) agar plate.
5. Incubate for at least 8 h at 30 °C to allow conjugation, plasmid transfer, and first homologous recombination.
6. Wash all cells off the plate with 2 mL LB broth, wash three times in LB broth, and resuspend in a final volume of 1 mL.
7. Dilute 1:10 and 1:100 in LB broth and plate on LB agar plates containing 50 μg/mL kanamycin, incubate overnight at 30 °C.

8. Pick ten colonies and place on LB + 10% (w/v) sucrose agar plates and LB agar plates containing 50 μg/mL kanamycin to screen for Km^r /Sac^s (*see* **Note 3**).
9. Incubate cells of two clones with the respective phenotype in 10 mL LB broth in a 100 mL flask at 30 °C with orbital shaking at 200 rpm for ~10 generations (min. 5 h, but not more than 8 h) to allow the second homologous recombination to occur.
10. Dilute 1:50 and 1:500 in LB broth and plate on LB + 10% (w/v) sucrose agar plates.
11. Pick 50 colonies and place on LB agar plates + 50 μg/mL kanamycin and LB agar plates + 10% (w/v) sucrose to screen for Km^s /Sac^r.
12. Identify the correct insertion clones by colony PCR (*see* **Note 4**).
13. Stable fusions can be determined using immunofluorescence standard procedure using antibodies raised against the stator proteins (if available) or dsRed (which are commercially available).

3.2 Motility Assays

The motility assays we commonly use to either evaluate motility of whole populations on soft agar plates or of single cells in liquid environments.

3.2.1 Soft Agar Assay

1. Prepare LB + 0.25% (w/v) agar, pour plates of at least 1 cm thickness when agar has reached temperature of 45–50 °C and let dry for at least 4 h.
2. Carefully place aliquots of 3 μL of cultures grown to exponential growth phase onto the agar surface and let sink in before moving (*see* **Note 5**).
3. Incubate for 12–24 h at 30 °C.
4. Images of plates are acquired by scanning and the radii are measured using ImageJ (Fiji package) [19].

3.2.2 Liquid Motility Assay

1. Culture *S. oneidensis* MR-1 strains in 10 mL LM100 broth in a 100 mL flask at 30 °C with orbital shaking at 200 rpm overnight.
2. Inoculate a fresh LM100 culture to an OD600 of 0.02. Grow to an OD600 of 0.2–0.25.
3. Add small droplets of silicone grease into the corner of a coverslip and carefully place it onto a microscope slide.
4. Carefully place 50 μL of culture close to the edge of the coverslip and allow it to flow under it. Lower the coverslip until the whole space is filled.
5. Immediately take a video of the swimming cells at a high frame rate (*see* **Note 6**).
6. Average swimming speed can be determined using the TrackMate plugin of ImageJ (Fiji package) [18, 19] (*see* **Note 7**).

3.3 FRAP

1. Cultivate *S. oneidensis* MR-1 cells in 10 mL LB broth in a 100 mL flask at 30 °C with orbital shaking at 200 rpm overnight.
2. Inoculate a fresh LB culture to an OD600 of 0.02. Grow to an OD600 of 0.2–0.25.
3. Stop protein production by the addition of chloramphenicol (final concentration: 10 μg/mL).
4. Prepare 1% agarose (w/v) pad by melting agarose in LM100 and immediately placing ~1 mL between a microscope slide and a coverslip spaced by a microscope slide on each side to yield a final thickness of ~1 mm. Let dry for 5–10 min before carefully removing the coverslip.
5. Cut agarose pad of 15 × 15 mm using the edge of the coverslip or a scalpel, place 5 μL of culture onto the pad, and cover with a 20 × 20 mm coverslip as soon as the liquid has evaporated.
6. Seal with VALAP heated to ~70 °C, add a droplet of microscope oil, and place on the microscope.
7. FRAP experiment starts by acquisition of a pre-bleach image (*see* **Note 8**).
8. Apply a single focused laser pulse of 30 ms to bleach individual polar stator clusters.
9. Acquire a post-bleach image immediately after bleaching the cluster.
10. Acquire a series of images at defined time points to determine fluorescence recovery until plateau is reached (*see* **Notes 8** and **9**).
11. The raw fluorescence intensities of the whole cell (*I*wc), the bleached cluster (*I*c), and the background outside the cell (*I*bg) are measured for each time point using ImageJ (Fiji package) [19]. Transfer raw values to excel.
12. Subtract the background fluorescence *I*bg from the whole cell intensity *I*wc for each time point (= adjusted whole cell intensity (aIwc)).
13. Subtract the background fluorescence *I*bg from the clusters intensity *I*c for each time point (= adjusted intensity of cluster fluorescence (aIc)).
14. Divide the adjusted cluster fluorescence aIc by the adjusted whole cell intensity aIwc to calculate the relative integrated fluorescence cluster intensity (*I*ifc).
15. Calculate the percentage of initial fluorescence loss by dividing the post-bleach *I*ifc of the bleached region immediately after bleaching by the pre-bleach *I*ifc, multiplying by 100 and subtracting this from 100. Typically, photobleaching should result in a 60–80% loss of fluorescence intensity in the bleached region (*see* **Note 9**).

16. Cluster fluorescence is double normalized to the pre- and post-bleach value. To this end, the post-bleach *I*ifc value is subtracted from the *I*ifc of each time point. Subsequently, each of these normalized values is divided by the normalized *I*ifc value of the pre-bleach time point.
17. Average the double normalized values of at least 15 cells for each time point and plot against the time.
18. Recovery rates were determined by fitting the data obtained for the bleached region to the single exponential function $y = a^*(1-\exp.(-x/b))$ where y is the fluorescence at time t, a is the maximum change in fluorescence during recovery, and x is the time in min, $1/b$ is the rate constant in $\min^{-1}$ (*see* **Note 10**).
19. Recovery half-times were calculated according to the equation $\tau 1/2 = \ln(2)^* b$.

4 Notes

1. For conjugation, the donor strain *E. coli* WM3064 is used in combination with a suicide plasmid (we use pNTPS138R6K [20]) which provides a kanamycin resistance as a selective marker and a *sacB* gene for counterselection. The vector is linearized in its multiple cloning site (*Eco*RV). By Gibson assembly [21], three PCR fragments are assembled into the vector: a ~550 bp long C-terminal fragment of the gene of the corresponding B-subunit which ends with the penultimate codon excluding the stop codon, the *mCherry* gene with its stop codon, and a ~550 bp fragment of the region downstream of the corresponding stator gene. Between *pomB* or *motB* gene, a linker is introduced using the reverse primer of the upstream fragment and/or the forward primer of the *mCherry* fragment. In *S. oneidensis* MR-1 this linker is at least three amino acids (GlyGlySer) long (*see* **Note 2**).
2. In our hands, the fluorescence tag renders the stator function severely impaired if fused to the stator without a linker. If stator function is critical to the experiment, longer flexible linkers may result in more functional stators. In an experiment following the study authored by Paulick et al. [15], up to 22 amino acids were tested in case of MotB and were found to impair motility significantly less than a shorter linker [22].
3. The counterselection based on *sacB* and the addition of sucrose is not completely impairing growth, the evaluation of the Km^r/Sac^s phenotype should therefore take place 10–15 h after picking the clones.
4. For the colony PCR, a small amount of cells (less than pinhead-sized) is mixed with 50 μL ddH2O in an eppendorf cup and

incubated at 100 °C for 5 min. 1 μL of this suspension is used as a template in a standard PCR using appropriate primers that are located ~100 bp up- and downstream of the ~550 bp regions flanking the *mCherry* gene.

5. Alternatively to liquid cultures, small amounts of cells from colonies can be transferred directly to the soft agar plate using a toothpick or pipet tip. In either case, individual strains should be at least 5 cm apart. Due to different properties of soft-agar plates, all strains to be compared need to be on the same plate!
6. Specification of video acquisition used in our group: DIC illumination, 63× or 100× objective lenses, 20 frames per second with each video containing at least 100 frames.
7. TrackMate plugin of ImageJ (Fiji package) [18, 19] can be used to determine the average speed of multiple cells per video. As many parameters depend on the microscope settings and pixel size of the cells, several parameters in TrackMate have to be adjusted accordingly by testing which values will yield the best results. Initially, TrackMate requires the user to determine the frame rate (Time interval box). Detector routinely used for *S. oneidensis* MR-1 tracks is the LoG detector. Using the preview function, the pixel value for the blob diameter and threshold have to be selected to recognize single cells. Tracker used is the "Simple LAP tracker" with a linking max distance of 10 pixel, gap-closing max distance of 10 pixel and gap-closing max frame gape set to 2 when acquiring videos using the 63× lense system. Filters set on tracks are "Number of spots in track" (at least 5), "track displacement" and "mean velocity" to eliminate diffusing cells moved by Brownian motion. "Minimal velocity" filter can additionally filter out cells that cease to swim during the video. Swimming speeds of all remaining tracks are annotated in the "Analysis" window.
8. There is no universal protocol for FRAP image acquisition as the settings always have to be adjusted to suit the fluorescently tagged protein and microscope features. To avoid increased bleaching by the pre- and post-bleach fluorescence image acquisition, we advise to determine the optimal number of frames per FRAP experiment, exposure time and illumination strength by testing various settings. For *S. oneidensis* MR-1 PomB-mCherry and MotB-mCherry usually a series of 12–15 images in total are acquired. To account the rapid recovery, the initial images are taken in shorter intervals, while the later images are acquired 1 or 2 min apart from each other.
9. Include only cells that reach at least 60% fluorescence loss by bleaching in the subsequent analysis. If bleaching depth frequently fails to achieve this value, laser intensity or bleaching duration may have to be adjusted.

10. The program used for curve fitting by our group is OriginPro (OriginLab Corporation). For curve fitting copy the averaged values except the pre-bleach time point with the corresponding time points (in minutes) into a new project in Origin. Plot data as scatter using the command in the menu bar. Since OriginPro does not provide the fit function mentioned above, it has to be defined first. To this end, select "Analysis/Nonlinear curve fit/open dialoge" and enter "user defined" as category and "new" as function. In the opening window a name can be assigned. Function model (explicit) and function type (expression) are preselected. Next, the independent variables (x), the dependent variables (y), and parameters (a, b) have to be defined. In the next part the function has to be entered ($y = a^*(1\text{-exp.}(-x/b))$). Save the function. Now it can be used to fit curved by selecting "analysis/nonlinear curve fit/category: user defined/Function" (name you entered).

References

1. Minamino T, Imada K (2015) The bacterial flagellar motor and its structural diversity. Trends Microbiol 23:267–274
2. Minamino T, Imada K, Namba K (2008) Molecular motors of the bacterial flagella. Curr Opin Struct Biol 18:693–701
3. Sowa Y, Berry RM (2008) Bacterial flagellar motor. Q Rev Biophys 41:103–132
4. Stock D, Namba K, Lee LK (2012) Nanorotors and self-assembling macromolecular machines: the torque ring of the bacterial flagellar motor. Curr Opin Biotechnol 23:545–554
5. Beeby M, Ribardo DA, Brennan CA et al (2016) Diverse high-torque bacterial flagellar motors assemble wider stator rings using a conserved protein scaffold. Proc Natl Acad Sci U S A 113:1917–1926
6. Leake MC, Chandler JH, Wadhams GH et al (2006) Stoichiometry and turnover in single, functioning membrane protein complexes. Nature 443:355–358
7. Reid SW, Leake MC, Chandler JH et al (2006) The maximum number of torque-generating units in the flagellar motor of *Escherichia coli* is at least 11. Proc Natl Acad Sci U S A 103:8066–8071
8. Kojima S (2015) Dynamism and regulation of the stator, the energy conversion complex of the bacterial flagellar motor. Curr Opin Microbiol 28:66–71
9. Fung DC, Berg HC (1995) Powering the flagellar motor of *Escherichia coli* with an external voltage source. Nature 375:809–812
10. Sowa Y, Rowe AD, Leake MC et al (2005) Direct observation of steps in rotation of the bacterial flagellar motor. Nature 437:916–919
11. Tipping MJ, Steel BC, Delalez NJ et al (2013) Quantification of flagellar motor stator dynamics through in vivo proton-motive force control. Mol Microbiol 87:338–347
12. Tipping MJ, Delalez NJ, Lim R et al (2013) Load-dependent assembly of the bacterial flagellar motor. MBio 4:e00551-13. doi:10.1128/mBio.00551-13
13. Lele PP, Hosu BG, Berg HC (2013) Dynamics of mechanosensing in the bacterial flagellar motor. Proc Natl Acad Sci U S A 110:11839–11844
14. Fukuoka H, Wada T, Kojima S et al (2009) Sodium-dependent dynamic assembly of membrane complexes in sodium-driven flagellar motors. Mol Microbiol 71:825–835
15. Paulick A, Delalez NJ, Brenzinger S et al (2015) Dual stator dynamics in the *Shewanella oneidensis* MR-1 flagellar motor. Mol Microbiol 96:993–1001
16. Paulick A, Koerdt A, Lassak J et al (2009) Two different stator systems drive a single polar flagellum in *Shewanella oneidensis* MR-1. Mol Microbiol 71:836–850
17. Thormann KM, Paulick A (2010) Tuning the flagellar motor. Microbiology 156:1275–1283
18. Meijering E, Dzyubachyk O, Smal I (2012) Methods for cell and particle tracking. Methods Enzymol 504:183–200

19. Schindelin J, Arganda-Carreras I, Frise E et al (2012) Fiji: an open-source platform for biological-image analysis. Nat Methods 9:676–682
20. Lassak J, Henche AL, Binnenkade L, Thormann KM (2010) ArcS, the cognate sensor kinase in an atypical Arc system of *Shewanella oneidensis* MR-1. Appl Environ Microbiol 76:3263–3274
21. Gibson DG, Young L, Chuang RY et al (2009) Enzymatic assembly of DNA molecules up to several hundred kilobases. Nat Methods 6:343–345
22. Brenzinger S, Dewenter L, Delalez NJ et al (2016) Mutations targeting the plug-domain of the Shewanella oneidensis proton-driven stator allow swimming at increased viscosity and under anaerobic conditions. Mol Microbiol 102:925–938

Chapter 24

Ion Selectivity of the Flagellar Motors Derived from the Alkaliphilic *Bacillus* and *Paenibacillus* Species

Yuka Takahashi and Masahiro Ito

Abstract

Many bacteria can swim using their flagella, which are filamentous organelles that extend from the cell surface. The flagellar motor is energized by either a proton (H^+) or sodium ion (Na^+) as the motive force. MotAB-type stators use protons, whereas MotPS- and PomAB-type stators use Na^+ as the coupling ions. Recently, alkaliphilic *Bacillus alcalophilus* was shown to use potassium ions (K^+) and rubidium ions (Rb^+) for flagellar rotation, and the flagellar motor from *Paenibacillus* sp. TCA-20 uses divalent cations such as magnesium ions (Mg^{2+}), calcium ions (Ca^{2+}), and strontium ions (Sr^{2+}) for coupling. In this chapter, we focus on how to identify the coupling ions for flagellar rotation of alkaliphilic *Bacillus* and *Paenibacillus* species.

Key words Flagellar motor, Stator, Alkaliphiles, *Bacillus*, *Paenibacillus*, Divalent cation

1 Introduction

The bacterial flagellar motor, which is embedded in the cell envelope, consists of three parts, the filament, hook, and basal body, and is generally powered by an electrochemical gradient of protons (H^+), sodium (Na^+), potassium (K^+), rubidium (Rb^+), magnesium (Mg^{2+}), calcium (Ca^{2+}), or strontium (Sr^{2+}) across the cytoplasmic membrane [1–5]. Torque is generated by the electrostatic interaction at the rotor (FliG) and stator interface. *Bacillus subtilis* and *Shewanella oneidensis* MR-1 employ H^+-coupled MotAB and Na^+-coupled MotPS/PomAB stators to generate the torque required for flagellar rotation [2, 6, 7].

Before 2008, the Mot complexes were believed to contain channels that used either H^+ or Na^+, with some bacteria having only one type and others having two distinct types with different ion-coupling [7, 8]. However, in 2008, alkaliphilic *B. clausii* KSM-K16 was identified as the first bacterium with a single stator-rotor unit that uses both H^+ and Na^+ for ion-coupling, based on external pH [9]. Subsequent findings have shown that alkaliphilic *B. alcalophilus* AV1934 uses Na^+, K^+, and Rb^+ as coupling ions for flagellar rotation

Tohru Minamino and Keiichi Namba (eds.), *The Bacterial Flagellum: Methods and Protocols*, Methods in Molecular Biology, vol. 1593, DOI 10.1007/978-1-4939-6927-2_24,

[1], suggesting that coupling ions other than H^+ and Na^+ can power the bacterial flagellar motor, including Ca^{2+}, which are abundant in nature. Subsequently, *Paenibacillus* sp. TCA20 has been shown to have a novel bacterial flagellar stator that can use Mg^{2+}, Ca^{2+}, and Sr^{2+} as coupling cations for flagellar rotation [3].

In this chapter, we introduce the methods for measuring motility and intracellular ion concentrations to elucidate the coupling ion selectivity of the flagellar motor stator. Here, we focus on this motility organ in alkaliphilic *Bacillus* and *Paenibacillus* species.

2 Materials

Ultra-pure water is used in the present study.

2.1 Motility Analysis

1. Tris medium (pH 7.7): Mix 3.63 g of Trizma base and 1.47 g of citric acid monohydrate in 930 mL of deionized water. After autoclaving, cool to 55 °C and add 10 mL of 10% (w/v) Yeast Extract, 10 mL of trace elements [10], and 50 mL of 1 M D-(+)-Glucose (*see* **Notes 1** and **2**).
2. Motility assay buffer: 30 mM Trizma base and 5 mM D-(+)-Glucose are mixed, adding the necessary amount of NaCl, KCl, or RbCl (*see* **Notes 1** and **2**).
3. Inhibitors of the flagellar motor: Proton-coupled and Na^+- and K^+-coupled flagellar motors are inhibited by the protonophore carbonyl cyanide m-chlorophenyl hydrazone (CCCP) and Ethylisopropyl amiloride 3-Amino-N-(aminoiminomethyl)-6-chloro-5-[ethyl(1-methylethyl)amino]-2-pyrazinecarboxamide) (EIPA), respectively. Both inhibitors are dissolved in dimethyl sulfoxide (DMSO). The prepared inhibitor solution is added to the appropriate concentration into the motility buffer.
4. Microscope with a microscope stage automatic temperature control system adjusted at 32 °C and recorded in high definition with a digital color camera.
5. The speed of each individual swimming cell was analyzed by 2D movement measurement capture using 2D–PTV software (Digimo Co., Ltd.).
6. The diameters of the colonies formed on the surface of soft agar medium were measured using a Vernier caliper.

2.2 Measurement of Intracellular K^+ Concentrations

1. LBK medium: LBK medium (pH 7.5) contains 10 g of Tryptone, 5 g of Yeast Extract, and 6 g of KCl per liter of deionized water. The pH is adjusted with 4 M KOH. Thereafter, the medium is sterilized.
2. A defined medium for *Escherichia coli* TK2420: The composition of a defined medium (pH 7.0) contains 4.9 g of Na_2HPO_4, 2.34 g of NaH_2PO_4, 0.29 g of Trisodium citrate, 1 g of

$(NH_4)_2SO_4$, 1 mL of 6 mM $FeSO_4$, 0.1 g of $MgSO_4$, and 1 mL of 1 M KH_2PO_4 per liter of deionized water. After autoclaving, cool to 55 °C and add 10 mL of 1 M glucose, 1 mL of 1000× thiamine (stock solution is 1 mg/mL), and various amounts of KCl. TK2420 will grow on 50 mM KCl but not on 5–10 mM KCl.

3. 0.5 M sucrose solution: The solution contains 171.1 g of sucrose per liter of deionized water. Thereafter, the solution is sterilized.
4. Growth was monitored by measuring the absorbance at A600 using a spectrophotometer.
5. The K^+ concentration in solution was determined using a flame photometer.

2.3 Measurement of Intracellular Mg^{2+} Concentrations

1. 2×TY medium: 2×TY medium contains 16 g of Tryptone and 10 g of Yeast Extract per liter of deionized water. Thereafter, the medium is sterilized.
2. 0.3 M sucrose solution: The solution contains 102.7 g of sucrose per liter of deionized water. Thereafter, the solution is sterilized.
3. The Mg^{2+} concentration in solution was determined by MG Metallogenics Mg^{2+} measurement LS (Metallogenics Co., Ltd., Chiba, Japan) using spectrophotometer.

3 Methods

3.1 Swimming Velocity Analysis in a Liquid Medium

1. Highly motile cells in the late logarithmic phase were harvested by filtration on OMNIPORE membrane filters (0.45 μm) and washed three times with 2 mL of motility analysis buffer (*see* **Notes 3** and **4**). For swimming speed assays with inhibitors, the cells were resuspended in motility assay buffer at the indicated values of pH and monovalent cation content for the assay. (CCCP is used as an inhibitor for a proton-coupled motor and EIPA is used as an inhibitor for a Na^+- and K^+-coupled motor)
2. Cells were suspended in 1 mL of the same buffer and incubated at growth temperature for 10 min.
3. For the measurement of swimming speed, cell motility was observed under a dark-field microscope. Videos of approximately 10 s were recorded from five suspensions (*see* **Note 5**).
4. The speed of each individual cell that was swimming (not tumbling) for more than 10 s was analyzed by 2D movement measurement capture (*see* **Notes 6** and 7).
5. The swimming speed of each cell is calculated at least 1 s or more of recorded data (*see* **Note 8**). All results are the average of three independent experiments where the speed of 30 different cells was measured.

3.2 Swarming Analysis on Soft Agar Medium

1. To observe motility, bacterial cells were inoculated on the surface of Tris media containing 0.25–0.3% Noble agar (Difco) (*see* **Note 9**).
2. Plates were incubated under static culture conditions using stationary phase bacteria at a temperature suitable for growth (*see* **Notes 10** and **11**).
3. After incubation, the diameters of the colonies formed on the surface of the Tris soft agar medium were measured using a Vernier caliper. All results are the average of three independent experiments.

3.3 Measurement of Intracellular K^+ Concentration

1. A potassium uptake-deficient (Δ(*kdpABC*) *trk*Δ1 Δ*trkA*) mutant strain of *E. coli* TK2420 was used as a host for these experiments. A plasmid carrying the cloned stator gene of interest was transformed into TK2420.
2. The transformant was grown in LBK medium at 37 °C overnight and then the culture was inoculated into 50 mL of TK2420 medium [11] containing 10 mM, 25 mM, or 50 mM KCl at an absorbance of 0.01 at 600 nm (A_{600}) and grown aerobically at 37 °C. Growth was monitored by measuring the absorbance at A_{600} (*see* **Notes 12–14**).
3. The cells were harvested by centrifugation (3000 × *g*, 10 min, 25 °C).
4. The cells were washed by resuspending them in a 300 mM sucrose solution and then harvested by centrifugation (3000 × *g*, 10 min, 25 °C).
5. The cells were resuspended in 5 mL of 0.5 M sucrose solution.
6. The protein concentration was measured by the Lowry method using 100 μL of the cell suspension (*see* **Notes 15** and **16**).
7. The rest of the suspension was harvested by centrifugation (3000 × *g*, 10 min, 25 °C) and resuspended in 10 mL of 10% (w/v) TCA solution.
8. The cell suspension was shaken at 200 rpm.
9. Cells were harvested by centrifugation (10,000 × *g*, 5 min, 25 °C).
10. The supernatants were transferred to a new test tube and then diluted by 10-fold or 100-fold.
11. The K^+ concentration of the stock and the 10-fold diluted and 100-fold diluted solutions were determined using a flame photometer calibrated with standard K^+ solutions of known concentration (*see* **Note 17**).
12. The protein concentration was determined by the Lowry method with lysozyme as a standard [12]. The intracellular concentration was calculated with an assumed cell volume of 3 μL/mg cell protein [13, 14].

3.4 Measurement of Intracellular Mg^{2+} Concentration

1. A major magnesium uptake- and flagellar stator-deficient (*lys3 trpC2* Δ*motAB*Δ*motPS* Δ*ykoK* Δ*yfjQ*) mutant strain of *B. subtilis* ΔABPSΔKQ was used as a host for these experiments. A plasmid carrying cloned stator genes of interest was transformed into ΔABPSΔKQ. Wild-type and ΔABPSΔKQ strains of *B. subtilis* were used for positive and negative controls.
2. The transformant was grown in LBK medium at 37 °C overnight and then the culture was inoculated into 50 mL of 2×TY medium containing 0 mM, 1 mM, 2.5 mM, 5 mM, or 10 mM $MgCl_2$ at an absorbance of 0.01 at 600 nm (A_{600}) and was grown aerobically at 37 °C. Growth was monitored by measuring the absorbance at A_{600} (*see* **Notes 12–14**).
3. The cells in the late logarithmic growth phase were harvested by centrifugation (3000 × *g*, 10 min, 25 °C).
4. The cells were resuspended in 0.3 M sucrose solution and then harvested by centrifugation (3000 × *g*, 10 min, 25 °C).
5. The cells were resuspended in 5 mL of 300 mM sucrose solution.
6. The protein concentration was measured by the Lowry method [12] using 100 μL of the cell suspension (*see* **Notes 15** and **16**).
7. The rest of the suspension was harvested by centrifugation (3000 × *g*, 10 min, 25 °C) and resuspended in 10 mL of 0.1 M HCl.
8. The cell suspension was shaken at 200 rpm.
9. The cells were harvested by centrifugation (10,000 × *g*, 5 min, 25 °C).
10. The supernatants were transferred to a new test tube and then diluted by 10-fold or 100-fold.
11. The Mg^{2+} concentration of the prepared sample of the stock and 10-fold diluted solution and a 100-fold diluted solution was measured by MG Metallogenics Mg^{2+} measurement LS (Metallogenics Co., Ltd., Chiba, Japan) that was calibrated with standard Mg^{2+} solutions of known concentrations. (*see* **Note 17**).
12. Intracellular Mg^{2+} was calculated by assuming that a *B. subtilis* cell was a cylinder (radius, 0.4 μm; length, 4 μm).

4 Notes

1. Tris medium was not subjected to pH adjustment. When preparing the pH in the alkaline range, both NaOH and KOH cannot be used for pH adjustment. Instead, N-Methyl D-glucamine was added to adjust the pH in the alkaline range.
2. Some bacteria, for example *B. subtilis*, show some inhibition of motility in a Tris buffer. If the tested bacteria do not show

vigorous motility in the Tris buffer, we recommend using other buffer suitable for optimum pH.

3. Some bacteria show slow or no motility after the collection of the cells because of the physical shock of the filtration and centrifugation. In this case, it is suitable to perform dilution with a wide mouth tip. If the bacterial concentration is low, then buffer substitution under microscopic observation is the best method.
4. When the turbidity of the late logarithmic cells (A_{600}) does not reach 1.0, it was necessary to concentrate the cells by two to five times because the motility observation required a sufficient amount of motile cells.
5. Motility observed under the microscope was performed using the hanging drop method. A number of bacteria were present in the intermediate layer, but the depth of focus changed quickly. Therefore, it was sometimes difficult to track each swimming cell. We recommend that swimming observation be performed by focusing near the slide glass or cover glass surface.
6. The tumbling frequency of *B. subtilis* and its derivatives is drastically increased by $CaCl_2$ and it is very difficult to measure the linear swimming velocity of each strain.
7. Bacteria that were swimming relatively straight were chosen because tracking cells that change direction resulted in variation of swimming speed.
8. The distribution of the swimming velocity for all of the cells was determined. Subsequently, 90 bacterial cells were selected to be used in the swimming velocity measurement based on the standard deviation of this distribution.
9. For best results, soft agar medium plates should be prepared freshly. Plates cannot be stored more than 1 week even if humidity is controlled.
10. The motility of the cells was observed on the surface of the soft agar medium.
11. The bacteria on the soft agar plate were spread in a relatively symmetrical circle. However, after a prolonged incubation, the spread of the bacteria was no longer circular. Therefore, this assay should be performed using stationary growth phase cells.
12. LBK medium contains KCl and is suitable for the culture of the TK2420 strain that is defective in the major potassium uptake system.
13. The function of the flagellar motor was extremely reduced in the stationary phase. Therefore, for the measurement of the intracellular concentration of the coupling ion of the flagellar motor, it is desirable to use cells in the late logarithmic growth phase.
14. It is possible to calculate the number of bacterial cells from the turbidity at the time of harvest.

15. Washing the cells with the sucrose solution removed the salt from the medium so that only the internal potassium concentration was measured. For consistency, the protein concentration was measured under the same conditions.
16. To calculate the number of cells, it was necessary to remove flagellar fiber and cilia from the cell body with a waring blender. Otherwise, the amount of protein content per cell is overestimated because of the large amount of flagellar fiber and cilia present on the cells.
17. Equipment and commercial kits differ in the ranges that they can measure ion concentration. Therefore, it was necessary to prepare the samples to a suitable concentration for the method of measurement.

Acknowledgment

We thank Dr. Arthur A. Guffanti for reading of the manuscript. This work was supported by a Grant-in-Aid for Scientific Research on Innovative Areas No. JP24117005 of the Ministry of Education, Culture, Sports, Science and Technology of Japan (MI).

References

1. Terahara N, Sano M, Ito M (2012) A *Bacillus* flagellar motor that can use both Na^+ and K^+ as a coupling ion is converted by a single mutation to use only Na^+. PLoS One 7:e46248
2. Minamino T, Imada K (2015) The bacterial flagellar motor and its structural diversity. Trends Microbiol 23:267–2674
3. Imazawa R, Takahashi Y, Aoki W, Sano M, Ito M (2016) A novel type bacterial flagellar motor that can use divalent cations as a coupling ion. Sci Rep 6:19773
4. Li N, Kojima S, Homma M (2011) Sodium-driven motor of the polar flagellum in marine bacteria *Vibrio*. Genes Cells 16:985–999
5. Ito M, Takahashi Y (2017) Nonconventional cation-coupled flagellar motors derived from the alkaliphilic *Bacillus* and *Paenibacillus* species, Extremophiles, 21:3–14
6. Fujinami S, Terahara N, Krulwich TA, Ito M (2009) Motility and chemotaxis in alkaliphilic *Bacillus* species. Future Microbiol 4:1137–1149
7. Paulick A, Delalez NJ, Brenzinger S, Steel BC, Berry RM, Armitage JP et al (2015) Dual stator dynamics in the *Shewanella oneidensis* MR-1 flagellar motor. Mol Microbiol 96:993–1001
8. Terahara N, Fujisawa M, Powers B, Henkin TM, Krulwich TA, Ito M (2006) An intergenic stem-loop mutation in the *Bacillus subtilis ccpA-motPS* operon increases *motPS* transcription and the MotPS contribution to motility. J Bacteriol 188:2701–2705
9. Terahara N, Krulwich TA, Ito M (2008) Mutations alter the sodium versus proton use of a *Bacillus clausii* flagellar motor and confer dual ion use on *Bacillus subtilis* motors. Proc Natl Acad Sci U S A 105:14359–14364
10. Cohen-Bazire G, Sistrom WR, Stanier RY (1957) Kinetic studies of pigment synthesis by non-sulfur purple bacteria. J Cell Physiol 49:25–68
11. Epstein W, Kim BS (1971) Potassium transport loci in *Escherichia coli* K-12. J Bacteriol 108:639–644
12. Lowry OH, Rosebrough NJ, Farr AL, Randall RJ (1951) Protein measurement with the Folin phenol reagent. J Biol Chem 193:265–275
13. Ito M, Cooperberg B, Krulwich TA (1997) Diverse genes of alkaliphilic *Bacillus firmus* OF4 that complement K^+-uptake-deficient *Escherichia coli* include an *ftsH* homologue. Extremophiles 1:22–28
14. Bakker E (1993) Cell K^+ and K^+ transport systems in prokaryotes. In: Bakker E (ed) Alkali cation transport systems in prokaryotes. CRC Press, Boca Raton, FL, pp 205–224

Chapter 25

Measurement of Free-Swimming Motility and Magnetotactic Behavior of Magnetococcus massalia Strain MO-1

Wei-Jia Zhang, Sheng-Da Zhang, and Long-Fei Wu

Abstract

Magnetococcus massalia strain MO-1 represents a group of fast-swimming marine magnetotactic coccoid-ovoid bacteria. They show polar magnetotaxis behavior in uniform magnetic field. MO-1 cells swim forward constantly with rare stop. When they meet obstacles, MO-1 cells could squeeze through or circumvent the obstacles. Here, we describe the methods for characterization of magnetotactic behaviors of MO-1 cells using adapted spectrophotometer and microscope mounted with magnetic fields.

Key words Magnetotactic bacteria, Motility, Magnetotaxis, Magneto-spectrophotometry, Magnetodrome

1 Introduction

1.1 Background

Magnetotactic bacteria (MTB) synthesize bacterial organelles, magnetosomes that consist of single-domain magnetic crystals enveloped by bio-membrane [1]. Magnetosomes arrange in chains and confer on the cells a magnetic moment, which allows them swimming along the geomagnetic field lines [1]. This behavior is referred to as magnetotaxis. Two kinds of magnetotactic behaviors are observed: axial or polar magnetotaxis [1]. The axial magnetotactic cells swim toward both directions along the magnetic field lines, whereas the polar magnetotactic cells swim always in one direction parallel (North-Seeking, NS) or antiparallel (South-Seeking, SS) with the magnetic field lines. Polar magnetotactic ovoid strain MO-1 is isolated from marine sediments. It grows at the oxic-anoxic transition zone (OATZ) in chemically stratified sediments or semisolid minimal media in laboratory [2]. MO-1 cells consist of dominant north-seeking population and minor south-seeking population when incubated under the geomagnetic field at North hemisphere [2, 3]. As a representative of marine magnetotactic coccoid-ovoid bacteria, MO-1 cells could swim at a speed of up to

Tohru Minamino and Keiichi Namba (eds.), *The Bacterial Flagellum: Methods and Protocols*, Methods in Molecular Biology, vol. 1593, DOI 10.1007/978-1-4939-6927-2_25,

300 μm/s, about 100 body-length per second [2]. North-seeking MO-1 cells swim consistently toward north pole and can squeeze through or southward for a short distance and then back northward to circumvent the obstacles [4]. This magnetotactic capacity appears to be an adaptation strategy to the marine sediments where the chemo-litho-autotrophic MO-1 cells should swim to obtain electron donors and receptors at different layers in the chemically stratified sediments [4]. The robust flagellar apparatus of MO-1 cells underpins the swimming capacity. Each MO-1 cell possesses two sheathed flagellar bundles on the long axis side of their ovoid body. Each flagellar bundle consists of 7 flagella and 24 fibrils arranged into seven intertwined hexagonal arrays [5]. Such architecture is currently observed for MO-1, its closely related *Magnetococcus marinus* MC-1 and other uncultivated magnetotactic cocci living in marine sediments [6, 7]. Magnetotactic cocci are the most abundant morphotype of magnetotactic bacteria found in both fresh water and marine sediments. In this section, we describe using two magnetic-optical approaches, Magneto-spectrophotometry assay [8] and Magnetodrome analysis [4] to determine the polarity and homogeneity of polar magnetotactic bacteria and to characterize their magnetotaxis behavior using MO-1 cells as an example.

1.2 Overview of Methods

1.2.1 Magneto-Spectrophotometer

A Magneto-spectrophotometer is a simple apparatus for measuring the magnetotaxis behavior of magnetotactic bacteria based on the changes in light absorbance resulted from cells moving across light path in response to application and reversal of magnetic fields [8, 9]. It is based on a Varian's Cary 50 spectrophotometer incorporated with a Xenon flash lamp that emits a super-concentrated beam, which is important for the measurement. Two pairs of coils made of copper wire were mounted around cuvette holder to generate two mutually perpendicular homogenous magnetic fields, in parallel or perpendicular to the light path of the spectrophotometer (Fig. 1) (*see* **Note 1**). When 0.5 V voltage was applied, currents of 0.042 A and 0.116 A, with current densities of 0.33 A/mm^2 and 0.35 A/mm^2, were generated in the parallel and perpendicular coils, respectively. The magnetic flux densities of the corresponding static magnetic fields were both 1.9 mT at 20 °C. This device can generate magnetic fields with magnetic flux densities ranging from 0 mT to up to 6 mT by controlling the supplied current density. An adjustable current power supply controls the electromagnetic system, switching to reverse current direction, and "OFF" or "ON" for the parallel magnetic field or the perpendicular magnetic field (*see* **Note 2**).

1.2.2 Magnetodrome System

A Magnetodrome device was designed and constructed by Petersen Instruments (Petersen Instruments, Munich, Bavaria, Germany). It consists of a Zeiss Axio Observer A1 microscope with a set of four electromagnets (Fig. 2a) generating homogenous magnetic fields up to 1500 Oe and controlled by MAGNETODROME software

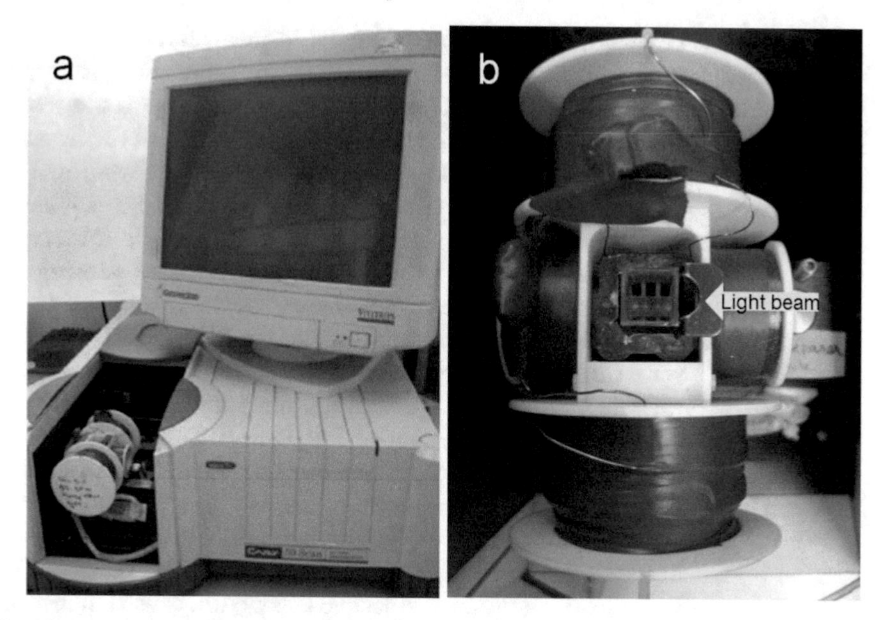

Fig. 1 The Magneto-spectrophotometer. Panel (**a**) shows the modified spectrophotometer, with the special cuvette holder inside. Panel (**b**) shows the details of the cuvette holder mounted with two pairs of coils. The yellow arrow indicates the direction of light beam

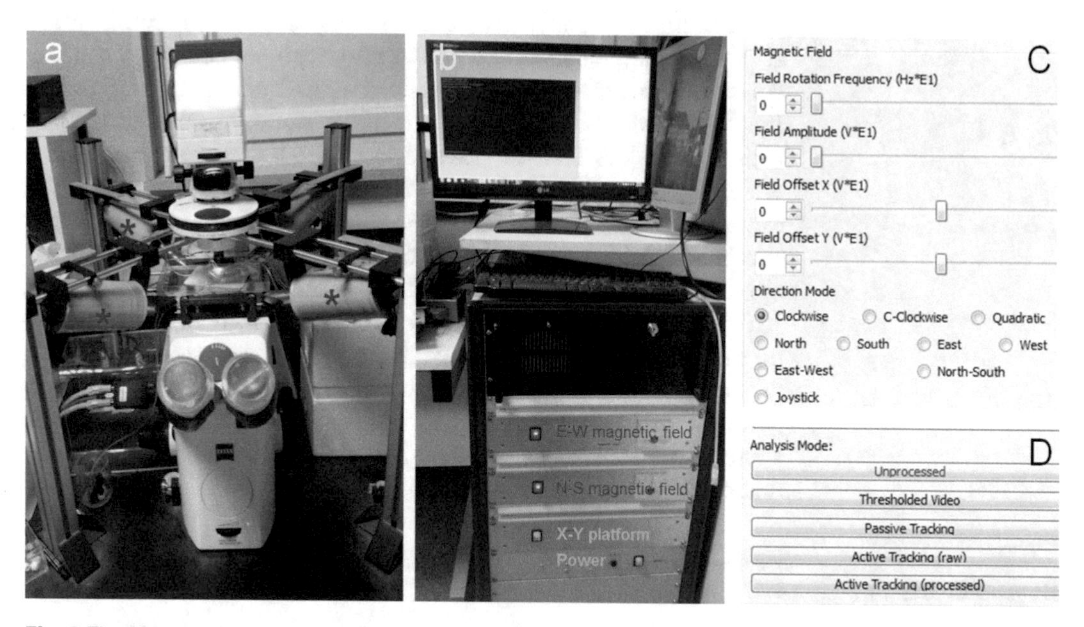

Fig. 2 The Magnetodrome system. Panel (**a**) shows Zeiss Axio Observer A1 microscope and the four electromagnets (stars). Panel (**b**) shows the MAGNETODROME controller with real-time display screen and the power control units of Magnetodrome system (Bottom with function marks). Panels (**c**) and (**d**) show the control panels of magnetic field and analysis mode of MAGNETODROME software, respectively

(Fig. 2b) (Petersen Instruments, Munich, Bavaria, Germany). The applied magnetic field parameters, including field strength, direction mode, and rotation frequency, could be set up within the MAGNETODROME software (Fig. 2c) and all the information is shown on the left top of the real-time display window (Fig. 2b). Field strength on 0 Gs means no magnetic field supplied by the electromagnets set. This condition is used in observing free-swimming behavior of the MO-1 cells under the geomagnetic field condition. Either "Passive tracking" or "Active tracking" could be chosen as an analysis mode (Fig. 2d) in MAGNETODROME software for swimming behavior recording and measure. A ".csv" format file is reported when the recording finished and further analysis could be down using data analysis software (*see* **Note 3**).

A fast-speed camera is necessary for the record of MO-1 swimming trajectories. A camera Gazelle of Point-Grey at 90 frames per second (fps) and a sCMOS camera of ANDOR up to 400 fps are used in tracking the MO-1 cells with Magnetodrome system (*see* **Note 4**). Long working distance objective lens of 40× is used in recording the swimming trajectories of MO-1 cells (*see* **Note 5**). The swimming behaviors are recorded and further analyzed using various software.

1.2.3 Micro-Channels

Besides analyzing the motility between slides and coverslips, two kinds of micro-channels are used. The first is the commercially available Ibidi micro-slide $VI^{0.4}$ with 0.4 mm depth and 30 μL of channel volume. The second is specifically designed and manufactured in PolyDimethylSiloxane (PDMS) using a standard soft-lithography technique [4]. A 130-μm-thick layer of SU-8 was spin-coated on a silicon wafer and soft-baked at 95 °C during 44 min. Then it was exposed to UV light using a MA150 mask aligner (Suss Microtec) and baked at 95 °C during 3 min. Uncross linked SU-8 was developed in PGMEA and the completed structure was hard baked at 110 °C. PDMS was casted on the mold and baked at 90 °C during 60 min. It was then pealed from the mold, cut and washed in a bath of isopropanol. Inlets and outlets were punched off with a razor sharp cutting tip. Finally, the PDMS device was sealed to a glass slide after a short exposure to O_2 plasma. The design and dimension of the microfluidic chip is detailed in Fig. 3.

2 Materials

2.1 Culture of Magnetotactic Bacterium MO-1

1. Artificial sea water: Weight 19.45 g NaCl, 5.90 g $MgCl_2$, 3.24 g Na_2SO_4, 1.80 g $CaCl_2$, and 0.55 g KCl, and dissolve in 1 L water.
2. Mix of vitamins: Weight following chemicals and dissolve in 100 mL water: 1 mg Biotin, 5 mg nicotinic acid, 0.4 mg Folic acid, 40 mg D-Ca-pantothenate, 40 mg pyridoxin HCl, 50 mg vitamin B-12, 900 mg thiamine HCl, 50 mg p-aminobenzoic acid, 5 mg riboflavin, and 5 mg lipoic acid. Filter through a

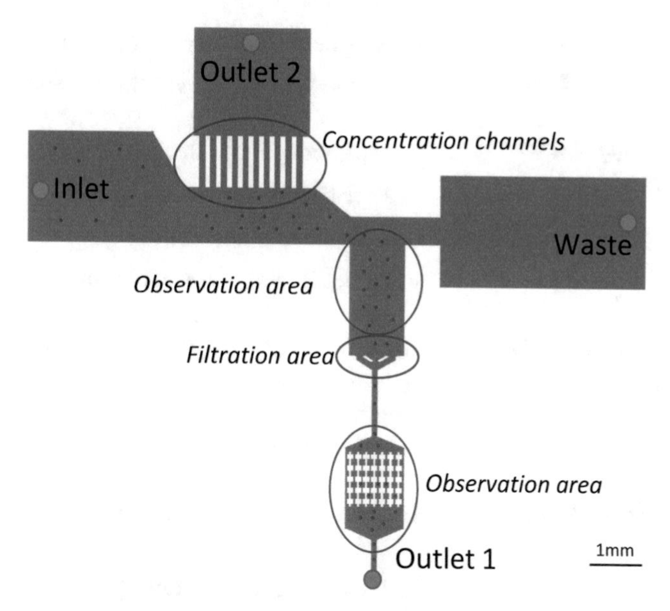

Fig. 3 Specially designed and constructed microchannel for study of MO-1 mobility. The channel is indicated by *dark gray color*. It has a depth of 130 μm and the *scale bar* is 1 mm. *Red dots* represent bacteria. Bacteria were injected using syringe pumps at Inlet. The cells entered the observation areas either with flow or being magnetically guided by the application of a magnetic field. The filtration area is used to block agarose (Courtesy of A.-M. Gué, LAAS, CNRS)

0.22 μm Millipore filter. Make up to 1 L with sterilized water and store at 4 °C, in a bottle wrapped with aluminum foil.

3. Mix of trace elements: Dissolve 1.5 g Nitrilotriacetic acid in around 500 mL water, adjust pH to 6.5, and then add following components: 0.18 g $ZnSO_4 \bullet 7H_2O$, 3 g $MgSO_4 \bullet 7H_2O$, 0.02 g $CuSO_4 \bullet 5H_2O$, 0.5 g $MnSO_4 \bullet 2H_2O$, 0.02 g $KAl(SO_4)_2 \bullet 12H_2O$, 0.1 g $FeSO_4 \bullet 7H_2O$, 0.01 g H_3BO_3, 1 g NaCl, 0.4 g $Na_2MoO_4 \bullet 2H_2O$, 0.18 g $CoSO_4 \bullet 7H_2O$, 0.01 g $NiCl_2 \bullet 6H_2O$, and 0.1 g $CaCl_2 \bullet 2H_2O$. Make up to 1 L with water and adjust the pH to 7 with KOH. Autoclave at 120 °C for 15 min and store at 4 °C.
4. Ferric quinate: Weight 0.45 g $FeCl_3 \bullet 6H_2O$ and 0.19 g quinic acid, transfer into 80 mL water preheated to 70 °C, continue incubating at 70 °C until fully dissolved. Autoclave at 120 °C for 15 min and store at 4 °C.
5. 1.86% K_2HPO_4 in water.
6. 40% Na thiosulfate in water.
7. Semisolid medium of MO-1: Dissolve 0.2 g NH_4Cl, 1.26 g $NaHCO_3$, 3 g HEPES, 0.6 g Cysteine-HCl, 0.35 g agarose in 1 L artificial seawater (*see* **Note 6**). Mix and adjust pH to 6.8 (*see* **Note 7**). Aliquot 80 mL of medium into 100 mL bottle with blue cap (*see* **Note 8**). Autoclave at 121 °C for 20 min.

2.2 Magneto-Spectrophotometer

1. Magneto-spectrophotometer.
2. Carbonyl cyanide m-chloro phenyl hydrazine (CCCP).

2.3 Magnetodrome System

1. Magnetodrome device.
2. Fast-speed camera.
3. Glass microscope slide.
4. 22 × 22 mm glass coverslip.

2.4 Manual Tracking with ImageJ

ImageJ (http://imagej.net/Fiji/Downloads) with particles tracking and measurement plugin MtrackJ (http://imagescience.org/meijering/software/mtrackj/) [10] are used in the motility manual tracking, measurement, and analysis.

2.5 Micro-Channels

1. Ibidi micro-slide VI$^{0.4}$ with 0.4 mm depth and 30 μL of channel volume.
2. PolyDimethylSiloxane (PDMS) devise [4].

3 Methods

3.1 Culture of Magnetotactic Bacteria MO-1

1. Prepare a mixture containing: 0.5 mL mix of vitamins, 2 mL of 0.01 M ferric quinic, 3 mL of 40% Na thiosulfate, 4 mL of 1.86% K_2HPO_4, and 5 mL of trace elements. Add 1.16 mL mixture above to each bottle of semisolid medium; mix up (*see* **Note 9**).
2. Carefully take MO-1 cells from the white cellular layer of a 3-days culture into a sterilized tube, avoid taking up too much medium (*see* **Note 10**).
3. Take 1 mL of MO-1 cells and inoculate the cells softly into a bottle of fresh medium, and place the bottles on the bench for a while, until a compact thin layer of MO-1 cells is formed at the oxic-anoxic transition zone. Carefully transfer the bottles into a dark cupboard far from high magnetic fields and cultivate for 3 days under room temperature (23–25 °C).
4. Incubation of the cultures in an electromagnetic field with the direction opposite to the geomagnetic field lines or in plastic tubes allow obtaining south-seeking MO-1 cells [3].

3.2 Magneto-Spectrophotometry (MSP) Assay

1. Take 1 mL of MO-1 culture. Measure the OD value at 600 nm and adjust to around 0.5 with artificial seawater if the concentration is too high.
2. Add MO-1 cells into the cuvette, mix up thoroughly with shaking or pipetting, and place the cuvette in the holder of magneto-spectrophotometer.
3. Turn on the electromagnetic system, set the magnetic field at "OFF."

4. Open the software of the spectrophotometer named "Kenetics," set the wavelength at 600 nm and the capture time of 10 min, and start recording the OD value.
5. After recording the OD value for around 20–30 s, set the electromagnetic system to apply a magnetic field of required strength perpendicular to the light path for 60 s (*see* **Note 11**). Change the current direction to apply an anti-perpendicular magnetic field and record for 60 s. Reverse the magnetic field again if needed.
6. When no magnetic field is applied, the MO-1 cells are homogenously dispersed in the cuvette (Fig. 4a, b, stage a). When a magnetic field perpendicular to the light beam is applied (Fig. 4a, the downward arrows), the OD value decreases as cells migrate along with the magnetic field northward away from the light path (Fig. 4a, b, stage b). In around 60 s, the decreasing of OD value slows down and the majority of MO-1 cells are accumulated at the north wall of the cuvette (Fig. 4a, b, stage c). Reverse the direction of magnetic field (Fig. 4a, the upward arrows), the MO-1 cells will swim toward the novel north wall. When the bulk of MO-1 cells cross the light path, the OD value reaches the maximum and then starts to decrease as MO-1 cells pass the light beam (Fig. 4a, b, stage d and e). As a control, the MO-1 cells treated with carbonyl cyanide m-chloro phenyl hydrazine (CCCP), an inhibitor of bacterial flagellar motor, are not mobile and the OD value does not change despite application and reversal of the magnetic field (Fig. 4a, black line).
7. The time gap between the reversal of the magnetic field and the point that OD value reaches the maximum implies the time needed for the bulk of cells to react to the magnetic field and swim from the cuvette wall to the light-beam path. When the strength of magnetic field is fixed, the time is mainly affected by two factors, the speed of cells to align along the magnetic field lines and the swimming velocity. By dividing the distance between the cuvette wall and the light beam by the time gap, we could estimate the swimming velocity of MO-1 cells. The sharpness of the absorbance peak reflects the homogeneity of the MO-1 cells.
8. The MSP assay could also be used to determine the polarity of MO-1 population. The light beam is asymmetrically positioned across the cuvette (Fig. 5a). Therefore, it takes longer time for cells to swim from one wall of the cuvette to the light beam (path b) than from the opposite wall of the cuvette (path a). Based on this characteristic, we could evaluate the magnetotaxis polarity of MO-1 population. MSP assays were performed with NS and SS MO-1 cells, respectively. The NS MO-1 cells swim in parallel to the magnetic field. When the perpendicular magnetic field is applied (the first arrow, downward), NS MO-1

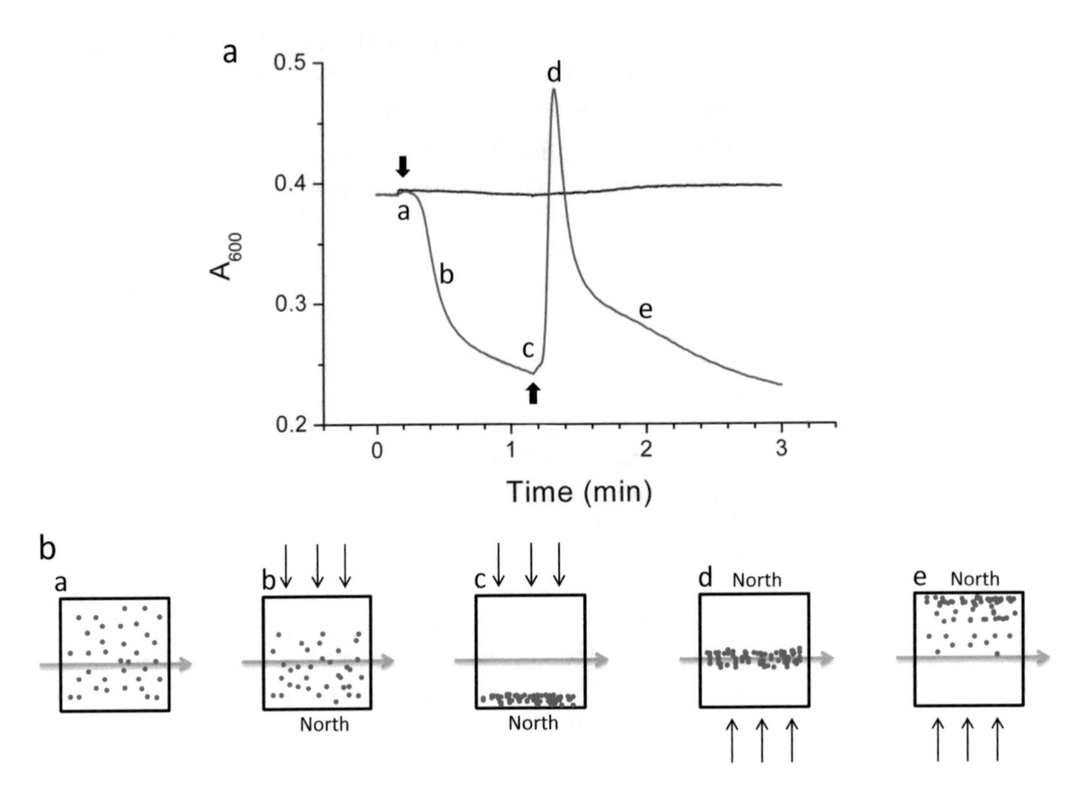

Fig. 4 Magneto-spectrophotometer characterization of motility of MO-1 cells. The *panel a* shows the absorbance curves of the mobile MO-1 cells (*red line*) and the immobile (*black line*) MO-1 cells treated with CCCP. The *arrows* indicate the changing of direction of the magnetic field. Different stages of the assay are indicated by "a" to "e," which is in correspondence with panel b. The *panel b* illustrates the position of MO-1 cells in the cuvette at different stages of the assay. The *orange arrow* indicates the sharp light beam of the magneto-spectrophotometer, the *black arrow* represents the direction of the magnetic field in the cuvette, and the *dots* represent MO-1 cells

cells accumulate at the north pole of the cuvette. Along with the reversal of magnetic field (the second arrow, upward), NS MO-1 cells swim toward the novel north wall and pass the light beam, leading to the asymmetric peak P1 (Fig. 5, b1). The subsequent reversal of the magnetic field (the third arrow, downward) results in the second peak, P2 (Fig. 5, b1). Notably, P2 is higher and sharper than P1. The reason is that, in reaction to the first reversal of magnetic field, NS MO-1 cells travel 2.7 mm from the cuvette wall to the light beam, while the travel distance after the second reversal of magnetic field is shorter than the first trajectory (1.3 mm versus 2.7 mm). Therefore, the cell swarm remained better grouped in the second case, and as consequence, the second peak (P2) is higher and sharper than P1. For SS MO-1 cells, the first peak of their MSP profile was higher and sharper than the second peak when a parallel-antiparallel-parallel sequential alternative magnetic field was applied in the MSP assay (Fig. 5, b2) (*see* **Note 12**).

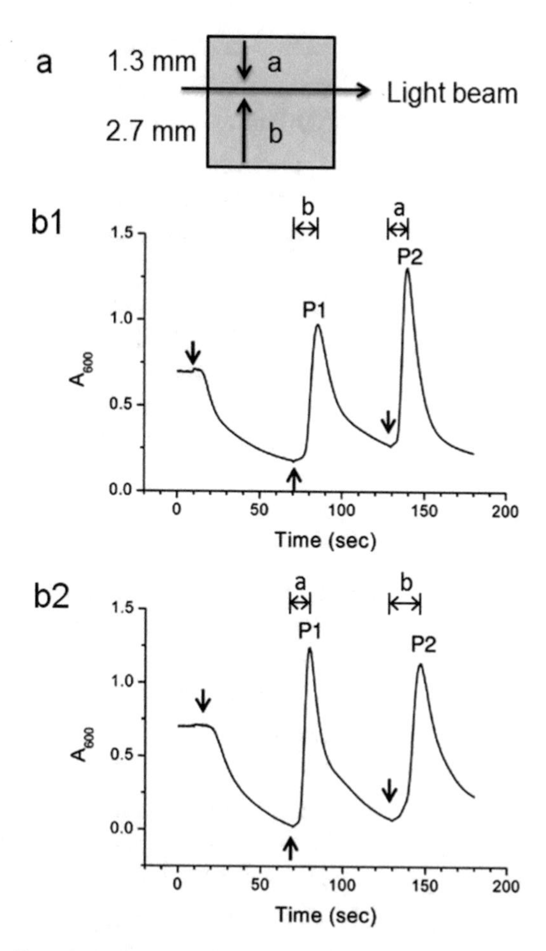

Fig. 5 Detection of magnetotaxis polarity of MO-1 population. *Panel a* shows the asymmetrically positioned light beam across the cuvette, and the distance from one wall and the opposite wall of the cuvette to the light beam. MSP assays were performed with north-seeking (*b1*) or south-seeking (*b2*) MO-1 cells. *Downward* and *upward arrows* indicate the direction of the applied 1.9-mT perpendicular uniform electromagnetic field with respect to the light beam of the spectrophotometer. The first and second absorbance peaks of cells swimming across the light beam in reaction to the reversal of the magnetic field are marked as *P1* and *P2*, and the swimming path is indicated as a or b, in correspondence to the diagramme in *panel a*

3.3 Measurement of Motility and Magnetotaxis Behavior with Magnetodrome System

1. Turn on the control of power, *x–y* platform, the S-N magnetic field, and E-W magnetic field on position "I."
2. Turn on the microscope, use 10× and 20× objective lenses to have a quick general inspection, then use 40× objective lens for the detailed analysis.
3. Carefully collect 1 mL of 2-day growth MO-1 cells (*see* **Note 13**) from the thin cellular layer to a sterilized tube.
4. Load 4 μL of collected MO-1 cells on a glass microscope slide. Cover the drop tightly with a 22 × 22 mm glass coverslip (*see* **Note 14**). Remove any air bubbles between the slide and the coverslip. Place the slide on the microscope slide holder.

5. Launch the MAGNETODROME software (*see* **Note 15**).
6. Set up the field strength on 0 Gs and record the free-swimming motility under geomagnetic field of MO-1.
7. Apply a uniform magnetic field of 1.5–10 Gs in a fixed direction (north, south, east, or west), clockwise or counter-clockwise rotary, quadratic, and north-south or east-west alternate mode.
8. Using either "Passive tracking" or "Active tracking" in MAGNETODROME software to record the motility data of the MO-1 cells. The former just shows the swimming tracks on the computer screen and movie files while the latter takes record of the particle positions. Unprocessed mode just makes a video of the swims, which can be analyzed with ImageJ, too.
9. When active processed mode is used, a CSV format file is automatically created. It contains the average value of velocity, swimming direction, and relative position of cells in each frame.
10. To simulate the magnetotaxis motility in marine sediments, pieces of agarose in the cultures and aggregated immobile cells can be used as obstacles to analyze the adjustment of MO-1 swimming behavior (Fig. 6).
11. Load MO-1 cell suspension between a slide and cover-slip and observe with Magnetodrome system.
12. Apply a uniform magnetic field of 1.5 Gs. Record the swim behavior at 50 fps and analyze it with ImageJ-MTrackJ plugin. The swim track is analyzed frame by frame.
13. MO-1 cells swim in parallel to the magnetic field lines (Fig. 6, a1). When encountering an obstacle and if it cannot go through the obstacle, a cell swims backward with a variable distance (Fig. 6, a2). Then the cell swims again toward north pole and squeezes through obstacles until being blocked again by another obstacle (Fig. 6, a3, white arrow). The cell changes swim direction and circumvents the obstacles (Fig. 6, a4).
14. Use 100× objective lens to obtain more details (Fig. 6, b1 and b2). Analyze in more detail about the motion by MTrackJ to calculate the instantaneous speed (Fig. 6c) and instantaneous angle (Fig. 6, D1 to D3) according to Meijering et al. [10] (*see* **Note 16**).
15. Although they swim constantly toward the north, MO-1 cells display a south-seeking motion when they encounter obstacles. To characterize the transitional south-seeking behavior of the population, inject MO-1 cells into micro-channels. When they encounter the channel wall, MO-1 swim antiparallel to the magnetic field lines to various distances, depending on the strength of the applied magnetic field. Active track the swim, analyze the distances of the south-seeking swim (Fig. 7).

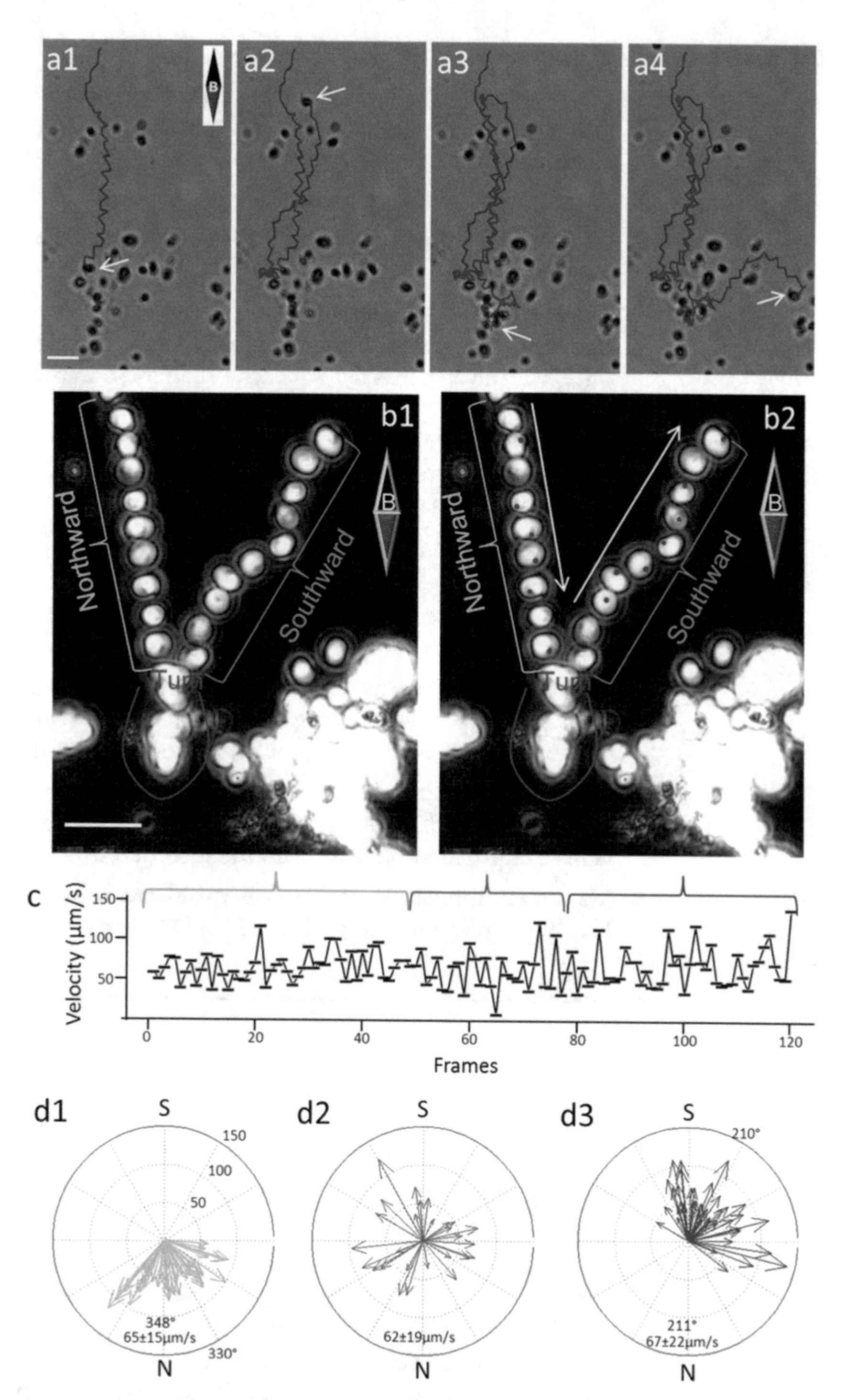

Fig. 6 Swimming behavior of MO-1 cells in applied uniform magnetic field. *Panels a1* to *a4* show the trajectories of a north-seeking cell (*white arrow*) to squeeze through (*a3*) or circumvent (*a4*) obstacles in a uniform magnetic field of 1.5Gs. *Panel b* shows motility details of a typical representative. A dark crack resulting from refringent intracellular granules can be seen in *b1*, and indicated in *red dots* in *b2*, which is used as a marker for the anterior of this cell. One frame has been extracted from every five to make the cellular morphology clearly observable. The instantaneous speed is shown in *b3* for the swimming along (*green*) or against (*blue*) the field lines, or turning to find a way out (*pink*). Instantaneous angles are plotted using MATLAB compass function

3.4 Manuel Tracking in ImageJ [10]

1. Open the record video file in ImageJ.
2. Click "Image"–"properties." Check and set up the all information of this video on this panel. It is crucial for the tracking and measurement in the following step.
3. Choose "Plugins"–"MTrackJ."
4. Click "Add" on the panel of MTrackJ to start tracking the motility of cells.
5. When the tracking finished, click "Measure." Two files are shown. One gives information of each cell in each frame; the other has the statistics data of each cell in its all tracked frames.

3.5 Measuring Magnetic Moment of Magnetotactic Cells

Magnetosome crystals confer on cells a net magnetic moment dipole. The magnetic moments of the cells can be calculated according to their magnetotactic behaviors.

1. North-South alternate field is applied; MO-1 cells make U-turns upon reversal of the field direction. Adjust reversal frequency with the swim velocity, and apply a magnetic field with a strength ranging from 1.5 to 10 Gs.
2. MO-1 swim behaviors are recorded either by active tracking (Fig. 8a) or passive tracking (Fig. 8b). For passive tracking, one frame from every 5, 10, or 15 frames are kept depending on the swim velocity of the MO-1 cells.
3. The swim speeds and diameter of the U-turn and magnetic field strength are obtained from the recorded data.
4. Magnetic moment of each cell is calculated using eq. (1) in Fig. 8 according to Esquivel and Lins De Barros [11].

4 Notes

1. In our machine, the light beam of the spectrophotometer does not go through the center of the cuvette, but 1.3 mm from one side and 2.7 mm from the opposite side.
2. Strong permanent magnets can be used instead of the mounted electro-magnetic fields.
3. In the literature, various types of microscopes mounted with electro-magnetic fields or magnets have been reported [12–14]. They serve for the same function of Magnetodrome by a similar operation concept. Once the data are obtained, characterization of magnetotactic behavior can be performed with ImageJ as described here.
4. The MAGNETODROME software is designed to show and record the cell motility only with the camera Gazelle of Point-Grey. If other types of camera are used, their specific recording software should be also used independently and MAGNETODROME software controls only the magnetic field parameters.

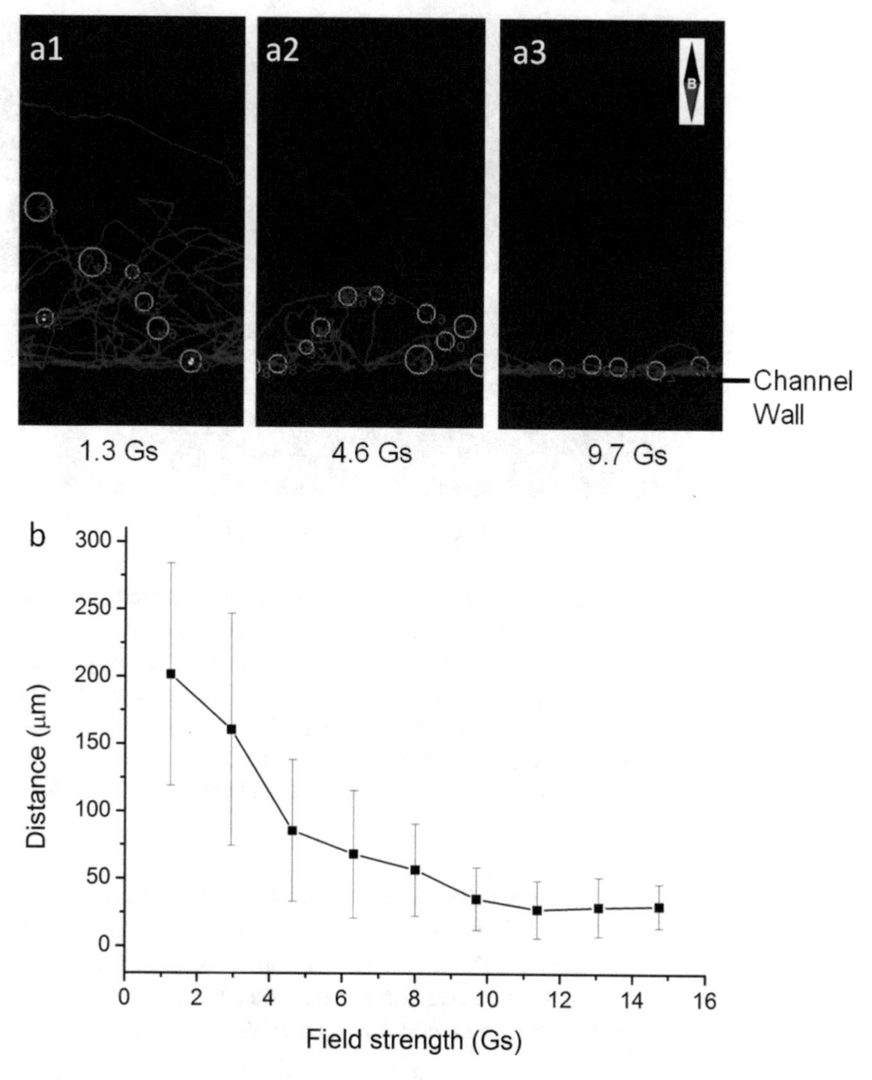

Fig. 7 Effect of magnetic field strength on the swimming against the field lines. *Panels a1* to *a3* are active tracking images showing the trajectories of north-seeking cells swimming away from the north wall of the micro-channels toward magnetic south. *Panel b* shows the plot of the average distance with standard deviation of the southward swimming from the north wall in different applied magnetic fields

5. Objective lenses of 10× and 20× are used for general overview and 40 x and 100× can be used in showing details of MO-1 motility during tracking.
6. 0.1% of resazurin could be added optionally as indicator of redox. Before autoclave, the resazurin is in light cyan color. After autoclave, the top layer of semisolid medium should be colorless, and the region below is in light pink. MO-1 cells grow at the oxic-anoxic transition zone.
7. The original pH of the semisolid medium of MO-1 should be around 6.6, and it takes about 3 mL 10 M NaOH to adjust to the right pH.

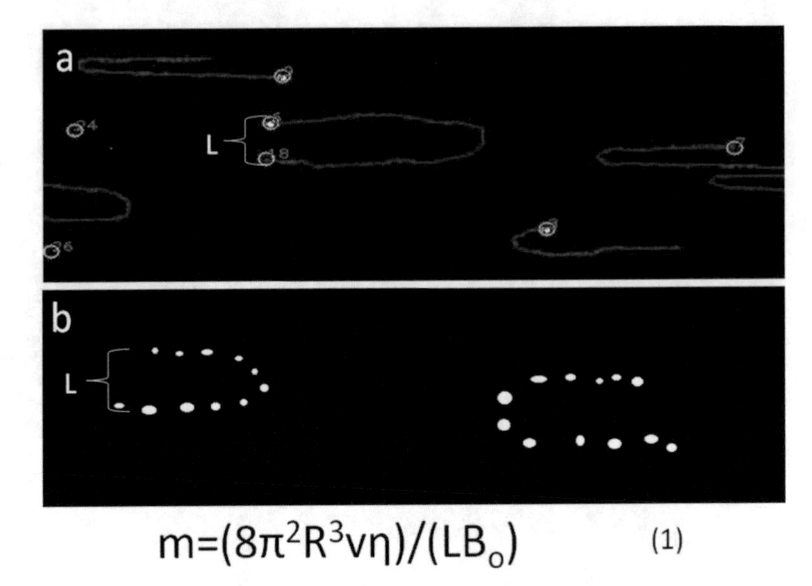

Fig. 8 Calculation of magnetic moment of magnetotactic bacteria. North–South alternate magnetic fields with various magnetic strengths are applied. The swim trajectories of cells are recorded by either active tracking (**a**) or passive tracking extract one from every ten frames (**b**). The swim speed (v) and the diameter of the U-turn are calculated, and the magnetic moment can be obtained using eq. (1). Where m is the magnetic moment, R is the radius of the cell, v is the swim speed, η is the viscosity of the media, L is the diameter of the U-turn, B_0 is the magnetic field

8. As there is 0.035% agarose in the semisolid medium, which is not solubilized before autoclave, special cares should be taken for the aliquot. The cylinder containing the medium is kept on a magnetic stirrer for all the time. We first transfer part of the medium to a beaker flask, then aliquot into bottles.
9. The mixture should be added when the medium is cooled down to room-temperature.
10. After 3 days growth, the band of MO-1 cells is at the surface of medium. MO-1 cells are taken with 1 mL pipette. In order to have a more concentrated MO-1 cells suspension, it is better not to take too much medium below the cell layer.
11. The time of recording is dependent on the swimming velocity, cellular status, and the strength of magnetic field. With fresh culture of MO-1, 60 s is enough for the majority of MO-1 cells to swim to one side of the cuvette under a magnetic field of 1.90 mT.
12. Under normal conditions in the geomagnetic field, the MO-1 cultures consist of mainly NS cells and few SS cells. Due to the asymmetric position of the light path, the time required for NS and SS cells reaching the light beam is different. Therefore, a major and a minor peak could appear sometimes.

13. The density of 2-day-old MO-1 cells could be observed directly. A 1:2 ~ 1:5 dilution of 3-day-old MO-1 cells could also be used. To keep the cell on good condition, all the dilution should be made with fresh liquid MO-1 medium.
14. The drop volume of MO-1 cells is according to the size of coverslips used. Too big volume leads to liquid flows between the coverslip and the slide. These flows are the main distributions in the swimming behavior observation. Alternatively, cells can be loaded in channels with 0.4 mm depth of Ibidi slides or specifically designed and constructed micro-channels (Fig. 3) and inspected by focusing on the middle of the channel depth to avoid boarder effect and minimize aero-taxis.
15. If there is a problem to start the software, check first if the camera is turned on.
16. Manual tracking with MTrackJ is more precise than automatic tracking with the MAGNETODROME. However, the workload is also heavier for manual tracking. If the result is based on the statistics of large amount of data, difference between the two tracking methods is not significant.

References

1. Bazylinski DA, Frankel RB (2004) Magnetosome formation in prokaryotes. Nat Rev Microbiol 2(3):217–230
2. Lefèvre CT, Bernadac A, Yu-Zhang K, Pradel N, Wu L-F (2009) Isolation and characterization of a magnetotactic bacterial culture from the Mediterranean Sea. Environ Microbiol 11(7):1646–1657
3. Zhang W-J, Chen C, Li Y, Song T, Wu L-F (2010) Configuration of redox gradient determines magnetotactic polarity of the marine bacteria MO-1. Environ Microbiol Rep 2:646–650
4. Zhang SD, Petersen N, Zhang WJ, Cargou S, Ruan J, Murat D, Santini CL, Song T, Kato T, Notareschi P, Li Y, Namba K, Gue AM, Wu LF (2014) Swimming behaviour and magnetotaxis function of the marine bacterium strain MO-1. Environ Microbiol Rep 6(1):14–20
5. Ruan J, Kato T, Santini C-L, Miyata T, Kawamoto A, Zhang W-J, Bernadac A, Wu L-F, Namba K (2012) Architecture of a flagellar apparatus in the fast-swimming magnetotactic bacterium MO-1. Proc Natl Acad Sci U S A 109(50):20643–20648
6. Bazylinski DA, Williams TJ, Lefevre CT, Berg RJ, Zhang CL, Bowser SS, Dean AJ, Beveridge TJ (2013) *Magnetococcus marinus* gen. nov., sp. nov., a marine, magnetotactic bacterium that represents a novel lineage (Magnetococcaceae fam. nov., Magnetococcales ord. nov.) at the base of the Alphaproteobacteria. Int J Syst Evol Microbiol 63(Pt 3):801–808
7. Zhou K, Pan H, Yue H, Xiao T, Wu L-F (2010) Architecture of flagellar apparatus of marine magnetotactic cocci from Qingdao. Marine Sci 34(12):88–92
8. Lefèvre CT, Song T, Yonnet JP, Wu L-F (2009) Characterization of bacterial magnetotactic behaviors by using a magnetospectrophotometry assay. Appl Environ Microbiol 75(12):3835–3841
9. Philippe N, Wu L-F (2010) An MCP-like protein interacts with the MamK cytoskeleton and is involved in magnetotaxis in *Magnetospirillum magneticum* AMB-1. J Mol Biol 400(3):309–322
10. Meijering E, Dzyubachyk O, Smal I (2012) Chapter nine - methods for cell and particle tracking. In: Conn PM (ed) Methods in enzymology, vol 504. Academic Press, New York, pp 183–200
11. Esquivel DM, Lins de Barros HGP (1986) Motion of magnetotactic microorganisms. J Exp Biol 121:153–163

12. Chen C-Y, Chen C-F, Yi Y, Chen L-J, Wu L-F, Song T (2014) Construction of a microrobot system using magnetotactic bacteria for the separation of *Staphylococcus aureus*. Biomed Microdevices 16(5):761–770
13. Ma Q, Chen C, Wei S, Chen C, Wu L-F, Song T (2012) Construction and operation of a microrobot based on magnetotactic bacteria in a microfluidic chip. Biomicrofluidics 6(2):024107–024112
14. Pan Y, Lin W, Li J, Wu W, Tian L, Deng C, Liu Q, Zhu R, Winklhofer M, Petersen N (2009) Reduced efficiency of magnetot axis in magnetotactic coccoid bacteria in higher than geomagnetic fields. Biophys J 97(4):986–991

INDEX

Tohru Minamino and Keiichi Namba (eds.), *The Bacterial Flagellum: Methods and Protocols*, Methods in Molecular Biology, vol. 1593, DOI 10.1007/978-1-4939-6927-2, © Springer Science+Business Media LLC 2017

D

E

F

G

H

Q

R

S

T

U

X

Printed in the United States
By Bookmasters